DIGITAL TWIN
COMMUNICATION TECHNOLOGIES IN TRANSPORTATION SYSTEMS
Insights from China

DIGITAL TWIN COMMUNICATION TECHNOLOGIES IN TRANSPORTATION SYSTEMS

Insights from China

Guanghui Zhao
Yu Yang
Guizhou University of Finance and Economics, China

World Scientific

NEW JERSEY • LONDON • SINGAPORE • BEIJING • SHANGHAI • HONG KONG • TAIPEI • CHENNAI • TOKYO

Published by

World Scientific Publishing Co. Pte. Ltd.
5 Toh Tuck Link, Singapore 596224
USA office: 27 Warren Street, Suite 401-402, Hackensack, NJ 07601
UK office: 57 Shelton Street, Covent Garden, London WC2H 9HE

Library of Congress Cataloging-in-Publication Data
Names: Zhao, Guanghui author | Yang, Yu author
Title: Digital Twin Communication Technologies in Transportation Systems : insights from China / Guanghui Zhao, Yu Yang, Guizhou University of Finance and Economics, China.
Description: New Jersey : World Scientific, [2026] | Includes bibliographical references and index.
Identifiers: LCCN 2025043540 | ISBN 9789819824618 hardcover | ISBN 9789819824625 ebook for institutions | ISBN 9789819824632 ebook for individuals
Subjects: LCSH: Transportation engineering--China--Technological innovations | Intelligent transportation systems | Digital twins (Computer simulation)--Industrial applications
Classification: LCC TA1145 .Z43 2026
LC record available at https://lccn.loc.gov/2025043540

British Library Cataloguing-in-Publication Data
A catalogue record for this book is available from the British Library.

For any available supplementary material, please visit
https://www.worldscientific.com/worldscibooks/10.1142/14627#t=suppl

Desk Editors: Soundararajan Raghuraman/Veronica Lee

Typeset by Stallion Press
Email: enquiries@stallionpress.com

Content Summary

Digital twin is a new digital technology in recent years. It makes full use of physical model, sensor update, operation history, and other data, integrates multi-disciplinary, multi-physical quantity, multi-scale, multi-probability simulation process, and realizes the visual replication of the whole life cycle of physical equipment. As the cutting-edge technology of modern surreality, digital twin has played an important role in various fields of public life, and its role in promoting transportation development will become more and more prominent with technological upgrading.

This book makes a detailed analysis and discussion from the following four aspects by taking the characteristics and functions of digital twin as the entry point: the impact on China's transportation development, the current situation of the integration with China's transportation technology, the way to accelerate the process of strengthening China with strong transportation network, as well as the trend of leading China's transportation development. In this book, various technical points of digital twin to promote transportation development are included, and moreover, a large number of cases including Guiyang and Guizhou are listed to analyze the key technologies and public policies in detail, and innovation points of the combination of digital twin and transportation technology are proposed. Through reading this book, more practitioners, managers, and entrepreneurs in the field of transportation are enabled to deeply learn that digital twin technology is not only an inevitable direction for the development of China's transportation power but also a key force for high-quality and high-speed development.

The digital twin is in line with key policies such as the "Outline for the Construction of Strengthening Country with Strong Transportation Network", the "Outline for the Development Plan of Digital Transportation", and the "Outline for the National Comprehensive Three-dimensional Transportation Network". In combination with the application trend in the process of intelligent transportation upgrading, this book puts forward the practical suggestions for the application of digital twin in the field of transportation in China, including strengthening the application depth of digital twin and creating a comprehensive three-dimensional transportation network with outstanding intelligence; moreover, starting from the perspective of public management, the technology used for reference and the laws, regulations, and policies to support the technology diffusion, and in combination with the practice of China's modern transportation development, the transformation of China's transportation development is promoted from traditional element driven to innovative technology driven, relying on digital forces to support and promote the country's modern development, and accelerating the construction of a world powerful country with strong transportation network.

Keywords: Strengthening country with strong transportation network, digital twin, big data transportation, transportation governance, transportation policy.

Introduction to the Authors

Professor Guanghui Zhao (born September 1976) is a Professor at Guizhou University of Finance and Economics, China. He is also a Researcher and Senior Visiting Fellow at the School of Engineering, University of Michigan, USA. In addition, he is the Founding Director of the Research Center for Modern Transport Development at the College of Management Cadres, Ministry of Transport, as well as a Researcher at the Research Center for Industrial and Academic Research Issues, Wuhan University.

Prof. Zhao has led numerous major projects, including National Social Science Fund of China (NSSFC) research projects commissioned by the Ministry of Transport, the China Postdoctoral Foundation Special Fund Project, Special Fund Projects from the China Postdoctoral Foundation, and over 30 China National Academy of Educational Sciences planning projects. He has authored over 100 academic papers in leading core journals and more than 20 treatises.

Prof. Zhao has published 10 monographs on transportation science and technology, including *Big Data Transportation Systems* (World Scientific), *Assessment and Policy Research on the Transportation Science and Technology Development in China* (World Scientific), *Redefining Traffic: How AI Leads the Change* (World Scientific) and 區塊鏈交通：開創未來交通新格局 (English title: *Blockchain in Transportation: Creating a New Landscape for Future Transportation*) (Posts & Telecommunications Press).

Associate Professor Yu Yang (born December 1986) is a business executive and academic with a strong expertise in manufacturing enterprise management. He currently serves as a Committee Member of the Industrial Committee at the Institute for Quantitative Finance, Xi'an Jiaotong-Liverpool University, China. Additionally, he is an Industry Mentor for professional degree postgraduates at Beibu Gulf University and a part-time Associate Professor at the university's School of Economics and Management. Beyond his academic roles, Yu Yang is the Co-founder of a leading domestic plant cell synthesis technology company in China and a Founding Partner of an early-stage commercial aerospace enterprise specializing in China's domestic key propulsion systems. His research interests encompass industrial operations management, corporate governance, and supply chain management.

Foreword I

In terms of promoting the development of China's transportation industry and the supply of public policies, as well as the research on Transportation Science and technology, Professor Zhao and I have some contacts and exchanges, and I very agree with his attitude of pursuing the integration of science and technology and public management. I have a new understanding of the author after reading the "Research on high-quality development of digital twin enabling 'strengthening country with strong transportation network'".

This is the first domestic research book that specializes in the analysis of digital twin to promote transportation development, which has a certain leading value. The foothold of this book is the digital twin of emerging technology in the era. It closely combines the current situation of China's transportation development, analyzes the development trend, and makes people have to pay tribute to its keen insight and rigorous logic.

I believe that there are two major sections in this book "Research on high-quality development of digital twin enabling 'strengthening country with strong transportation network'" worth readers' review: The first is the application and achievement of digital twin technology in transportation developed countries, and the second is Professor Zhao Guanghui's analysis on the promotion of digital twin to the high-quality development of China's transportation. The contents of these two major sectors make people aware that the transportation technology at the forefront of the times can actually bring such a huge force to the development of a country. Only by recognizing this situation in a timely manner can we ride the wind of the development of the times and achieve something. Digital twin will become a powerful tool in China's strategy of becoming a strong country with strong transportation network. The digital road cast by it indicates the direction and trend of China's transportation development in the future. In the digital era where everything can be digitalized, the development of digital twin constantly changes the quality of life of the public. It not only realizes the real replication and evolution of the real world in the digital space but also completes the twin connection of the real things in the virtual space under the background of the Internet of Things. This technology greatly promotes the improvement and development, upgrading, and iteration of things since it can provide dynamic demonstration and corresponding changes of the state information of real things. After the digital twin landing in China's transportation field, it has gradually become an important technology for China's transportation development. Digital twin is not only the technical support for China's transportation development but also the most common application field in the global transportation industry.

From this book, I can clearly see the integration path of digital twin technology and China's transportation industry and also agree with Professor Zhao's views and strategies on "digital twin enabling high-quality development of strengthening country with strong transportation network". This book helps us to understand the development of China's transportation industry and also clarifies the gap between China and developed countries. In terms of the current situation of the development of China's transportation industry, the gap between the technical level of some fields and that of developed countries has not shortened even though the overall situation is great, which requires the effective and correct use of digital twin technology, and more Chinese transportation personnel to condense digital quality and deeply cultivate the technical strength of China's transportation.

In my point of view, since the book "Research on high-quality development of digital twin enabling 'strengthening country with strong transportation network'" can bring practical value and can be regarded as a guide for career development, it will certainly be favored by numerous transportation-related people. For example, in the section "New momentum for transportation development in the digital era" in the book, the key technologies of transportation development are analyzed from multiple aspects, and the momentum, the demand for momentum, and the direction of momentum are specified in combination with the development views of the industry. This will not only help readers to have a more concrete understanding of the integrated development of digital power and transportation but also guide relevant people to discover important opportunities in this trend.

I had an in-depth discussion with Professor Zhao Guanghui after reading the sample manuscript of "Research on high-quality development of digital twin enabling 'strengthening country with strong transportation network'", in which I benefited a lot. I found that the author's objective and frank academic attitude is the main reason why the views in the book are convincing. The author explains the deficiencies in the application of digital twin in China and analyzes the substantial gap between China and the developed countries in a more careful way. It makes people feel unwilling, and at the same time, it makes people not only help but also stimulate a kind of power, which should be the accountability and responsibility of China's transportation people.

In summary, in the process of reading "Research on high-quality development of digital twin enabling 'strengthening country with strong transportation network'", I was struck by a lot, and Prof. Zhao's deep and simple expression also brought me a lot of inspiration. Therefore, I am very happy to recommend this book to all readers but also hope that China's transportation people can carefully read this excellent book. I believe this book will set off a "digital twin" development boom in China's transportation field, creating a trend and numerous opportunities for China's transportation development.

Xu Xiaolin
(Former Dean, Professor, and Doctoral Tutor of the
School of Public Administration,
Huazhong University of Science and Technology,
Member of the Public Administration Discipline Evaluation Group
of the Academic Degrees Committee of the State Council)

Foreword II

Dr. Zhao Guanghui is my visiting researcher at the University of Michigan. I agree with his persistence in research on big data transportation and big data technology, especially the spirit of the integration of intelligent manufacturing and transportation technology in the past 10 years, and always encourage him to write down his own ideas in the field of transportation that he enjoys. I was very pleased when I got the sample manuscript of "Research on high-quality development of digital twin enabling 'strengthening country with strong transportation network'". Since, in the years of research on intelligent manufacturing technology, including transportation technology, I have seen the improvement of this digital technology on the transportation development of Western countries many times, but after contacting various information from the world, I found that there are few books on the in-depth research of this technology in the field of transportation in China. Dr. Zhao Guanghui is a visiting researcher under my guidance at the University of Michigan. It can be said that the publication of this book fills a big gap in the field of transportation technology in China. I firmly believe that this book will also be favored by many friends in the field of transportation.

I found that Dr. Zhao Guanghui's view is very unique after reading "Research on high-quality development of digital twin enabling 'strengthening country with strong transportation network'". The analysis of digital twin in this book not only combines the industry demand and the trend of the times but also relates to the economic and social development and technical value. In the meantime, the author also interprets various strategies for the development of digital transportation to ensure the consistency and practicability of the views in this book and national policies. At the same time, this book has a certain depth of scientific research. Dr. Zhao Guanghui has learned from the domestic and foreign advanced experience in the analysis of the "high-quality development of digital twin enabling strengthening country with strong transportation network", led by the proper way of transportation developed countries, and combined with the actual situation of China's transportation development, resulting in a road to the landing of digital twin.

Generally speaking, this book is a transportation science and technology research work with rigorous writing style and distinctive views. This book introduces the specific application of a large number of domestic and foreign digital twin in the field of transportation and carefully analyzes and explains the application method of this digital technology so that readers can clearly understand the correct method of learning and using this technology.

The digital and intelligent development of China's transportation industry has become an inevitable trend. Therefore, how to obtain excellent development results is the problem that contemporary transportation people are thinking about. The answer to this question seems to be a partial one. It uses a new transportation technology to explain a development trend. However, all kinds of views have strong landing features and a large number of facts have been used to support them. I can't help but secretly praise Dr. Zhao Guanghui's thinking is so meticulous and clear, and the expression of the concept of "reel silk from cocoons" has pointed out a development path in the field of people's livelihood. The "Action Plan for New Infrastructure Construction in the Field of Transportation (2021–2025)" issued by the Chinese government clearly states: "Adhering to innovation driven and intelligent development, we will take digitalization, networking and intellectualization as the main line, organize and promote a number of key projects of new transportation infrastructure, create an influential model of new transportation infrastructure, create an innovative development environment, promote the development of new infrastructure from point to area, improve the efficiency, expand functions increase momentum of transportation, and continuously enhance the sense of gain, sense of happiness and sense of safety of the people". In this book "Research on high-quality development of digital twin enabling 'strengthening country with strong transportation network'", I saw this development trend and also found an effective way to enhance the "sense of gain, sense of happiness and sense of safety" of the people through transportation development. Therefore, I think this book will become a good reading work for transportation people. This work with broad vision, practical basis, and thorough analysis of the development trend of modern transportation digitalization and intellectualization will definitely benefit readers. This work of "making way clear" will also set off a new wave of China's transportation technology development, leading China's transportation industry to obtain a huge breakthrough.

In our era, everyone has a story, and turning the ideas they like into stories that can be shared, moved, and meaningful to others is a source of power for academic research. Dr. Zhao Guanghui is expected to introduce the latest digital twin practice of China's transportation to the world with a vivid story, spread the voice of China's transportation power, and enable the world to understand that Rome is not built in a day, and "strengthening China with strong transportation network" is not built in a day.

Ni Jun
(Tenured Professor of the School of Engineering of the University of Michigan,
Chairman of the Future Manufacturing Committee of the World Economic Forum,
Winner of International Science and Technology Cooperation of the People's Republic of China, as well as President, Professor, and Doctoral Supervisor of the School of Future Science and Technology of Shanghai Jiaotong University)

Preface

In 2011, the term "digital twin" was born in the research sharing of the US Air Force Research Laboratory, which represents the beginning of a digital power to change the world and change the era. In 2021, digital twin has spread all over the world and become the core development force of various industries in the world. It has accelerated the pace of world development and created remarkable achievements one by one by making the real world into a binary system, sampling, quantifying, and coding real things, and creating all elements in the virtual space.

In the process of digital twin changing the development of the world, its application in the transportation industry is the most in-depth, and the development brought by it to the transportation field is the most prominent. Nowadays, digital twin has not only become an important sign of intelligent development in the transportation developed countries, such as the United States, Japan, and Germany, but also driven the economic progress of these countries. China's transportation has also begun to enter the digital and intelligent upgrading stage since the Ministry of Transport issued the "Outline for the Development Plan of Digital Transportation" in 2019. The digital twin is then deeply integrated with the development of China's transportation industry, leading China's transportation gradually to a world-class level.

We have formed this book after three years of unremitting efforts with the support of the monograph fund of Guizhou University of Finance and Economics. On the basis of this trend of China's transportation development, this book comprehensively analyzes and elaborates the strategies, approaches, and modes of digital twin enabling China's transportation high-quality development. This book is divided into 4 articles with a total of 22 chapters. The trend of digital twin driving the rapid development of global transportation is described; the application methods and results of international digital twin transportation are analyzed; after that, it explains the path of digital twin to help China to achieve the goal of "strengthening country with strong transportation network" in detail and finally summarizes the direction and key points of the future development of digital twin.

This book can be regarded as a white paper on the integration of digital twin in China's transportation field and can also be regarded as a guide for the direction change of the transportation industry in the digital era. I believe that practitioners in the field of transportation, managers in the field of transportation, relevant entrepreneurs in the field of transportation, and readers interested in transportation technology will benefit a lot in the process of reading this book and fully feel the strong pace of China's transportation industry supported by digital power.

The author has collected a large number of classic cases of domestic and foreign digital twin and transportation integration in order to improve the readability of the content of this book; the author read a large number of policies related to the development of digital transportation and analyzed and interpreted the core points of the policy in order to ensure the correctness of this book; the author disassembled each application mode of digital twin in detail and carried out a summary and analysis of relevance based on the actual situation in China in order to enhance the practicality of the knowledge points in this book. It is believed that the publication of this book can promote the development of China's transportation industry, can make up for the knowledge gap of the qualitative change of digital twin in China's transportation development, and let the readers learn about the theoretical basis of the integration of digital twin and transportation.

In a word, the compilation and publication of this book is not only an analysis of the path to achieve the strategic goal of strengthening China with strong transportation network but also an interpretation of the global cutting-edge digital technology. It is believed that this book can bring more surprises to readers and contribute to the development of transportation in China.

Finally, due to the limited level of the author, there are inevitably deficiencies and omissions in this book. The author urges readers and friends to criticize and correct it. The author is very grateful.

Contents

Introduction

1. The Origin of Digital Twin Transportation Development Theory

Combining the theory of digital twin and transportation development, the first step is to recognize that the concept of digital twin is a new concept originated from the industry, and its official definition is more complicated, that is, fully utilizing physical model, sensor update, operation history, and other data, integrating multi-science, multi-physical quantity, multi-scale, multi-probability simulation process, and completing the mapping in virtual space. Thus, it reflects the whole lifecycle process of the corresponding physical equipment. Simply put, digital twin is to create a digital version of a "clone" based on a device or system. Its conceptual level combines the highlights of the transportation development theory and lies in its three characteristics of "whole lifecycle process", "real time", and "two way". The whole lifecycle refers to the whole cycle of product design, development, manufacturing, service, maintenance, and even scrap recycling, which is limited to not only helping make products better but also help users use products better; real time refers to the comprehensive real-time or quasi real-time connection between the noumenon and the twin, which is not completely independent; two way refers to the data flow between the ontology and the twin which can be two-way, not only can the ontology output data to the digital twin but also the twin can feed back information to the ontology. The management of transportation level can take further actions and intervention to the noumenon according to the information fed back by the twin. From the concept extension, it can be seen that the digital twin theory, by virtue of twin urban transportation, the application of digital twin in other aspects, the application cases of digital twin, the challenges and development trends faced by digital twin technology, and the industrial policies of digital economy, to a certain extent, is and will change the underlying logic of the transportation industry, and moreover, as an important technical point representing the future transportation field, it continuously enables the high-quality development of transportation to master its technical framework (Chen Gen, 2020).

The combination of digital twin and transportation development theory also comes from the integration of digital twin and the concept of smart city. 5G ushered in the era of Internet of Everything, enabling a closer connection between the cloud and the end. This means that more data will be collected and aggregated. These data can help build a stronger digital twin. For example, a digital twin city. Further, the transportation system in digital twin cities is the first to bear the brunt. Under the background of smart city, the operation status of infrastructure (water, electricity, gas, as well as transportation, etc.) and the allocation of municipal resources (police, medical, as well as fire, etc.) will be collected through sensors, cameras, digital subsystems, and transmitted to the cloud through Internet of Things technologies including 5G. Based on these data and urban models, urban transportation managers construct digital twin to manage the transportation status of the city in minutes and seconds more efficiently. The digital twin transportation theory provides reference for the industries related to smart cities, and provides a new idea, new model, new approach and new direction for the construction of smart cities in various regions. The precise mapping and twinning of the two worlds of digital urban transportation and physical urban transportation load full transportation data resources on the digital urban model to achieve data driven panoramic visualization and dynamic and efficient management. At the same time, transportation adjustment decisions and schemes are simulated in virtual space and executed in the real world to serve the real world. The combination of digital twin transportation and smart city will have a more acute perception of urban pulse, a more comprehensive data integration ability, a more thorough insight into urban operation rules, and a stronger prediction and simulation ability. In the future, a new form of highly intelligent urban transportation development will be formed with virtual reality coordination, in-depth learning, self-optimization, and endogenous development. Digital twin will bring greater returns in the field of transportation under the integration of smart cities, and of course, the deployment of urban digital twin will be more difficult (Gao Yanli and Chen Cai, 2019).

The development of digital twin transportation development theory originated from the macro policy background of domestic large-scale infrastructure construction, and the two are closely related and associated. Infrastructure engineering is an important application area of digital twin. Especially for China, the "infrastructure mania", the introduction of digital twin is more significant, and behind the infrastructure projects, it is closely related to transportation and its theoretical support (Du Mingfang and Xing Chunxiao, 2021). The performance of the specific theory extension is that before the construction of infrastructure such as expressways and bridges, the digital modeling of the project is completed, and then the project is simulated in the virtual digital space to evaluate the structure and bearing capacity of the transportation engineering, and the flow data can also be imported to evaluate whether the project can meet the needs after it is put into use. It is also possible to assess whether the project can bear the pressure of special situations in the maintenance phase, as well as to monitor the potential accident risks after the delivery of the project. Digital twin transportation not only is the upgrading of technical means but also changes the collaborative mode of infrastructure projects from design, construction to later operation and maintenance under the trend of large-scale infrastructure construction; it solves the problem that data changes cannot be fed back in the past and makes the whole transportation status based on a data integration platform to get the latest data update in real time (An Xiaopeng, 2021). The academic community generally believes that the next ten years will be the installation period of digital infrastructure. The imagination space of the new infrastructure driven by digital twin in the transportation field is very broad in the era of accelerated iteration of digital transportation, and it will be the next wind vent of the industry.

The digital twin transportation theory also ushered in the accelerated transformation of related theories based on the background of Industry 4.0. The core requirements of Industry 4.0 in the field of transportation are digitalization. The digital twin gives full play to a series of digital solutions such as model-based transportation load solutions, model-based classification and modular management solutions, model-based software whole lifecycle management solutions, model-based transportation cost management solutions, model-based innovation and verification solutions for transportation reconstruction and model-based closed-loop compound comprehensive solutions; relying on massive data, it integrates multi-disciplinary digital technologies such as virtual reality, simulation, big data, Internet of Things, artificial intelligence, and blockchain, and records, simulates, and connects in real time through two-way mapping, dynamic interaction, and real-time connection between the real world and virtual space, predicts the operation track of the whole lifecycle of the transportation object, and realizes the allocation of massive data information and resources in the digital transportation system advocated by the Industrial 4.0 policy (Christian Manzai and Linus Schlaugipna, 2021). The advanced information, communication, electronic control, and computer processing technology will be effectively integrated and applied to the transportation system to improve the operation efficiency of the transportation system, in the event that the intelligent transportation system driven by Industry 4.0 is effectively combined with digital twin technology. At present, the relevant theoretical research on intelligent transportation system is mainly applied to urban transportation and expressway. However, the strategy of "Internet +" and coordinated development of various transportation modes has been proposed with the increasing development of China's transportation industry, such as smart city rail transit, smart high-speed rail, smart aviation, and smart ports (Ceng Yue and Zhu Liangcai, 2020). The theory of digital twin transportation development under Industry 4.0 needs more demonstration and practical cases to prove in order to meet the development needs of science and technology and industry.

In general, digital twin is based on the work of Professor Michael Grieves of the University of Michigan, a concept that has since gained widespread currency and is still used today. The theory and method of digital twin itself have gone through decades of development. Since the 1960s, when the Apollo landing program of the National Aeronautics and Space Administration (NASA) began, digital twin technology has been widely applied in various industries. However, the origin of the theory of digital twin combined with the field of transportation has been widely discussed and mentioned in the past 5 years. Its theoretical evolution has experienced the breaking circle of digital twin concept from the industrial field, gradually known and recognized by the domestic academic community along the Dongfeng of the industrial 4.0 concept and finally realized the initial integration of theory and practice relying on the 5G, the concept of smart city, and the background of domestic infrastructure. The relevant research sub-divided in the field of digital twin transportation is still relatively thin objectively at present, and the academic research with higher academic value and penetrating into this subdivided concept for further exploration is relatively scarce. Although some unique literature references can be found in the limited academic disputes at present, the overall research space can be said to have been opened with China attaching more and more importance to the digital economy with China, and attaching importance to the combination of science and technology and the competition at the people's livelihood level and quickly opening up, the traceability of the above digital twin transportation development theory is only the starting point, and the future may be expected.

2. Digital Twin Transportation Leads the Modern Development of Transportation

Digital twin transportation is becoming an important application scenario of digital twin and continuously reflects its core value. Its overall direction is to continuously enable the 5G, digital, and intelligent transportation under the premise of realizing the modernization of transportation, anchor a broad imagination space, and gradually overcome the realization in the end.

In fact, digital twin transportation is to solve a series of decision-making problems encountered in transportation engineering by using digital twin technology to help the modernization of transportation. In essence, digital twin transportation is a digital file of transportation systems that should help optimize the history and current behavior of physical objects or processes in transportation. The digital model is based on large-scale, cumulative, real-time, real transportation world data measurement across a series of dimensions. Therefore, digital twin transportation must rely on high-precision transportation detector data. Digital twin transportation is a technical mean to create a virtual entity (including transportation infrastructure, transportation participants (motor vehicles, non-motor vehicles, and pedestrians) and transportation environment) of various transportation physical entities in the transportation system in a digital way means to simulate, verify, predict, and control the whole life-cycle process and the whole traffic operation process of all kinds of traffic physical entities with the help of historical data, real-time data, and operation methods (Song Xiaofeng, 2020). Therefore, the real digital twin technology can not only accurately express the real transportation world in a digital way but also simulate/analyze/virtually test the problems or unknown problems in the real transportation world. In addition, it can go back to the past, solve problems, predict the future, reduce failures, and realize interaction with the real transportation system. The efficiency of transportation system in the real world will be improved to a new height, so as to realize the integration of modernization and digital development of transportation.

In response to the "Opinions on Promoting the Modernization of Transportation Governance System and Governance Capacity", the goal of transportation modernization is highly compared with the application goal of digital twin transportation technology. The first is to establish and improve the legal system of transportation. On the basis of the important instructions of general secretary Xi Jinping on "rule of law and reform are the two wings of the bird and the two wheels of the car", the implementation of comprehensive rectification of transportation governance issues in accordance with the law, especially the rule of law innovation and secondary clarification under the trend of digital transportation needs to be supported by digital technology. The second is corresponding to the three governance dimensions of government, market, and society to improve the transportation administrative management system, improve the transportation market governance system, and improve the transportation social collaborative governance system. Digital twin mainly focuses on technical output and technical support. The third is to establish and improve the policy system for high-quality development of transportation infrastructure, improve the policy system for transportation travel protection, and establish and improve the modern logistics supply chain system in response to the three major responsibilities and tasks of infrastructure construction, travel service, and freight logistics. The technology behind these systems relies on digital twinning to make each other work. The unity of system and technology is reflected in the application level of digital twin transportation. The fourth is to improve the transportation safety and emergency management system, improve the transportation technology innovation system, improve the transportation green development system, and improve the transportation open cooperation system corresponding to the governance concept and value orientation of the four aspects of safety, wisdom, green, and open. Digital twin will take the lead in solving the two key points of safety and wisdom, and the two aspects of green and open need the extension of related technologies. The fifth is to improve the high-quality transportation talent system and consolidate the basic guarantee for the modernization of industry governance. An important direction of transportation talent supplement is to fill the digital talent gap and to recruit some high-tech talents who are familiar with the digital twin technology into the transportation industry, which will be an important assessment and consideration indicator for the modernization of transportation development in various regions. The sixth is to adhere to and strengthen the party's leadership in the modernization of transportation governance, implement a comprehensive and strict party governance system, and improve the authoritative and efficient system implementation mechanism. Digital transportation management has a more efficient solution to this problem. The application of digital twin technology can achieve the efficient realization of party-managed transportation through two platforms: virtual and real (Geng Yanbin *et al.*, 2019).

In the process of digital twin combined with the modernization of transportation, it should also be objectively judged that it is a long process that needs time and technology investment accumulation. Some problems of the existing related products show the misunderstanding and the wrong direction of the digital twin to help the modernization of transportation. First, good looking interface does not exactly mean digital twin transportation. The beautiful graphic display interface is created by 3D or even VR/AR and other audio-visual means to present the beautiful scenery of "cars in the middle of the painting, people walking in the scenery", as long as the vehicles are three-dimensional, the characters are voluptuous, the buildings are beautiful, and the scenery is roamed, it can be called twin. Second, simulation does not exactly mean digital twin. The problem is that the product claims to use real-time transportation flow data, which can completely reproduce the motion track and motion characteristics of all transportation objects in the video and can maintain synchronous reproduction with all physical entities within the video field of vision. It is believed that being able to reproduce in real time is modern digital twin. Third, scene controllable does not exactly mean digital twin. It is often publicized that real-time simulation scene reproduction can be achieved, and transportation operation parameters and scene environment can be modified according to different requirements. This situation can be controlled in addition to viewing, and it is easier to produce the false impression that such technology has realized the transportation digital twin (Zhang Jingtao, 2020). The development of digital twin technology is bound to lead the development process of transportation modernization, but we should also recognize the progress of the realization of objective technology. However, we should neither overstate the value of digital twin in the transportation industry nor underestimate the huge development potential and broad technological reality it contains.

In general, digital twin transportation can only be said to lead the way at present, and there is still a long way to go to truly achieve technological change. If we want to achieve full fidelity mapping of transportation system, just like achieving the same effect as *Avatar*, in terms of the current digital technology level, it is difficult to achieve the picture drawn by digital twin transportation in the next 20 years. At present, more attempts of transportation modernization are still at the stage of transportation simulation technology, that is, the application of transportation simulation hardware and transportation simulation software through simulation experiments, with the help of certain numerical calculations and problem-solving, the model technology reflecting transportation system behavior or process is a method to simulate the transportation flow by transforming the transportation model which includes some certainty rules and some transportation mechanisms into software. The modernization of transportation requires transportation simulation technology as the core technology to create and operate digital twin transportation, which is the basis for data interaction and integration of digital twin transportation (Wang Qiang, 2020). On this basis, the digital twin transportation relies on and integrates other new technologies, and is online together with the perception data to ensure its authenticity, real time, and closed loop, so as to achieve a breakthrough in the real digital twin 1.0 stage of the domestic transportation industry, and further look forward to the digital twin technology to explore more diversified and digital technology implementation and management level iteration for the transportation field.

3. Global Trend of Digital Twin Transportation Development

Looking at the global transportation development and management situation represented by digital twin, it is not difficult to see that the information transportation technology represented by the developed countries of transportation is forming a convergence effect, that is, the United States, Japan, Europe, and other countries value more about it, and the technology iteration has been far ahead of other countries. However, it is difficult to significantly change the objective situation of backward core technology in the short term.

In contrast, the representative information systems of the United States include TRAVTEK and FASTTRAC. TRAVTEK is composed of transportation management center, information and service center, and vehicles equipped with navigation devices for the purpose of real-time route guidance and service information system practicality. The transportation management center is responsible for the collection, management, and provision of road transportation information, as well as the information management necessary for the system operation; the vehicle navigation device is composed of three functions: vehicle position determination, route selection, as well as interface. FASTTRAC is an intelligent transportation system (ITS) project which combines advanced transportation management system (ATMS) and advanced transportation information system (ATIS) technology. It plans to conduct tests to unify experimental vehicles and information control methods, i.e. to optimize signal control and green signal according to waiting time measured by vehicles (Hu Xiaoli, 2020). Japan attaches great importance to the technical implementation of twin virtual system in transportation control and management. The VICS center collects transportation information from both transportation managers and road managers through the Japan road transportation communication center. There are five types of information provided by VICS: transportation congestion information, required time information, transportation obstacle information, transportation control information, and parking lot information. ATIS is an advanced transportation information service system. Its communication media are telephone lines (wireless, wired). Users of transportation information can receive multi-media map information and text information as required through the navigation device loaded on the vehicle or their own home and office computers (Li Zhe, 2019). Europe is represented by EUROSCOUT, which is a dynamic route guidance system with infrared beacon as the media. The infrared communication between vehicles and beacon is two-way, and the vehicle becomes a probe to transmit transportation information data such as travel time, queuing time, and OD information to the central guidance computer, forming a two-way interconnection of dynamic and static combination, physical road conditions, and virtual computer simulation (Zhu Guozhen, 2019).

From a global perspective, Siemens of Germany is currently the best in the field of digital twin transportation applications. Among them, Aimsun, a software company, is the latest member of the Siemens

Mobility family. It is a global transportation modeling company, building digital models of transportation networks and simulating the movement of people and goods. This identifies weaknesses in the operation of the network and allows the solution to be tested to optimize transportation. Not only can local intersections be tested, as Aimsun's solutions not only simulate road transportation but also assess a variety of travel patterns and identify and try out potential solutions. Siemens Mobility has successfully simulated the transportation flow in the transportation control center in real time to forecast the upcoming transportation disruption and determine the best short-term mitigation strategy to minimize congestion. Centers in the United States, the United Kingdom, France, and Australia are already using the Aimsun solution to make their cities smarter. These centers use simulation applications to continuously monitor and process real-time data, simulating the movement of vehicles on a network of any size, from a single trunk road to a metropolitan area. The system operates so fast that transportation controllers can virtually test various scenarios and know which scenario is most effective to implement in the real world (Feng Bo, 2019).

The digital twin development and construction of transportation is essentially a virtual copy of its physical assets, infrastructure, vehicles, and interactions. At present, with the gradual attention and assessment of the prospect of digital twin technology by various countries, it has been used to develop and test transportation engineering solutions. However, in general, we are just beginning to realize its potential. Only under the global vision, Siemens Mobility in Germany is currently at the forefront, and China still has a broad time window to achieve corner overtaking. The premise is to recognize the technical gap with developed countries and catch up with and surpass them in an orderly manner.

4. Digital Twin Transportation Improves Transportation Technology Innovation

Why does transportation need digital twin? How does digital twin enter into the vision of information construction of transportation quickly? Transportation congestion, driving difficulties, parking difficulties, public travel inaccuracy, and other problems not only make ordinary transportation participants' headaches but also have been plague by the key livelihood issues of transportation management departments. Most of the transportation participants have also experienced such things. For example, on a road that is congested, they just walk through a congested intersection under the command of the transportation police, and within a few minutes of driving, they run to another section of the road that is not usually congested, but becomes seriously congested due to the influence of other road dredging. This reflects the main problems existing in the current transportation management: the application of discrete, information island, relatively single and fragmented treatment of transportation problems, is difficult to take the integrity of transportation into account. Intelligent transportation emerged here. In the process of project construction and operation, we pay more attention to the interconnection of transportation data and system, and emphasize the quality and effect of the overall solution. The digital twin with real-time and closed-loop features has entered the transportation field, which just makes up for the shortcomings of transportation management and control (Wu Kewei, 2019).

With the development of digital twin and the continuous penetration of Internet technology enterprises, digital twin has gradually penetrated into various segments of the transportation industry: project construction planning, urban transportation, expressway, vehicle road collaboration, as well as automatic driving, etc. At the level of urban transportation construction and upgrading, in the digital twin space, we can use the collected data to carry out model construction, conduct simulation according to the corresponding needs, automatically calculate the density of urban road network, trunk network density, per capita road area and other contents, analyze and evaluate the situation of road network, and automatically optimize the solution to obtain the optimal effect, so as to improve the road transportation efficiency. Transportation intersections are the concentrated places of urban transportation accidents, bringing together a variety of transportation participants and urban transportation problems. Therefore, the transportation intersection is one of the scenarios with the most transportation perception equipment. Due to this, as the main strategic attack place for players in the current transportation industry, Internet technology giants such as Baidu and Huawei have successively released relevant solutions and system architecture such as smart intersection, hologram intersection, and hologram road network, trying to take the lead in some applications of digital twin technology, and laying a foundation for the further development of vehicle-road cooperation and automatic driving which are more dependent on digital twin technology. Perception hardware manufacturers have also launched new products and solutions, trying to seize the market share and market voice driven by the expansion of digital twin. Relevant departments make overall planning through the construction of transportation data center and transportation brain in digital space, and strive to achieve the global control of urban transportation (Zheng Guorong, 2020).

Vehicle road coordination and automatic driving are the image embodiment of digital twin technology development and application innovation. Through the digital twin technology, the transportation data collected by the perception equipment is combined with the high-precision map, and the automatic driving digital model is built. The real world 1:1 is restored to the digital scene in the digital twin, and the operation law of the physical world is restored, so as to meet the training needs of the artificial intelligence algorithm in the intelligent driving scene, thus improving the efficiency and safety of training. Digital twin relies on the breakthrough of AI and other technologies in the professional field. Therefore, Internet technology giants are the main construction promoters in the field of vehicle road collaboration and automatic driving. Baidu, Tencent, and other technology companies have entered the construction of this field one after another and continue to increase investment (Wu Zhaohui *et al.*, 2021).

At the same time, it should be pointed out that it is necessary to treat the effective benefits of China's digital twin transportation to the level of transportation technology innovation in a dialectic way because it is currently driven by technology, while there are still many problems to be solved. First, the construction and development are not clear, the application scenarios are relatively single and not deep enough, and the purpose of construction promotion is lack. Each transportation sub-division field has launched project planning and construction based on digital twin. However, in the initial planning and design, it still stays in solving the transportation problems in a single scene, and the transportation problems in specific application scenarios are not solved deeply enough, lacking the overall planning of road transportation and without purpose for the final presentation of the construction effect. In some regions, the concept and construction direction of digital twin are not clear, and they follow the construction of digital twin projects in a blind manner, thus breaking away from the actual business needs of the projects, and the digital twin transportation remains in name only (Wu Fubiao, 2020). The second is the lack of clear construction standards and norms. The construction of digital twin is a comprehensive project covering the entire industry field. However, the demand for responsible parties behind the project is different due to the real world and different fields. The repeated construction of the same region often occurs, and the project construction standards and specifications applying data and system architecture are not unified. In the subsequent collaborative processing and integrated application of the project, the issue of who shall prevail for unification appears (Xiao Jinghua *et al.*, 2021). The third is that the key technology has the risk of neck sticking, which requires innovation and breakthrough. Digital twin was born in the explosive development of many advanced science and technology, which depends on the rapid development of a variety of perception means. However, the development and integration of many key technologies involved in the current digital twin, such as new surveying and mapping, logo perception, collaborative computing, full element expression, and simulation, need to be strengthened. The maturity of massive data loading technology, cloud edge computing collaborative technology, and simulation technology also needs to be improved, and artificial intelligence and edge computing also lack the ability to rapidly analyze and process dynamic data (Huang Haisong *et al.*, 2020). The fourth is that the interactive model of twin scene and dynamic transportation scene is not mature enough. The application of digital twin in the field of transportation is still at the primary level, and the algorithms and models in various professional fields need to be further developed; the maturity is not high, so the interaction between twin scenes and actual dynamic transportation is not enough, the value of digital space simulation and situation prediction is far from release, and many applications eventually become traditional information construction projects (Tao Fei *et al.*, 2018).

5. Digital Twin Transportation Changes Transportation Life Cycle

With the new iteration of digital technology, customers' demand for product experience is increasingly diversified. As a comprehensive demonstration of the development of industrial Internet and intelligent manufacturing, digital twin has raised its expectations to an unprecedented level. It is expected that the application of digital twin technology will change the whole lifecycle of transportation, help managers, travelers, and market participants to better create and use products, and bring efficiency improvement and cost reduction.

The whole lifecycle reform of transportation can rely on the practice of digital twin core technology, use three-dimensional digital technology to build a shared model with intelligent tools, and make the data information of each stage in the whole lifecycle of transportation integrated (comprehensive integration of software and hardware systems), fused (integration and integration of application data), linked (real-time linkage of all professional sub-systems, twin interaction), and visualized (real visual, interactive visual). Its essence is to integrate the data resources of the existing information systems of various departments of transportation, deeply integrate the application of 5G, big data, cloud computing, AI, integrated communication, and other cutting-edge technologies, and organically combine the information, technology, equipment, and transportation management needs, covering various business areas such as the overall transportation situation monitoring, regional transportation monitoring, hub situation monitoring, transportation population distribution analysis, travel demand analysis, and duty residence contact analysis, it has enabled user business applications to achieve "intelligent perception, intelligent disposal, intelligent assessment and intelligent improvement", effectively improving the efficiency of cross department decision-making and resource coordination (Dong Hua, 2019).

In recent years, with the rapid development of economic and social construction in China, the technological iteration at the transportation level is realizing the upgrading and transformation from the transportation lifecycle of 1.0–2.0, and from the traditional transportation dualism to the diversified transformation and expansion of transportation in the new era. During the period from 2010 to 2018, the scale of China's intelligent transportation market has grown at a compound growth rate of 23.33% year by year. In 2018, the market scale of China's intelligent transportation management system industry reached RMB 72 billion. As of 2018, more than 200 large and medium-sized cities in China have established urban transportation command centers, and the scale of urban intelligent transportation investment has reached RMB 45 billion. As an important development direction of intelligent transportation, digital twin is also gradually reflecting its significance and value of changing the whole lifecycle construction of transportation industry. Based on the practical problems faced by the traditional transportation in the past, such as the short verification cycle of the management system, the urgent upgrade cycle, and the frequent update of the road management and control rules, the digital twin technology with pilot intelligence can help solve the key problems, such as the management platform and the scheduling coordination resources, and help reduce the cost and increase the efficiency of China's transportation in the new era, providing an overall solution for the whole lifecycle of transportation (Luo Yu, 2019).

Digital twin emphasizes the interconnection of virtual environment and real data. In the digital era, everything can be described in the form of data and is parameterized and modeled according to the inherent rules. In the application of intelligent transportation, digital twin takes data as the basis and model as the core, injects data and model into the interior of the three-dimensional reconstructed urban transportation network, carries out macro, meso, and micro simulation, and constructs the digital twin transportation engine, which can be applied to the urban static auxiliary planning and urban dynamic road performance, or the evaluation of transportation plans for specific scenarios. For example, in the urban transportation scene, more than 128 targets are tracked with the use of over-the-horizon radar fusing strong twin perception scheme, the use of long focus cameras, fish eye cameras, and MMW radar, a variety of types of motor vehicles and non-motor vehicles detection is carried out, transportation simulation and prediction are made by relying on the ability of large-scale parallel acceleration of Tencent cloud, the transportation flow data is visualized by using game engine technology, and the transportation flow data is imported into the three-dimensional reconstruction environment to form a transportation thermal map, which can intuitively show the transportation conditions of the whole city and run through the whole cycle of transportation management (Zhang Long, 2018). From virtual simulation to digital twin, each link and each terminal in the transportation can realize the flow of data and help the optimization of transportation construction, management, operation, and service. With the integration of future and automatic driving technology, digital twin will likely continue to provide digital toolbox for intelligent vehicles and even intelligent transportation industry to achieve a more efficient and convenient transportation in the new century.

China's "Fourteenth Five-Year Plan for National Economic and Social Development of the People's Republic of China and the Outline of Long-term Goals for 2035" clearly emphasizes the need to accelerate the construction of strengthening country with strong transportation network. The programmatic requirements put forward are that we need to closely follow the goal of building a new development pattern, focus on building a modern high-quality comprehensive three-dimensional transportation network, and strive to make transportation a solid support for the coordinated development of the modern industrial system, an important link for the mutual promotion of internal and external economic cycles, a guarantee cornerstone for the safety and stability of the supply chain of the industry chain, and help build a solid chassis of the national economic cycle. The development of various modes of transportation must be driven by innovation and technology instead of traditional elements, so as to build a transportation power that satisfies the people and has strong guarantee and is at the forefront of the world (Zhang Yanli *et al.*, 2020). With the increasingly complex domestic transportation scenarios, people are increasingly calling for the safety, convenience, and environmental protection of daily travel. It will become more urgent for intelligent transportation to achieve global management, synchronous visibility, and virtual reality interaction. The whole lifecycle of digital twin-enabled intelligent transportation is bound to be a new direction of future transportation development.

The transportation power strategy cannot be separated from the support of emerging scientific and technological forces. We have "Internet +", big data, blockchain, and artificial intelligence technology and are now ushering in digital twin technology. This book discusses the innovative logic of the high-quality development of digital twin transportation, which will boost the national transportation strategy to move toward a higher, broader, and deeper development space and provide more satisfied transportation and social services for the people.

Chapter 1

The Emerging Technology to Boost the Development of the Era to a Higher and Deeper Space

In the digital era where everything can be digitalized, the development of digital technology is constantly changing the quality of life of the public. In recent years, a new digital technology has created a virtual space which is completely the same as the real world by sampling, quantifying, and coding the real things through the real world's dualization. In this virtual space, the development and changes of the real world can be demonstrated and predicted, and the development of the real world to a higher and deeper space will be visualized accordingly. The technology to complete this feat is called "digital twin".

1.1 Overview of Digital Twin Technology

Digital twin is a hot cutting-edge technology in recent years. It realizes the real replication and evolution of the real world in digital space and completes the digital twin connection of real things in virtual space under the background of Internet of Things. This technology greatly promotes the improvement and development, upgrading and iteration of things since it can provide dynamic demonstration and corresponding changes of the state information of real things.

As of 2021, digital twin has demonstrated great value in seven areas of public life and vigorously promoted the development of the times:

(1) Smart home management.
(2) Remote control of intelligent devices.
(3) Industrial equipment monitoring and command field.
(4) Smart city construction and management.
(5) Public health testing and management.
(6) Exploration of unknown areas of digital technology.
(7) Detection and management of human brain activity.

According to the prediction of industrial Internet experts, in 2022, 85% of the Internet of Things platforms will use digital twin technology to achieve intelligent monitoring, and digital twin will become the core technology of the development of the Internet of Things in the next 10 years.

1.1.1 *Origin and meaning of digital twin technology*

The idea of digital twin was first proposed by Professor Michael Greaves of the University of Michigan in the United States. However, he initially named the digital twin "Information Mirroring Model" and then evolved into "digital twin" with the continuous maturity of technical ideas. Digital twin also has two other names, namely "digital twin" and "digital mapping", which are defined as digital modeling and replication of real things.

Digital twin is based on the development of MBD technology. Some enterprises need to establish a large number of physical models and digital models based on engineering systems when implementing large-scale projects. The development of these modeling technologies has laid a solid foundation for digital twin.

1.1.2 *Principle of digital twin technology*

In March 2011, the United States Air Force Research Laboratory clearly proposed the technical name of "digital twin" in a speech, which was the first appearance of "digital twin". At the same time, National Aeronautics and Space Administration (NASA) also began to focus on digital twin technology and applied it to the field of aerospace equipment development. However, in the recent 10 years of development, the contribution of digital twin to the National Aeronautics and Space Administration has not been significant but has been fully developed by the United States Department of Defense.

In 2012, the National Aeronautics and Space Administration took the lead in providing a comprehensive explanation of the concept of digital twin. National Aeronautics and Space Administration explains that "digital twin refers to the integration of multi-disciplinary and multi-scale simulation processes by making full use of physical models, sensors, operating history and other data, which serves as a mirror image of physical products in virtual space and reflects the whole lifecycle process of corresponding physical products".

With the development of digital technology, the definition of digital twin has become more and more clear. Nowadays, digital twin has been recognized as a simulation process that integrates multi-disciplinary, multi-physical quantity, multi-scale, and multi-probability by using physical models, sensor updates, operation history, and other data. It can complete the mapping of real things in virtual space and reflect the whole lifecycle process of real equipment in virtual space. Combined with the simulation application of new generation information technology in virtual space, digital twin can analyze and predict the optimal results of the operation and application of various real physical equipment, so as to drive the improvement and improvement of real-world physical equipment.

It can be said that the essence of digital twin is to map real-world things in digital space, which will show great value in future development.

1.1.3 *Basic components of digital twin technology*

In 2011, Professor Michael Greaves clearly put forward three key components of digital twin in his book "Almost Perfect: Driving Innovation and Lean Products through PLM": ① physical products in real space; ② virtual products in digital space; ③ the data and information interface between real space and digital space. Among the three basic components of digital twin, the data and information interface between real space and digital space has become the key to determining the level of technology.

In 2016, Siemens group stood at the Siemens industrial forum from the perspective of "digital twin" and explained the basic components of digital twin from different perspectives. Siemens Group believes that the three key components of digital twin are: ① product digital twin; ② digital twin of production process; ③ device digital twin. Through this interpretation, digital twin can completely and truly show the entire enterprise in the digital space.

Soon after, Professor Zhuang Cunbo of Beijing Institute of Technology in China also explained the composition of digital twin in detail from the perspective of products. Professor Zhuang Cunbo believes that from the perspective of product design and research and development, digital twin mainly includes product design data, product process data, product manufacturing data, product service data, as well as product retirement and scrap data and other components.

In fact, the three explanations of digital twin are reasonable and academic. Taking Siemens Group's interpretation of digital twin components as an example, digital twin can directly and comprehensively replicate any production enterprise in digital space, and the whole lifecycle development of enterprises can also be clearly reflected in digital space.

Either way, the basic composition of digital twin will not be separated from the three basic elements of real space, digital space, and the connection of the two spaces.

1.1.4 *Research status of digital twin technology*

Since Professor Michael Greaves proposed the concept of digital twin in 2003, this technology has been used for the replication and transformation of physical products in virtual space. The main aspect of its application is the abstract development of physical products after application. As data can show various abstract developments in a concrete way, the real conditions of various products provided by digital twin in the virtual space are simulated, which greatly accelerates the development of core production technology in various industries.

The first to apply digital twin to the field of intelligent manufacturing is the National Aeronautics and Space Administration. In the U.S. Apollo space program, the National Aeronautics and Space Administration used digital twin technology to conduct simulation and operation analysis of the space vehicle, so as to obtain the real flight status data of the aircraft. The ground control personnel made various command decisions through data sorting.

In 2012, the National Institute of Standards and Technology of the United States put forward two concepts of digital twin technology application from different perspectives: MBD (model-based definition) and MBE (model-based enterprise), from which digital twin began to run through all aspects of product design, processing, maintenance of various industries, and deeply combined with all aspects of product manufacturing process.

In 2015, CPS (cyber-physical system) technology was officially put forward with the rapid development of network technology, and digital twin technology was also promoted to a new height by the manufacturing industry in the world. Especially in the intelligent manufacturing enterprises, digital twin has generated the best way for the enterprise's actual development and digital simulation, and also accelerated the integrated development of information technology and manufacturing industry. Digital twin began to optimize the implementation of the enterprise engineering and reduce the cost of enterprise engineering system construction.

In 2019, digital twin fully entered the development field of high-end transportation equipment in China and has shown high application value in aerospace, intelligent vehicle research and development, waterway management, and other fields.

In 2020, China Electronics Technology Standardization Research Institute of the Ministry of Industry and Information Technology of the People's Republic of China officially released the "White Paper on the Application of Digital Twin", in which the definition and characteristics of digital twin are elaborated in detail, and the research status and application fields of digital twin technology are analyzed. The analysis results show that digital twin has penetrated into six areas, such as China's intelligent manufacturing, smart city, intelligent transportation, as well as smart energy, and moreover, the technical level is constantly improving.

1.2 Development Status and Prospect of Digital Twin Technology

With the accelerated development of the new generation of information technology, the trend of social digitalization, networking, and intellectualization is becoming more and more obvious. In this process, numerous industries have shown new models and new business forms. Digital twin technology, as a core force, has gradually become the focus of attention and research from all walks of life, and its future development prospects are very broad.

In 2020, hundreds of leading enterprises in the world included digital twin into the core technology research ranks. Markets and Markets, the second largest market research consulting company in the world, forecasts that the market of digital twin technology will exceed US$15.7 billion by 2023 and will continue to rise at an annual growth rate of nearly 40%. Various phenomena indicate that digital twin technology will become the key force for the development of many industries in the future. Now, digital twin is creeping into the heart of every industry.

1.2.1 *Development trend of digital twin technology*

In 2020, Deloitte, a British professional service institution with more than 170 years of development, issued the "Technology Trend 2020 Report", which is an annual report on global technology trend issued by Deloitte company. Before that, the first 10 technical reports issued by Deloitte company accurately predicted the development trend of advanced technology in the world.

In the "Technology Trend 2020 Report", Deloitte company clearly pointed out that in the next 18~24 months, digital twin will become an important technology development trend in all major industries in the future. This forecast is derived from the analysis of hundreds of technical experts of Deloitte company on the world's top enterprises and industry leaders, which also includes the exchange and discussion between Deloitte company and well-known venture capital companies and technology suppliers.

Deloitte pointed out that in the future, digital twin is a force that cannot be ignored in the aggregation of macro forces that will affect the development of the world. It may cause disruptive changes in five aspects in the short term and stimulate key emerging trends in a number of areas.

According to the "Technology Trend 2020 Report", the development trend of digital twin technology can be divided into five aspects.

1.2.1.1 *Connect the real world and digital space*

Digital models have begun to optimize the production processes and related services of modern industries since the birth of digital technology. With the continuous improvement of digital technology, this optimization ability is constantly tending to be meticulous and comprehensive. After the birth of digital twin technology, its complex simulation and modeling capabilities have improved the operability of the digital model in an all-round way. Its main feature is to build a simulation platform in the digital space through IoT sensors and power system so that the digital model can be more dynamic and presented in front of the public.

Through the effective use of digital twin technology, major industries can improve production efficiency and optimize the maintenance system and supply chain. The most outstanding application effect of modern digital twin technology is to effectively ease the transportation congestion and improve the utilization rate of transportation resources.

It can be found that this technology can become a data driver for multiple industries and play an important role in model design, technological innovation, and other aspects as digital twin technology deepens in various fields. At the same time, because digital twin has obvious effect on the soundness and acceleration of the ecosystem of various industries, it also shows great potential.

1.2.1.2 *The awakening of digital architecture*

As the core force of the development of the times, technology has always been the focus of the development of various industries. Many people believe that digital twin has disrupted the law of technological development in the industry, but in fact, digital twin merely reordered the logic of technological development in various industries and eliminated some unnecessary trial and error links.

With the promotion of digital twin, a large number of technology architects have emerged in various industries, and this IT position has improved the market competitiveness of enterprises. In future development, the strategic position of the industry technology structure will be greatly enhanced by digital twin, or even grow into the core force of the industry.

1.2.1.3 *Predicted real development*

Digital twin is still in the auxiliary position of the core technology of various industries, but its biggest feature is to achieve digital simulation and digital simulation of various industries. In future development, digital twin can be deeply integrated with the core technologies of various industries, so as to achieve real and comprehensive prediction of industry development and accelerate the development of the industry.

1.2.1.4 *Creating the field of human experience*

The artificial intelligence began to cover all areas of public life with the development of AI technology. However, the current AI technology generally lacks "emotion", and the core of the technology is still the objective calculation of numbers. After the emergence of digital twin technology, the proportion of "emotional computing" began to increase, and the degree of emotional simulation was also rising. In future development, the emotion of AI technology will become more and more abundant, which will also depend on the support of digital twin.

1.2.1.5 *Deepening the integration of IT and finance*

The digital development of finance has become the mainstream trend. IT finance will be more accurate in exploring the future after the birth of digital twin, thus improving the security and stability of financial operation. Moreover, IT finance will be more agile in the future under the support of digital twin, bringing more competitive advantages to major enterprises.

1.2.2 *Practical value of digital twin technology*

In short, digital twin technology is a bridge between the physical world and the digital space. Digital twin carries out real-time data analysis of all physical elements to form digital copies which are presented in the virtual space for understanding, analysis, and operation. Over the years of development, digital twin technology has shown great value and was listed as one of the top 10 strategic technology trends in the field of Internet in 2019. With more and more fields being penetrated by digital twin, the actual value of this technology has gradually been demonstrated.

For example, the modern medical industry can monitor and obtain various physical indicators of patients in real time through digital models, coordinate equipment and medical personnel through sensors, and formulate a diagnosis and treatment process that more conforms to the specific situation of patients, so as to implement diagnosis and treatment measures more timely and accurately.

In 2020, according to the report of a hospital applying digital twin technology in California, the utilization rate of the emergency room of the hospital doubled, and the utilization rate of other resources of the hospital was also significantly improved. The comprehensive management cost of the hospital was 90% lower than that of 2019. In addition, with the support of digital twin technology, the emergency incident rate of the hospital's healthcare network decreased by 61%.

For example, digital twin has been widely used in the field of transportation management in China, whose main value is to optimize the urban transportation network and improve the utilization rate of transportation resources.

In general, the actual value of digital twin has been reflected in various fields, and the value is still being explored. Its value is mainly reflected in the following three aspects.

1.2.2.1 *Cost reduction*

Digital twin technology can significantly reduce research and development costs in the process of new product design in various fields. Through digital simulation of product prototypes, it can complete multiple iterations in virtual space, thus saving time cost, human cost, and trial and error cost in the real world and reducing various financial risks, making the industry development safer and more convenient.

1.2.2.2 *Improving efficiency*

Digital twin can complete the preview of enterprise operation in digital space, and then analyze the shortcomings and disadvantages of the operation mode. The whole lifecycle simulation of products is helpful for enterprises to optimize their business strategies and improve their operating efficiency and profit margin.

1.2.2.3 *Forecast maintenance*

Predictive maintenance is the greatest value of digital twin technology in various fields currently. Especially, digital twin realizes remote transportation monitoring through simulation of virtual copies and reduces the occurrence rate of things with the help of sensors in the field of transportation. The value of digital twin is not only reflected in the economic field but also used to improve the effective use of transportation resources and personal safety.

In a word, digital twin has improved the speed of world development, optimized the development process, and perceived various development problems in advance in the process of continuous development and evolution. This technology is worth exploring, and it's getting more valuable.

1.2.3 *New things brought by digital twin*

The development of digital twin technology in the past 10 years has not only changed the development pace of many industries but also brought a variety of new things to China. In other words, many industries show new forms of business and new scene because of the application of digital twin technology.

At present, the new things brought by digital twin are mainly shown in the following aspects:

(1) **New model:** The digital modeling of physical objects in the real world in virtual space becomes more and more smooth under the support of digital twin technology, and complex simulation schemes are clearly demonstrated in digital space. It becomes clear step by step from detection to evolution, from forming to material tracing, the emergence of this new model has helped a number of industries and fields to improve the depth of insight and practical reliability.
(2) **New data:** The multi-dimensional deployment of digital twin sensors has enhanced the controllability and timeliness of modern monitoring systems, especially the Internet of Things sensors placed in the equipment or within the supply chain and industrial chain, which has generated new data, constructed a new simulation system, and realized the real simulation and accurate simulation of the development of things.
(3) **Interconnection and interoperability:** In comparison with traditional digital technology, the digital twin based on monitoring and simulation has increased and enhanced the implementation and operation capabilities. This technological upgrade is mainly due to the enhancement of industrial communication standards of Internet of Things sensors. The emergence of this capability provides important technical support for the integration of multi-domain platforms.
(4) **Artificial intelligence visualization:** The combination of digital twin, AR, AI, and VR technologies has realized the visual transformation of artificial intelligence, the upgrade of artificial intelligence

has been demonstrated in a concrete way, and the "intelligence" has been greatly improved.

(5) **New technology platform:** Digital twin not only saves the construction and operation costs of each big data platform but also improves the processing capacity and load capacity of the platform. As a result, the research and development costs of various cloud platforms are getting lower and lower. A large number of new high-end platforms with complete functions are starting to appear, and the exploration boundaries of various industries and fields are continuously broadened.

1.2.4 *Future digital models and out-of-model applications*

From the current development situation of digital twin, it can be seen that digital twin will gradually penetrate into various industries around the world in the future, and it is believed that it will become the main technical support for various industries to make database decisions, revise products, upgrade services, and innovate business models in the near future. Especially in the transportation, manufacturing, public utilities, and other intensive industries, digital twin will lead the development as a pioneer, and the early practitioners will also show more advantages and characteristics.

From the perspective of long-term development, the digital models of various fields in the future will integrate the industry ecology and build a huge basic database, and the digital space will present a whole lifecycle preview model.

For example, after all cities in China have built a digital model, China's comprehensive three-dimensional transportation network can complete the whole lifecycle preview in digital space, the management and construction of transportation resources by the transportation department will be more effective, and the reasonable utilization rate of transportation will double.

This digital model can provide a more insightful decision-making basis for human beings, help human beings to more accurately predict future development, and to think about ways to overcome future suffering and cope with long-term challenges in the digital space. From this point, we can see that the contemporary efforts to digital twin are valuable and rewarding.

In addition to digital models, various areas of the world will also be positively affected. For example, in logistics, manufacturing, and other fields, the degree of mechanization will also continue to increase. With the new generation of information technology, each link of logistics and manufacturing can be tracked and monitored in real time, and the process can be adjusted according to the actual situation at any time to optimize the distribution and manufacturing effect.

Not only that, but the service field outside the model will also be greatly improved, which is reflected in not only the specific services but also the pricing analysis and channel optimization of the digital model for the products, which are the main performance of improving the services, and the application outside the model will also open up interest space for all walks of life.

For example, the monitoring enterprises can adjust the product production pace according to the digital twin preview and analysis, optimize the distribution strategy of regional markets, improve the timeliness and security of meeting customer needs, and open up a new monetization strategy for the development of enterprises, which is also the application mode outside the digital model.

1.3 Digital Space Created by Digital Twin

From September to November 2021, the “Third China Industrial Internet Competition - Industrial Internet + Digital Twin Professional Competition”, sponsored by the Ministry of Industry and Information Technology of the People’s Republic of China and jointly organized by the China Academy of Information and Communications Technology and the People’s Government of Wuhu City, was successfully launched. The technical competition aimed at promoting the integration and innovation of “Industrial Internet + Digital Twin” shows that China has paid more and more attention to the training of digital twin technical talents, and the layout of digital twin in the field of Internet in China has become more and more extensive.

In fact, the digital twin technology not only integrates perfectly with the industrial Internet, but also connects with the industrial chain and value chain of various industries in China, and promotes the transformation and upgrading of digital and intelligent industries.

The digital space created by the digital twin completely replicates the physical object, realizes the simulation mapping in the virtual space, and then realizes the comprehensive decision-making and operation optimization of the physical object through the integration of data and model.

In short, digital twin is to build a perceived and visual twin in the digital space. The emergence of this twin has strongly promoted the development of various industries.

1.3.1 *Industrial application of digital twin*

The industry in which digital twin technology was first applied is industrial manufacturing industry. With the development of the times, industrial manufacturing, transportation, Internet, and other fields have gradually developed into the main battlefield of digital twin.

In the industrial field, the R & D and design of products need repeated attempts and optimization, which requires a lot of manpower and material resources. Industrial designers can see the external changes of products and complete the dynamic observation of products in the virtual space constructed by digital twin, thus saving a lot of time, manpower and financial costs, and improving product research and development and production efficiency.

For example, most of the contemporary brand automobile manufacturers will build the motion model of parts and components in the digital space, especially the changes in the internal parts and components of the engine and the lines, which determine the quality and competitiveness of the automobile. The development of this space reduces the production and research and development costs of these enterprises.

Digital twin can also describe the production environment of the industrial chain in the virtual space and conduct targeted simulation on the production process in the production process of the industrial industry, so as to reduce the occurrence of production exceptions and production accidents.

Digital twin also plays an important role in the maintenance of the industrial industry. For example, the application of digital twin technology by the National Aeronautics and Space Administration is to improve the maintenance of the fighter. The global large-scale production enterprises will also use digital twin technology to optimize the use of equipment and extend the service life of equipment.

It can be seen that the digital twin has brought obvious cost saving and efficiency improvement to the industrial industry, especially for the transformation of traditional industrial manufacturing to modern and intelligent industry, and this technology will continue to play a more important role in the industrial industry.

1.3.2 *Application of digital twin in entertainment industry*

On 10 March 2021, Roblox, a children's game website company located in San Mateo, California, the United States, was officially listed. The stock market price increased from $45 to $70, becoming a very prominent listed brand in the field of entertainment games in recent years.

The game developed by Roblox has a major feature which allows players to create their own 3D world. The company will provide players with real-time responses to the game and basic settings for personalized data analysis. This mode of creating alternative game experience through digital tools has won the favor of modern young users, and has given them the development posture of "concept shares of the universe", which is based on the digital twin.

The emergence of digital twin makes the former sci-fi concept more and more real. This technology continuously obscures the boundary between reality and virtual. Through the real demonstration of digital space, it makes the virtual and reality increasingly integrated, and continuously promotes the process of "virtual world authenticity", which further drives the development of numerous entertainment enterprises.

For example, driven by Roblox, industry giants such as Microsoft and NVIDIA also took advantage of this trend to fully tap the value of the "metaverse" and launched their own application framework of the meta universe based on digital twin, "In the future, digital space or virtual space will be thousands of times larger than the physical world, and factories and buildings will have a digital twin to simulate and track their physical versions", added Jensen Huang, CEO of NVIDIA. This shows that the integration of entertainment industry and digital twin technology is imperative, and the concept of virtual space began to spread to the real world.

In fact, as early as 2015, digital twin technology had achieved remarkable response in the entertainment industry. In 2015, the digital twin technology was fully applied in the "If a Wish Could Be Made: Teresa Teng 20th Anniversary Virtual Person Memorial Concert" held at the Taipei Arena. Teresa Teng, a virtual space on the stage, sang a number of classic songs, and more than 7,000 audience were mesmerized. In this scene of a combination of the virtual and the real, the boundary between the digital space and the real world seems to completely disappear. Fei Yuqing, a guest on the stage, even sang with Teresa Teng in the virtual space on the same stage.

It has to be said that the application of digital twin technology in the entertainment industry has achieved remarkable results. Its main application is to eliminate the boundaries between the virtual world and the real world. In the process of improving the sense of mass entertainment experience, the value of digital twin technology is also growing.

1.3.3 *Transportation application of digital twin*

Digital twin has been widely used in the field of transportation in China. For example, in 2021, Guiyang pilot digital twin transportation system first became the real-time focus in the field of transportation. This technology plays an important role in the construction of intelligent transportation and the development of smart cities in China.

At present, the application of digital twin in the field of transportation is mainly reflected in the detection and management of people, vehicles, and roads. Through the modeling and analysis of virtual data, the deduction and operation of transportation development are carried out, and the improvement and construction of a comprehensive three-dimensional transportation network are optimized.

Taking the digital twin transportation system pilot in Guiyang as an example, it is through the combination of monitoring equipment and MMW radar equipment that vehicle and pedestrian hologram perception is carried out, and the overall mapping of urban transportation system is completed in digital space, and real-time analysis and tracking are carried out for the model data of digital space reality, so as to finally solve the problems of transportation resource waste and signal system simplification, and moreover, it also improves the predictability of transportation time and response speed.

From this, it can be seen that the stable and rapid development of China's transportation field in the future, as well as the guarantee of people's daily travel safety and convenience, cannot be separated from the support of digital twin technology, which will also become a new direction of the development of intelligent transportation and intelligent transportation.

At present, the application of digital twin technology in the field of transportation mainly shows three characteristics:

(1) The digital twin technology realizes the synchronous visual management of the transportation network, model development deduction, and data-driven decision-making, which improves and enhances the safety and stability of the transportation network.

(2) Digital twin technology has realized the real simulation of complex transportation environment and accelerated the landing of intelligent driving and automatic driving technology. Based on real environment data modeling, the digital twin technology can improve the response ability of intelligent driving and automatic driving to the external environment and emergencies in the virtual space, so as to accelerate the safe landing of intelligent driving and automatic driving.

(3) Digital twin technology provides a global perspective for transportation management. The digital twin technology accurately quantifies the daily situation of urban transportation by the real simulation of the complex urban transportation network, and the urban transportation management department can realize the real-time dynamic management and overall management of urban transportation in accordance with the data changes.

Digital twin technology is gradually becoming the cutting-edge trend of China's transportation development and has provided a steady stream of power for the strategy of transportation power. In the process of intelligent transportation construction, digital twin technology provides more new ideas and new ways. It is believed that in the near future, China's digital twin technology will form a perfect operating system in the transportation field.

1.3.4 *Technology application of digital twin*

The digital technology has gradually developed into a new type of productivity after the advent of the digital era, especially in the field of science and technology, AI, VR, AR, and other technologies have brought technological changes and created a new market space. China's Internet market has shown a vigorous development trend in this process, and digital twin technology has also been embodied as an important production element, which effectively provides the fitness of scientific and technological products and users and has become the focus of the development of modern digital economy.

The application of digital twin in the field of science and technology is mainly to promote the development of science and technology and meet the higher needs of the public. The digital twin technology not only completes the simulation of the physical world but also shows the future world to human in advance through the development of virtual space. It takes the physical world as the prototype technology, rehearses the future development in advance, and visualizes various bold imaginations.

For example, workers in the future society will gradually change from human to machine, social production will be developed from meeting demand to "meeting demand + creating demand", the labor process has changed from human led to data led, and the pace and process of social development will continue to break through human cognition.

These predictions of future social development are presented in digital space, and human life experience is constantly improving under the guidance of this trend.

Generally speaking, the application of digital twin in the field of science and technology is mainly to predict the future and development. This technology is conducive to the overall layout of social development and the formulation of national development strategies. The overall layout of social resources will be more strategic and forward-looking supported by high-end scientific and technological forces.

1.4 Standard System Constructed by Digital Twin

Digital twin, as a key technology to promote the digital and intelligent development of the times, has attracted much attention in various fields in the world. In the process of implementation of digital twin research, the standard system constructed by it determines the difference of technical effect.

The digital twin is mainly divided into five plates, namely physical plate, data plate, model plate, preview plate, as well as decision-making service plate, from the perspective of digital twin architecture. The content of these five major plates determines the application direction of digital twin and the actual application effect.

1.4.1 *Construction of digital twin physical plate*

In the standard system of digital twin technology construction, the construction of physical plate is the basic plate that supports the foundation and plays an important role. In this plate, digital twin makes digital analysis of physical objects in the real world, i.e. physical entities, and makes clear the rules and logic of their internal operation logic or operation process.

This process seems to be the process of digital twin reproduction of the physical world, but it can determine the impact of digital space on the physical world. For example, digital twin builds the physical segment of parts in the manufacturing industry, and the slight size change of a certain part of the parts will affect a series of problems, such as production, assembly, and running in. If there is a problem in the construction of this segment, it is easy to lead to the butterfly effect of the industrial chain of a certain enterprise and cause significant losses to the enterprise.

At present, large and well-known enterprises such as General Electric, Siemens, and ABB have used digital twin technology to accelerate industrial development and save research and development, production, and operation costs. These enterprises will increase investment in detailed data analysis during the construction of the digital twin physical plate. For example, General Electric controls the physical precision to micron during the construction of the physical plate, which saves a lot of costs for the subsequent production of General Company.

1.4.2 *Construction of digital twin data plate*

The collection of physical data is mainly through various types of sensors. Digital twin uses these inherent data to complete the construction of physical plates under the careful collection of multi-mode and multi-type sensors. After the completion of the physical plate construction, digital twin can convert the inherent physical data into virtual space.

The construction of data plate determines the scope of digital twin. Taking Siemens in Germany as an example, Siemens has built a virtual production workshop and a digital processing factory through digital twin technology. The data of the virtual space is completely synchronized with the data of the physical entities in the real world. Therefore, the optimization of the production process and the upgrading of the product functions of Siemens are all completed in the virtual space. After that, Siemens copied the change data of the virtual space to the real space and completed the industrial upgrading iteration.

In addition, the construction of digital segment pays more attention to the real-time and variability of data. Taking General Company of the United States as an example, General Company deeply integrated its-built industrial Internet platform Predix with digital twin technology and monitored the real-time data of aircraft engines through various sensor equipment. These data also clearly perceived the failure and maintenance requirements of engines.

It can be seen that the digital twin data plate construction focuses on the comprehensive replication of physical data, as well as the real-time perception and change monitoring of physical data.

1.4.3 *Construction of digital twin model plate*

The model built by digital twin in virtual space is a combination of the mechanism model of physical plate and the driving model of digital plate. It is based on physical entity data but also contains dynamic digital changes, which represents that the digital model plate has the ability to develop and adjust.

Many people believe that the construction of digital model is the digital simulation of physical entities, but this statement is not accurate because the digital simulation is only for the external simulation of physical entities, but the construction of digital model also has the ability of divergence and development. That is to say, the model constructed by digital twin is not only real, but also virtual and has the variability of virtual space.

For example, during the outbreak of COVID-19 in 2020, Central-South Architectural Design Institute was ordered to make a perfect copy of Beijing Xiaotangshan Hospital to Wuhan by using digital twin technology. It can be said that Leishenshan Hospital is the twin brother of Xiaotangshan Hospital. The Leishenshan Hospital, which was built in 10 days, is world-famous and its speed has surprised the world. However, in fact, the modeling in digital space had been completed before the completion of the construction of Xiaotangshan Hospital. The building information modeling team of Central-South Architectural Design Institute perfectly copied the physical information and digital information of Xiaotangshan Hospital and carried out detailed modeling in virtual space, so the hospital can be put into use immediately after the completion of construction, and its functionality is completely consistent with that of Xiaotangshan Hospital.

Generally speaking, the construction of the digital model plate is the process of organic combination of the physical plate and the digital plate. Digital twin technology gives the development and change ability to the combination in the virtual space and makes it practical in the process of combination.

1.4.4 *Construction of digital twin preview plate*

Nowadays, digital twin has been deeply used in various fields around the world; especially in the super era of science and technology, the shadow of digital twin can be seen everywhere. For example, in “Avengers 4”, the world’s first box office in 2019, the role of digital twin has been fully demonstrated.

In “Avengers 4”, Iron Man’s intelligent assistant “Friday” can make combat suggestions for Iron Man based on the status of the armor and the situation of the war, which is the role of digital twin preview. It can be seen that the digital twin preview is a kind of development forecast of physical things in virtual space, which has real basis and high accuracy and authenticity.

The construction of the preview plate is very important in the standard system of digital twin construction. It is the real dynamic display of the digital model and also the main basis for human analysis and decision-making of the physical prototype. After the digital twin technology completes the model construction in the virtual space, the digital model can display the changes of the physical entities under certain trends according to the inherent physical information, the real-time information collected by the sensors, and the overall structure provided by the digital model, simulate the behavior of the physical entities, accurately evaluate the state of the physical entities in change, and predict the future development, which is the preview plate of digital twin.

The preview plate of digital twin has the characteristics of high authenticity and high reduction, which is very accurate for the development of physical entities. Therefore, this plate is also an important plate of the practical application of digital twin technology.

1.4.5 *Construction of digital twin decision service segment*

After the digital twin technology constructs the preview plate, the digital twin can conduct future trend analysis for the model and the preview, thus improving the performance of the physical entity, optimizing the operation of the physical entity, and completing the synchronous control of the virtual space. This functional plate is the decision-making service segment of the digital twin.

It can be said that the function of digital twin is mainly demonstrated through this plate, which is also a plate for complex operation of digital twin. The decision-making service plate's construction can help human achieve the directional development of physical entities in the digital space and can expand, replace, or modify the replicates at will. This complex operation saves a lot of costs for the optimization of physical entities in the real space, and improves the functionality of physical entities.

Taking the engine model of General Company as an example, the function improvement of the engine of General Company needs numerous tests and modifications, and all of these processes are completed in the virtual space. General Company has saved a lot of research and development and testing costs through the application of digital twin technology, improved product quality, and obtained all-round protection for the brand.

After the directional operation in the virtual space, the digital twin can improve the adaptability of physical entities to the specific environment through the adjustment of various sectors or enhance a certain ability for pertinence. The presentation of these optimized data is the decision-making service brought by the digital twin for physical entities, and also the significance of the application of the digital twin technology.

1.5 Digital Twin Creates the "Digital Road" for Strengthening Country with Strong Transportation Network

Digital twin has shown tremendous value in the process of promoting the development of the times, and its driving force on the national economy is very obvious. Especially in the field of transportation in China, digital twin has become an important support for the upgrading and qualitative change of comprehensive three-dimensional transportation network.

China's transportation has undergone many changes driven by the digital twin, and the overall wisdom, convenience, and effectiveness have been greatly improved since 2020. It can be seen that the digital twin will become a powerful tool in China's strategy of becoming a strong country with strong transportation network. The digital road cast by it indicates the direction and trend of China's transportation development in the future.

1.5.1 *Digital twin intelligent transportation*

Li Yi, CEO of Beijing 51 World Digital Twin Technology Co., Ltd., said, "It is the significance of digital twin to enable things that need high cost or are difficult to achieve in the real world to be implemented quickly in the virtual space". This sentence fully highlights the value of digital twin technology in the field of transportation.

The scale of China's transportation network is among the top in the world, and its three-dimensional nature is extremely prominent. The trend of transportation development is to meet the travel needs of China's large population. The overall optimization and upgrading of the transportation network needs to consume ultra-high costs in the huge transportation network, and the digital twin technology just solves this problem so that the bad transportation environment and complex transportation network in real life can be perfectly copied in the virtual space, and hundreds of tests have been completed at a very low cost so that each route of the transportation network in China and each resource find the best way to use, so that the land, sea and air transportation all run out of the optimal solution.

At the same time, China's huge transportation network has also achieved real-time monitoring with the help of a large number of sensors, and dynamic data such as vehicle conditions and road conditions have been updated simultaneously, making transportation control more convenient, efficient, and simple.

It is clearly stated in the strategy of strengthening China with strong transportation network that the development goal of strengthening country with strong transportation network will be basically realized by 2035, in which digital transportation is the cornerstone of a large comprehensive three-dimensional transportation network, and the fundamental purpose of the application of digital technology is to improve the safety and convenience of transportation operation and support the rapid development of the transportation industry.

Thus, as an important technology for cost saving, consumption reduction, and quality improvement of transportation development, digital twin will gradually integrate with the transportation network and constantly change the ecology of transportation development and upgrade China's huge transportation network and transportation system as digital brain and intelligent brain so that the transportation development will truly enter the digital era and intelligent era.

1.5.2 *Digital twin building digital new city rail*

Digital twin makes transportation more smooth and safe in the process of intelligent transformation of China's transportation problems. Moreover, digital twin technology not only exists in virtual space but also appears in the public in the form of digital city rail, smart port, digital airport, etc., which is constantly improving the mass travel experience.

Among the many new transportation facilities built on the basis of digital twin technology, digital Metro is unique. This new transportation system, which is laid in the city and has much higher functionality and timeliness than traditional urban rail transit, fully meets the needs of rapid urban development, and makes urban transportation from informatization to intelligence and smart.

For example, the Metro Management System of many first-tier cities in China has deeply applied digital twin technology. Through the analysis of metro operation, security, and environmental control, the Metro Management Department integrates the actual information of the design and construction stage and various problems in the operation and maintenance process after the completion of the virtual space modeling and optimizes the metro service plan. At the same time, with the assistance of various sensors, it realizes the real-time interaction of data in physical space and digital space, so as to make the metro's intelligence and smart characteristics constantly improve.

In general, the digital new town rail built by digital twin technology inherits the characteristics of traditional rail transit, such as fast, public transportation, and electrification, and has been upgraded to be green, efficient, and intelligent on the basis of meeting the urban travel needs. At the same time, the construction cost and space use of the digital new town rail have also been optimized according to the actual situation of the city, and various improvements and innovations have been made in the details of material structure, safety protection, and convenience facilities during the development process.

It can be said that digital new town rail is the main trend of the development of rail transit in China. It effectively relieves the transportation pressure of urban roads, realizes the green development of transportation, improves the urban transportation network in a three-dimensional way by diversified travel modes, accelerates the travel efficiency of the public, and also accelerates the communication between cities.

1.5.3 *Digital twin gives birth to new digital infrastructure*

Digital twin technology and new transportation infrastructure have created a lot of sparks in the process of digital twin promoting the strategic development of a strong transportation country. The new infrastructure of transportation focuses on the application of a new generation of information technology and digital technology, while digital twin emphasizes the interconnection of digital space and the Internet in the real world, and digitalizes and models the physical things, so the two can be deeply combined and promote each other.

Especially in the various infrastructure construction of intelligent transportation, in order to ensure the functionality of intelligent transportation construction, digital twin has built various digital models based on various transportation data and injected data and models into the smart three-dimensional transportation network to conduct macro, meso, and micro simulation of the intelligent transportation network. This mode ensures the practical application of various functions of intelligent transportation and ensures that the transportation development meets the needs of urban dynamic and static changes or adapts to specific environment and conditions.

For example, in recent years, in the expansion project of the Chongqing expressway, the real modeling of permissions has been completed in the feasibility study stage of the project. This digital model is a physical overlay, expansion, or demolition based on the original expressway. The digital model provides technical support for the upgrading and iteration of Chongqing Expressway infrastructure, and it also improves the decision-making efficiency of the Chongqing transportation department in the process of highway expansion.

For example, during the construction of the Beijing–Harbin expressway, digital twin technology has been applied to the infrastructure construction of the Harbin section. The digital twin technology provides a visual presentation of the construction plan for the construction unit and provides a virtual audit platform for various infrastructure facilities, realizes the deep integration of construction and management, and saves a lot of construction funds for the construction party.

It can be seen that the construction mode and construction foundation will change due to digital twin in the process of future transportation construction, and the use effect, operation effect, as well as service, security, and intelligence of infrastructure in the virtual space can be reflected.

In other words, the future construction of new transportation infrastructure is mainly decided by digital twin.

1.5.4 *Digital twin leads the new direction of digital*

The development of intelligent transportation in China has undergone multiple iterations; however, each upgrade is completed under the promotion of new technologies. After the arrival of the digital twin transportation era, the relationship between cars, roads, and people will be more close, and the public travel experience and transportation travel services will also present a new development direction.

In future development, digital twin technology will lead China's transportation industry to complete the big migration of smart architecture, and the three-dimensional transportation network will gradually become scenario-based and intelligent, and the ability of transportation resources to cope with a complex environment will also be greatly improved.

Under the framework of data + model + preview, digital twin can open up broad development space for the transportation field.

For example, digital twin is creating a virtual space where trial and error can be made continuously and without limit. In the past, numerous collision experiments in the process of automobile design and wind tunnel experiments of tens of thousands of hours in the process of airport construction can be completed in virtual space. The emergence of zero-cost trial and error space greatly shortens the climbing period of transportation equipment quality.

For another example, in 2021, China's Shenzhou 12 successfully completed the manned space flight, which is also inseparable from the great exploration of the space industry with the support of digital twin. Before Shenzhou 12 set out, its digital model had completed many tasks in the virtual universe and responded to various emergencies. Various data provided in the virtual universe represented a new height and direction of China's aerospace industry.

Nowadays, China's transportation development has entered the stage of accelerating the construction of a strong transportation country. Digital twin is also giving full play to its advantages and forging a visual, safe, efficient, and high-quality digital development path for transportation development. We believe that in the future development, digital twin technology will lead the development of the transportation industry with more digital posture and digital mode so that the transportation level in China has been raised to a new height.

Case: 3D real scene platform of Sichuan–Tibet Railway constructed by digital twin

Sichuan–Tibet railway is the second access road to Tibet built in China after the Qinghai–Tibet railway. The Railway Construction Department of China overcame a variety of construction problems during the construction process, such as complex topography and geology, adverse weather conditions, and fragile ecological environment, and completed the railway project with the most difficult construction so far.

Since the completion of the construction of the Sichuan–Tibet railway, digital, intelligent and information upgrade have become the main work of the operation and maintenance of the Sichuan–Tibet railway, which has attracted the attention of many transportation research teams in the country. Among them, Professor Zhu Qing's team of Southwest Jiaotong University has conducted a lot of research on the application of digital twin in the Sichuan–Tibet railway.

In August 2020, Professor Zhu Qing released a report on the "Key Technologies of the Real 3D Space Information Platform of the Digital Twin Sichuan–Tibet Railway". In the report, Professor Zhu Qing detailed the important role of the digital twin in the construction, operation, and maintenance of the Sichuan–Tibet railway. The digital twin technology describes the whole elements, the whole lifecycle, and the relationship among them of the Sichuan–Tibet Railway as a whole. Through the digital display of various information in the real world, the construction difficulties, operation and maintenance difficulties, and environmental impact of the Sichuan–Tibet railway have been intuitively demonstrated.

With the support of digital twin, Sichuan–Tibet railway has realized the refined and efficient overall management of the whole lifecycle from planning, design, and construction to post operation and maintenance. The construction information and operation and maintenance information in the whole process are transparent and traceable. The construction status and operation and maintenance status of the Sichuan–Tibet railway can be dynamically displayed. It can be said that the application of digital twin in Sichuan–Tibet railway fully demonstrates the height and value of this technology. The digital platform and information platform constructed by digital twin have built a standardized and high-performance command system for Sichuan–Tibet railway, supporting the digital control of the whole lifecycle under the complex and sensitive physical space environment.

The efficient application of digital twin in Sichuan–Tibet railway has far-reaching significance for the digital construction and intelligent development of China's railway. Moreover, the smart space opened up by this technology for the railway industry is also amazing.

Chapter Summary

Digital twin technology is not so much integrating with all walks of life in the world as creating a new world. In the scene world constructed by digital twin, everything in the real world can be clearly perceived, and all walks of life can be simulated. Digital twin takes full advantage of its own advantages to present a visual, efficient, and high-quality digital development road for the world and promotes all walks of life to achieve breakthroughs and achieve new heights.

Chapter 2

Digital Twin Has Become a Key Driving Force for China's Transportation Development

In recent years, the comprehensive and three-dimensional transportation network has gradually taken shape with the rapid development of China's transportation industry. However, with the rapid development of China's economy, the accelerated pace of urban life, and the explosive growth of car ownership, the transportation system still bears great pressure. The transportation needs to be upgraded to intelligent and digital directions in order to fully meet the transportation needs of the masses in China and improve the travel experience. The digital twin technology gradually highlights the ultra-high value and accelerates the high-quality development of China's transportation as a key driving force in this process.

2.1 Digital Application: Leading the General Direction of Transportation Development

Transportation is not only a strategic industry with strong foundation and long-term benefits in China but also an important measure to promote the stable growth of China's economy and benefit people's livelihood. Since China has set the strategic goal of strengthening country with strong transportation network, it has become an important strategic layout of transportation development to strengthen the development effect with digital force. Digital application has started to dominate the general direction of transportation development in recent years. Digital application has shown the characteristics of growth and innovation though there is still a gap in the scale and extent of China's transportation digital application compared with that of developed countries, and it has played a leading role in the upgrading of intelligent transportation.

In 2019, the Ministry of Transport of China promulgated the "Outline for the Development Plan of Digital Transportation", which clearly states in the "basic principles":

"Led by innovation and empowered by data. With data as the key element, the company will empower transportation and related industries, promote the linkage innovation of models, business types, products and services, improve the quality of travel and logistics services, and make the digital dividend benefit the people and enhance their sense of gain.

Joint construction, sharing and integrated development. We should give full play to the institutional advantages of overall planning and coordinated promotion, and promote the coordinated development among government, enterprises, industries and departments. We should give full play to the role of the market as the main body, scientifically allocate various resource elements, and build a digital transportation industry ecology with cross-border integration and common creation and sharing. Moreover, we should promote the efficient connection of various modes of transportation through data chain, and promote the two-way transformation and utilization of data between government and enterprises.

Prevent risks and ensure safety. We should take into account innovation and security development, prevent and resolve information security risks brought by digital transformation, improve network security and data security capabilities, and protect public security and national interests.

Dare to explore, pilot first. We should adhere to the world vision, international standards and Chinese characteristics, adopt an open and inclusive attitude, adapt to the trend of technological development, take pilot projects as an important means, gather resources such as technology, intelligence and industry, and gradually form the "China Experience" and "China Program" of digital transportation development through typical guidance".

From this point, we can see that "digital power" is an important force to build China's transportation and modern transportation, and also the main trend of China's transportation modernization and globalization.

2.1.1 *General development direction of digital transportation in China*

The "Outline for the Development Plan of Digital Transportation" issued by the Ministry of Transport of China clearly states:

"By 2025, a new step will be made in the digital upgrading of all elements and full cycle of transportation infrastructure and transportation equipment, and the digital acquisition system and network transmission system will be basically formed". Transportation has become the main civil industry of Beidou navigation. The public network such as the fifth generation of mobile communication (5G) and the new generation of satellite communication system have been initially applied in the industry. The application level of transportation big data has been greatly improved, travel information services have been fully covered, the logistics service platform and integration have entered a new stage, and the industry governance and public service capabilities have been significantly improved. The transportation and automobile, electronics, software, communication, Internet services and other industries are deeply integrated, and the new business types and new technology application level remain advanced in the world.

"The transportation infrastructure will be digitalized in all elements and cycles, and the transportation control network integrating heaven and earth will be basically formed by 2035. The on-demand instant travel service will be widely used. Moreover, China will become a major international standard setter or participant in the field of digital transportation, and the overall competitiveness of the digital transportation industry will be leading in the world".

The future development of digital transportation includes the following: "building a digital collection system, building a network transmission system, building an intelligent application system, cultivating an industrial ecosystem, improving the network and data security system, improving the standard system, and improving the support and guarantee system".

From the above, it can be seen that the future development direction of China's digital transportation is to build a digital model of comprehensive three-dimensional transportation network, promote the digital transformation of the transportation industry ecology, and emphasize on the safety of the transportation network and the soundness of the transportation digital system on the basis of the digitalization of transportation equipment and transportation equipment.

The digital development of transportation has become a tangible development trend. The original intention of the digital development of transportation is to improve the development effect of transportation and ensure that the construction of transportation network meets the needs of social development. The application of digital twin technology is constantly deepening in the process of the application of digital technology in transportation, which brings more prominent help to the transportation industry and creates greater value in the field of transportation.

2.1.2 *Current situation and prospect of digital transportation development in China*

In 2018, the Ministry of Transport of China officially issued the "Digital Transportation Development Report (2017)"; the report clearly stated that digital power provides strong support for the development of strengthening a country with strong transportation network and has become an important strategic deployment in the construction of Digital China. Good digital development results have been achieved in terms of industry governance, and moreover, the intelligence of roads, railways, ports, and airports has been effectively improved.

In 2021, the "Digital China Development Report (2020)" issued by the Internet Information Office of China also clearly stated that digital transportation has spawned a lot of new technologies, demonstrated new business types and new hotspots, focused on practical application in development which has accelerated the development of the transportation industry, focused on policies to promote the livelihood industry and poverty alleviation projects in China, and focused on international standards, so as to lead China's transportation to a high-quality and international high-speed. Nowadays, digital transportation has become an important growth pole of China's digital economy and a representative industry in China's digital field.

From the development results of China's digital transportation, it can be seen that in the future, digital transportation will continue to upgrade the core technology and build a closed-loop system with virtual and real interaction under the support of digital twin. At the same time, the comprehensive digital process of China's transportation industry will also accelerate, and digital efficiency and precision will bring more surprises to the public.

2.1.2.1 *Upgrading core technology to build a closed-loop system with virtual real interaction*

At present, digital technology has added the perception ability, real-time monitoring and scheduling ability, human-computer interaction ability, as well as data collection, feedback, and control ability to China's transportation development. It can be predicted that the virtual reality interactive closed-loop system will be built in the field of digital transportation in China in the future combined with the digital development results of developed countries. In this system, digital power will play a greater role, transportation will develop simultaneously in the virtual space and the real world, and even the development of the real world will be led by the virtual space.

2.1.2.2 *Digital efficiency and precision will bring more surprises to the public*

With the upgrading of digital transportation technology, the digital efficiency and precision of the transportation industry will continue to improve and bring more surprises to the public. Taking the digital twin technology as an example, the simulation effect brought by the digital twin technology optimizes the use of urban transportation resources, relieves the pressure of urban transportation, and greatly improves the sense of public travel and travel experience.

2.1.3 *Digital twin brings new opportunities for intelligent transportation development*

As a new digital technology, digital twin technology has shown great value in various industries and fields and is also of great significance in the process of smart city construction. From the current situation of China's transportation development, it is not difficult to see that digital twin has brought stability, security, and convenience to the increasingly complex transportation industry in China and has brought many new opportunities for the development of intelligent transportation:

(1) The deep combination of digital twin technology and 5G technology has activated new kinetic energy of intelligent transportation. With the support of 5G technology, the construction of China's transportation

network has been increasingly improved, and the interaction and operability of digital transportation have been continuously improved. The industry development has entered a new phase. For example, the combination of 5G technology and digital twin has promoted the development of the automatic driving industry. The new industry format has increased the technical demand and talent demand, and the industry development space is also expanding.

(2) Digital twin brings new opportunities for the manufacturing, production, and maintenance of intelligent transportation equipment. Digital twin not only improves the intelligence of transportation equipment but also optimizes the production process and process of transportation equipment, thus accelerating the production efficiency and quality of equipment. In the meantime, the intelligent equipment that completes the process performance test in the virtual space, the maintenance work in the use process is more convenient and effective, and the overall development of the intelligent transportation equipment industry will also enter a new stage.

(3) Digital twin enhances the effectiveness of intelligent transportation control. With the support of digital twin, the transportation control department can flexibly and actively optimize the configuration of the transportation system according to the virtual space simulation of the transportation network and the real-time data collection in the real world, so as to improve the overall efficiency of intelligent transportation. For example, when a transportation accident occurs, the corresponding emergency plan shall be formulated according to the actual situation of the accident to ensure the safety and efficiency of the urban transportation network.

(4) The development of refined logistics and transportation industry. Digital twin can accurately sense the cargo conditions under different environments, temperatures, and vibration states through the sensor installation of logistics transportation equipment and ensure the transportation safety of special goods, high-precision instruments, and other goods. At the same time, the logistics network can build more diversified and more flexible distribution mode, formulate corresponding distribution plans according to the cyclical and seasonal characteristics of logistics business, optimize the allocation of logistics transportation resources, and improve the efficiency of logistics distribution with the support of digital twin technology.

As the representative of modern digital technology, digital twin is constantly improving the development effect of the transportation industry and promoting the realization of the strategy of strengthening the country with strong transportation network. This technology will become the key driving force of China's transportation upgrade and open up a new situation for China's transportation industry development.

2.2 Digital Transportation: Key Driving Force of Smart Travel

In September 2019, the State Council officially promulgated the "Outline for the Construction of Strengthening Country with Strong Transportation Network", which clearly stated that in the new era, China's Congress "promotes the development of data resources, empowers transportation" and "constructs a comprehensive transportation big data center system". In order to implement this development goal, promote the development effect of the transportation industry, and improve the comprehensive service level of transportation, the Ministry of Transport subsequently issued the "Action Outline for Promoting the Development of Comprehensive Transportation Big Data (2020–2025)".

According to the "Action Outline for Promoting the Development of Comprehensive Transportation Big Data (2020–2025)", China will vigorously promote the research and development of digital twin technology of transportation vehicles, in order to enhance the intelligent travel effect of transportation network, improve the experience of mass transportation, and take digital twin as the key driving force for future development, improving the implementation effect of the strategic goal of strengthening country with strong transportation network.

2.2.1 *Digital transportation is an important driving force for smart travel*

Digital transportation is an important area of China's digital economy development strategy. The development results of digital transportation directly affect the quality of life of the public and the overall operational efficiency of the society in China. In recent years, China's transportation development has gradually shifted from sharing to intelligent travel, and the overall digitalization of transportation is also increasing with the promotion of digital twin.

For example, at the "China International Digital Economy Expo" held in September 2021, more than 400 enterprises across the country demonstrated the cutting-edge and core technology products of digital technology in China, among which, the technology products in the transportation field became a major highlight of the Expo, and the latest scientific research results in many transportation fields let the public feel a new experience of smart travel in the future.

Future smart rail: qualitative change travel experience on virtual rail

In recent years, China's rail transit has shown a new development situation. With the support of digital twin technology, smart rail express system, urban rail vehicles, and intercity rail system can not only apply to different transportation environments but also give full play to the advantages of rail transit, integrate into the core of smart city, and provide efficient solutions for the development of urban public transport. For example, with the assistance of digital twin technology, rail transit can achieve intelligent forward-looking travel and test run on the established virtual track in advance without laying rail. The operation preview of this virtual space can find the construction difficulties and laying problems of rail transit in advance, which not only shortens the construction period but also advances the service life. The construction cost is also much lower than the construction cost of traditional rail transit.

The "digital brain" of urban rail transit has comprehensively enriched the operation mode of rail transit, effectively alleviated urban congestion, met the travel needs of the public, promoted the development of the strategy of strengthening country with strong transportation network, and upgraded the travel experience of the public.

Intelligent driving: innovative travel mode

In 2021, after the COVID-19 epidemic, the travel frequency of the public in China began to grow rapidly, and intelligent driving and automatic driving began to become the focus of public attention. In order to meet the needs of public travel, major automobile enterprises have launched new models with intelligent driving technology. Taking the Great Wall ORA Good Cat equipped with L4 level automatic driving technology as an example, the vehicle can complete continuous curve driving, lane change driving, and avoid obstacles under the situation that the driver does not need to touch the steering wheel, accelerator, and brake. Although it has not fully realized unmanned driving, this vehicle model using GNSS RTK high-precision positioning camera to identify and avoid obstacles demonstrates technological innovation in the field of intelligent driving in China.

The development of intelligent driving technology also needs the support of digital power, which is also a major area of digital transportation development. In future development, digital technology will combine with more market demands to improve the application scope of intelligent driving technology. Unmanned driving bus and unmanned driving cleaning will present the trend of industrialization, and China's transportation mode will be subverted and innovated.

Digital model of integrated three-dimensional transportation network: exploring the development path of transportation

Digital technology not only improves the mass travel experience in the field of rail transit and intelligent driving but also plays an important role in upgrading and optimizing the comprehensive three-dimensional transportation network in China. For example, the digital twin technology has built a digital model of urban transportation in the virtual space, improving the utilization rate of transportation resources and the efficiency of public travel. In future development, with the enrichment of digital models of China's urban transportation network, China's comprehensive three-dimensional transportation network will build an accurate and comprehensive digital model by connecting points and forming a surface. The overall development will be improved again, and the public travel experience will be upgraded again.

2.2.2 *China fully promotes the research and development of digital twin technology for vehicles*

The Ministry of transport clearly stated in the "Action Outline for Promoting the Development of Comprehensive Transportation Big Data (2020–2025)": "promote the research and development of digital twin technology of various types of transportation infrastructure and vehicles, accelerate the technical innovation of building information model (BIM) in various fields of transportation, and form application products with independent intellectual property rights. Study and formulate the address planning of the sixth version of Internet Protocol (IPv6) in the transportation industry, and promote the research, development and application of the fifth generation mobile communication technology and satellite communication information network in various fields of transportation. Carry out the research and development and application of key technologies of big data under the comprehensive transportation system". Thus, the research and development of digital twin technology for vehicles will become an important development task in the future development of the transportation industry. In this process, digital twin technology will also show higher application value.

2.2.2.1 *Research and development of digital twin technology in the field of land vehicles*

At present, digital twin has achieved good development results in the automobile field, supporting a number of technological innovations in road vehicles. Modern intelligent driving technology and unmanned driving technology are inseparable from the promotion of digital twin.

In addition to road vehicles, the technology promotion brought by digital twin for railway vehicles is also very obvious. For example, in 2021, DB Schenker, a wholly owned subsidiary of the Deutsche Bahn Group, announced the joint development of a digital twin system for complete trains with the Swiss railway company Stadler, which is expected to be operational by the end of 2021. In the R&D project, DB Schenker can periodically demonstrate the technical achievements of the new digital twin system. With the assistance of this system, the railway vehicles can improve the operation reliability and transportation capacity, prevent the occurrence of accidents, and reduce the number of times the railway vehicles interrupt the transportation; Mr. Milkovich, chief executive officer of Stadler Germany, said, "by creating digital twin technology for complete vehicles, we can obtain relevant data as a vehicle manufacturer and as DB Schenker. In this way, we can ensure proactive maintenance and continuous optimization of aging management. This helps both parties to continuously improve vehicle availability and minimize the damage caused by failures".

It can be seen from the research and development of digital twin technology of DB Schenker that digital twin technology is an inevitable way to improve the efficiency and safety of railway vehicles and is also the focus of future technology research of railway vehicles.

2.2.2.2 *Technology research and development of digital twin in the field of air vehicles*

In 2017, China successfully made its first flight of C919, a single-channel large trunk passenger aircraft developed in full accordance with the international advanced airworthiness standards. The successful research and development of this large passenger aircraft with China's full independent intellectual property rights represents that China has completely broken the air transportation situation monopolized by Boeing and Airbus, and China has officially become a country in the world that has the ability to develop large aircraft other than the United States, France, Germany, Russia, Brazil, Canada, and the United Kingdom.

The successful development of C919 has great significance for the aviation development in China. The emergence of this new large air carrier is also the result of the combination of digital twin and aviation technology. In the research and development process of C919, the aircraft structure design, aerodynamics application, system performance test, thermal energy management, and other aspects have all passed the digital twin simulation test. The electrical system, hydraulic system, environmental control system, flight control, fuel system, and landing system of C919 have been upgraded numerous times in digital space, so C919 has been successfully developed and successfully completed the first flight.

In future development, the application of digital twin in the research and development process of air transport tools will be more in-depth, and the intelligence of air transport tools will be improved, the safety of air transport tools will be enhanced, and the overall strength of China's air transport industry will be enhanced.

2.2.2.3 *Research and development of digital twin in the field of water vehicles*

The digital twin technology has played an important role in the intelligent ship management system currently. The digital operation system constructed by the digital twin has thoroughly solved the problems of ship remote monitoring and fault detection, which also improves the management efficiency, instead of only reducing the management cost in the process of ship operation. In particular, the ship model built by digital twin in virtual space can remotely and visually monitor the operation status and various operation parameters of the ship, so as to ensure the operation safety of the ship to the maximum extent.

This shows that the digital twin technology has a good role in promoting the health management of large-scale water vehicles. In the future, the application and research and development of digital twin in this field will be more in-depth, and the intelligence and safety of large-scale water vehicles will also be continuously improved.

2.3 Digital Innovation: Creating a New Era of Transportation

The digital twin of transportation can be regarded as the synchronous interactive simulation of transportation entities, new infrastructure, operating individuals, and operating systems in virtual space. This digital twin now has become the main force of testing and development in the field of transportation, and moreover, the huge development potential of digital twin also influences the overall development of the transportation industry.

In fact, since China formulated the development goal of "strengthening country with strong transportation network" in 2019, digital twin technology has penetrated into all aspects of transportation development. For example, the construction of China's high-speed road network requires digital twin to carry out transportation flow modeling in virtual space, to simulate the overall framework and infrastructure functions, and to provide construction parameters and implementation plans that are cost-effective and in line with the actual construction characteristics.

As of 2021, China's transportation has entered a new stage of digital construction, network development, and intelligent upgrading. The level of digital upgrading and transformation of transportation is also increasing with the support of various policies. The application upsurge caused by digital twin has generated new trends and new business types in the industry.

2.3.1 *"Construction of new infrastructure and new urbanization and major projects such as transportation and water conservancy" planning transportation development pattern*

On May 22, 2020, the State Council clearly stated in the "Government Work Report" that "Construction of new infrastructure and new urbanization and major projects such as transportation and water conservancy" (also known as 2-new and 1 major) will become the focus of China's construction and development in the future. The so-called "2-new" refers to new infrastructure and new urbanization, and "1 major" refers to major projects such as transportation and water conservancy. After the development focus of "Construction of new infrastructure and new urbanization and major projects such as transportation and water conservancy" is determined, China's urban construction, water conservancy, and transportation engineering construction will receive more policy support, and the overall development will show a trend of high standard and high quality.

We can understand from the future development situation of transportation that technological innovation is essential in the development process, which exactly needs the support of digital technology and new generation of information technology. For example, in the development of transportation industry in recent years, the upgrading and application of three-dimensional geographic information system has guaranteed the construction and management of a number of important transportation projects in China. In the meantime, the Ministry of Transport has also clearly stated in a number of documents on transportation development that in the future, transportation design, transportation planning, transportation operation, and maintenance management need to use digital technology in depth, improve the intelligent level of the transportation industry, and improve the construction, planning, and management capacity of transportation.

Generally speaking, digital twin has opened up a new situation for the development of transportation and reconstructed the development pattern of the industry. In future development, the proportion of traditional development model will continue to reduce, and the digital development model led by informatization and intelligence will become the core of the industry. The transportation industry will embark on a visual, analytical, and manageable smart development path.

2.3.2 *Full application of digital twin in transportation engineering construction*

The construction of transportation engineering in China will speed up again, and the application of digital twin in transportation engineering will be more extensive under the development pattern of "Construction of new infrastructure and new urbanization and major projects such as transportation and water conservancy". We can see that in future development, digital twin has two main application directions in transportation engineering construction in combination with the technical characteristics of digital twin.

2.3.2.1 *Application of digital twin in transportation engineering construction*

Digital twin technology can run through the whole process of transportation engineering construction, and digital twin can play an important role in each link from planning and design to construction and construction, and then to operation and maintenance.

First, digital twin can build a visual and all element construction model based on construction projects to ensure the integrity of the project functions; second, the all element digital model is convenient for the engineering design and implementation personnel to conduct convenient analysis and improvement; third, the combination of digital twin and Internet of Things can design the refined management system of construction site and construction distance to ensure the project quality.

2.3.2.2 *Building a full element virtual model of modern transportation facilities*

The intelligence of China's transportation network depends on the degree of digitalization and intelligent construction of transportation facilities. Moreover, based on the business needs of transportation facilities, digital twin can build a smart model of refined management. This model not only carries the travel requirements of the transportation facilities but also reflects the whole life cycle of the transportation facilities in the simulation preview. As a result, the transportation facility model constructed by digital twin is also called the total factor virtual model.

The all element virtual model constructed by the digital twin is the main way for the intelligent development of China's transportation because this model can simulate the construction environment, operation environment, operation situation, and other full scenarios of the transportation facilities, and then physically simulate the transportation facilities, integrate them with the external environment and transportation vehicles, and then carry out digital upgrade for the operation data; a set of improvement plans to improve the intelligence is calculated.

2.3.3 *Transportation assistance of digital twin*

The emergence of digital twin has supported the effective development of China's transportation sector. It integrates not only the existing data resources of the transportation industry but also deeply cutting-edge technology applications such as 5G, big data, cloud computing, as well as AI. In the normal operation of China's transportation system, information, technology, and equipment can be organically combined according to the actual needs, thus enhancing the overall management ability of transportation.

Taking urban transportation management as an example, digital twin technology can be used to monitor the overall situation of urban transportation, regional transportation, and hub situation. The density distribution of urban travelers and travel demand analysis can realize intelligent perception, intelligent disposal, and intelligent improvement. This technology effectively improves the decision-making ability and resource coordination efficiency of the transportation department.

In general, the auxiliary effect of digital twin technology on China's transportation development can be shown in the following aspects: ① overall transportation situation monitoring, ② transportation population distribution monitoring, ③ transportation demand analysis, ④ transportation network structure analysis, ⑤ regional and hub transportation situation monitoring, ⑥ transportation structure optimization analysis, ⑦ transportation intelligence warning, ⑧ intelligent disposal of transportation conditions, ⑨ intelligent optimization of transportation network, ⑩ integrated modeling of all factor transportation network, ⑪ the fusion of multi-source transportation data, and ⑫ full scale and full element 3D twin simulation.

It can be said that the construction efficiency, operation efficiency, safety, and intelligence of China's transportation can be greatly improved with the help of digital twin, which is also the main reason why digital twin is so popular in the transportation field.

2.3.4 *Digital twin helps transportation facilities realize intelligent construction*

Intelligent development of transportation industry is the main trend in the future. Intelligent construction of transportation infrastructure based on digital twin technology is also a clear task proposed in the "14th Five-Year Plan". With the aid of digital twin, both environmental data information and dynamic information of the construction site can be visually displayed in virtual space. This dynamic and static combination and virtual combination simulation can combine various complex information in the construction process with refined management and intelligent upgrading, so as to meet the dynamic construction management based on the needs of transportation infrastructure business and control of the real situation of the construction environment.

For example, in the digital model constructed by digital twin, the precision of each component and facility can be truly reflected from the whole project to the smallest one. Meanwhile, the virtual model constructed by digital twin also supports the fine management, analysis, and calculation function of a single target and can carry out the associated management of complicated engineering data, including drawings, schemes, and establishment process. In the visual model, the construction information is optimized point by point, and the construction quality, efficiency, as well as the operation and maintenance effect in the later period can be greatly improved.

2.3.5 *Digital twin supports online operation, maintenance, and management of transportation*

Digital twin can improve not only the quality and intelligence of transportation engineering construction but also the maintenance effect of operation and maintenance management of transportation network. The transportation facilities constructed based on digital twin technology can build a more efficient maintenance system. Each key facility and each key component of the transportation network can be monitored in real time, and an operation management file can be established. The storage and analysis of these data can help carry out dynamic analysis of transportation facilities and formulate a more reasonable maintenance plan.

Digital twin can optimize the scene of the facility operation in the operation and maintenance management of the transportation network, support the optimization management of the large-scale transportation network, check and analyze the actual real-time information of the transportation facilities components with the help of split sensors, and then carry out dynamic rendering in the virtual space so that the management personnel of the transportation network can more intuitively manage and maintain the transportation network in a more timely manner. In addition, with the help of the digital twin function, management personnel can more accurately predict the impact of environmental changes on transportation and achieve intelligent management and control of transportation facilities' operation, management, and maintenance.

2.4 Digital Twin: Boosting the Development of Intelligent Transportation

China's transportation network has developed from a closed network to a wide network in the recent 30 years of development, which not only changes the efficiency of mass transportation and travel but also benefits the economy and promotes the continuous improvement of people's living standards. However, the direction of China's transportation development has begun to change in recent years. China's transportation development has started to promote the improvement of the quality of life of the public, and the wisdom of the transportation industry has become increasingly prominent on the basis of the changes in the relationship between people, vehicles, roads, and the environment.

With the vigorous development of transportation infrastructure, China's transportation network has begun to achieve a new leap forward. The application of digital twin technology has injected a strong force into the development of intelligent transportation, gradually changing the development status of artificial management, post reflection, and repeated correction of transportation, which is of great significance to the overall development of all major industries in China.

2.4.1 *Digital twin: new development direction of intelligent transportation*

The development speed of intelligent transportation has been at the forefront of the world even though there is still a gap between China and the developed countries in terms of the intelligent development of transportation. Driven by a number of policies in China, we pay attention to the creation of intelligent attributes from the construction of transportation infrastructure, to the design of the overall transportation network, and then to the operation and maintenance management at a later stage. By 2021, China had become one of the countries with not only the fastest speed of transportation infrastructure construction in the world but also the fastest growth in transportation industry and intelligent transformation of transportation.

By combining and comparing the development status of developed countries, we can conclude that the main direction of current international transportation development is as listed in the following three aspects.

2.4.1.1 *Intelligentization and digitalization of transportation network*

The intelligent and digital development of the transportation network refers to the enhancement of the intelligent attribute of the transportation network and the digital control at the same time. Many developed countries in Europe and the United States have basically realized the intelligent and digital development of the transportation network currently. Moreover, the controllability and efficiency of the transportation network become more prominent, and public travel satisfaction is also increasing under this development mode.

At present, China's transportation network is also in the process of intelligent and digital development. Taking the digital twin transportation system pilot operated in Guizhou as an example, Guizhou Province realized the hologram perception of transportation elements such as motor vehicles, non-motor vehicles, and pedestrians by integrating the MMW Radar with video monitoring, and analyzed and tracked according to the real-time perception data. The problems such as waste of transportation resources and rigid system functions have been effectively solved, and the travel experience of mass transportation in Guizhou has been comprehensively upgraded.

2.4.1.2 *Prediction perception of transportation accident*

In the past transportation management systems, the prediction and perception of transportation accidents mostly relied on the experience of accident handling. After the integration of the transportation network and the digital twin technology, the transportation management system can predict the possible collision in a certain period of time in advance according to the operation status of the vehicles. This real simulation of all elements realizes the multi-state prediction and identification of transportation accidents and effectively improves the safety of the transportation network.

2.4.1.3 *Intelligent management system of vehicle road collaboration*

After the combination of digital technology and wireless communication technology, we can build an intelligent management system for driving coordination with the help of sensors of transportation equipment, which is also the focus of modern transportation development. In the intelligent management system of vehicle road collaboration, transportation safety and transportation efficiency can be greatly improved, while urban pollution can be reduced. As a result, the intelligent management system of vehicle-road cooperation is an important basis for building a high-quality and environmentally friendly transportation system.

From the main direction of modern transportation development, it can be seen that the development of intelligent transportation cannot be separated from the support of digital power, and digital twin is helping intelligent transportation enter a new development situation.

2.4.2 ***Implementation of digital twin in the field of intelligent transportation***

In recent years, there are still some pain points to be solved even though China's transportation development has made considerable progress. The landing mode of digital twin can just be regarded as the process to solve the pain point of modern transportation network. The deep application of this technology makes up for the deficiency of intelligent transportation in all aspects and highlights the intelligence of transportation network:

(1) The digital twin improves the real-time data collection, realizes the synchronous operation of the transportation network, constructs the deduction model of the transportation network, and creates a data-driven decision-making management mode, which is also the first step of the digital twin in the implementation of intelligent transportation.
(2) Digital twin enhances the intelligence of modern driving functions. The digital twin, which is modeled with all elements of real data, can effectively improve the safety and stability of intelligent driving and accelerate the landing of unmanned driving technology, which is the second step of the digital twin in the field of intelligent transportation.
(3) The digital twin technology, which plans in detail from a global perspective and finds the optimal solution for transportation management, has been applied to major cities in China. It also solves the difficulties

of transportation development according to the changes of transportation flow, quantifying the development data of smart city, instead of only effectively handling the complex situation of urban transportation. It can be said that digital twin is not only a control tool of intelligent transportation, but also a virtual portrait of smart city, which is the third step of digital twin in the field of intelligent transportation.

2.4.3 *New kinetic energy generated by "5G + digital twin"*

Digital twin has been effectively combined with artificial intelligence and new generation information technology in the development of transportation, but the most significant effect is 5G technology. 5G network has effectively improved the response speed of the transportation network with the upgrading of information technology. After the combination with digital twin, it gives birth to the new kinetic energy of the development of China's transportation industry.

First of all, the "5G + Digital twin" mode enables vehicles to realize real-time data interaction in high-speed movement. The effectiveness of vehicle road collaboration is ensured, the development of automatic driving technology is promoted, and the landing of remote driving technology is accelerated by this network feature.

Second, 5G promotes the characteristics of Internet of Things to enhance the modeling accuracy of digital twin, coordinates the development relationship of "people, vehicles, roads, and environment", greatly enriches the database of transportation departments, and makes the digital and intelligent characteristics of intelligent transportation more prominent.

All in all, the digital twin can show greater value with the support of 5G technology, which can ensure the overall management of China's transportation network, the interaction between the virtual and the real, and push China's intelligent transportation to the forefront of the times, create new ideas and new paths for China's transportation development, and create an international first-class comprehensive three-dimensional transportation network.

Case: iFLYTEK joined hands with Beijing 51 WORLD to build a "Silk Road Smart Port"

2021 is the 10th anniversary of the operation of China–Europe trains. In the 10 years of development, this China business road, known as "steel camel caravan", can not only run out of China's speed but also gradually develop from a line to a network. As of 2021, China–Europe trains have opened a total of 7377 trains and delivered 707,000 TEUs, supporting the sustained and stable growth of China's economy.

Behind the remarkable achievements of China Europe Express, there is also an efficient and intelligent operation mode in addition to its unique transportation advantages. Urumqi Land Port Assembly Center (Xinjiang) is the international marshalling hub nearest to Central Asia and Europe in the China section of the China–Europe block train west passage. It is the last station before China's China–Europe trains leave the country and the first marshalling station before entering Europe. Therefore, Urumqi Land Port Assembly Center (Xinjiang) can become the core hub of China–Europe trains.

In Urumqi Land Port Assembly Center (Xinjiang), there is a smart platform called "Silk Road Smart Port", which is a management platform based on digital twin technology, artificial intelligence technology, and other cutting-edge technologies. "Silk Road Smart Port" effectively improves the operation efficiency of Urumqi Land Port Assembly Center and also improves the security of China–Europe trains.

This platform is jointly built by iFLYTEK and Beijing 51 WORLD company, mainly based on digital twin technology, and is mainly used for full digital expression of the Urumqi Land Port Assembly Center (Xinjiang). Under the technical support of digital twin, the "Silk Road Smart Port" has built a digital model with the assembly center and the multimodal transportation center as the main body. The Urumqi Land Port Assembly Center (Xinjiang) has optimized the train organization mode, reasonably allocated transportation resources, and innovated the "assembly and transportation" mode under the simulation and preview of the digital model, and what's more, it has also completed the intelligent upgrade of the land port assembly center in Urumqi, Xinjiang.

The outstanding features of "Silk Road Smart Port" are mainly manifested in the following two aspects.

(1) **Full element simulation:** The reason why "Silk Road Intelligent Port" is known as the intelligent steward of the land port assembly center in Urumqi of Xinjiang is that it can complete the full element simulation of the assembly center, conduct overall analysis on the simulation data, provide the best and most timely decision-making information, and then carry out intelligent management on

the industrial operation, station operation, as well as safety management, etc. of the assembly center.

(2) **Realize unmanned operation:** After the combination of digital twin and 5G technology, Urumqi (Xinjiang) Land Port Assembly Center has realized efficient "unmanned operation". There is a set of 24 hours non-stop automatic operation system of unmanned trucks in the Land Port Assembly Center. These unmanned vehicles can be configured according to the information in the station fed back by the digital model to complete various operations in the unmanned mode. Meanwhile, the real-time status of unmanned vehicles can also be fully demonstrated in the digital model, which makes the management of the land port assembly center in Urumqi, Xinjiang, more reasonable and timely, and makes the transit of China–Europe trains more efficient and safe.

In general, the digital twin platform jointly built by iFLYTEK and Beijing 51 WORLD company has greatly improved the management and operation effect of Urumqi (Xinjiang) Land Port Assembly Center and accelerated the intelligent development of China–Europe trains. In the future, this model can be replicated and applied in other trunk lines of China's railways, and the value of digital twin will also be fully demonstrated.

Chapter Summary

As of 2021, China's intelligent transportation development has made remarkable achievements, which depends on the key drivers of digital twin technology. From the development of global transportation digitalization, it can be seen that digital twin will become the core support for future transportation, which will run through all aspects of transportation planning, design, construction, operation, and maintenance, and promote China to develop into a world power with outstanding intelligence and significant transportation economic growth.

Chapter 3

China's Transportation Construction Enters the Digital Twin Era

Since the Ministry of Transport of China issued the "Outline for the Development Plan of Digital Transportation" in 2019, proposing two-phased development goals of 2025 and 2035, a rapidly changing digital transportation era has arrived for us. As the main mode of global transportation development, digital transportation is continuously stimulating the potential of China's transportation resources and drawing a beautiful blueprint for China's transportation quality change. As a representative of digital technology, the digital twin has been accelerating the trend of digital and intelligent transportation in China, gradually becoming an important engine for China's economic growth.

3.1 Transportation Network Security Needs to Be Improved in the Digital Era

Digital technology has brought about great changes to China's economic development, which are hard to meet in 100 years. After simplifying the complex structure of various fields, it has initiated great changes in China's scientific, technological, and industrial development. In the digital era, all things can be connected and all things can be programmed. Digitalization and networking have become the main ways to drive the progress of the world.

However, after various fields such as energy, finance and transportation are linked to digital technologies, the development speed has doubled and improved, however security has been accompanied by a number of hidden dangers. As Zhou Hongyi, CEO of 360 Group, said at the Eighth Internet Security Conference, "entering the digital twin era, cyber attack has more influence than nuclear weapons".

In the future, information network security will become an important foundation to ensure the rapid development of various industries in the digital era. The transportation network with a large layout and large objectives needs to improve its own security, so as to play an important role in the strategy of strengthening the country with a strong transportation network, completing the established development goals and boosting China's economic growth.

3.1.1 *The form of cyber war in the digital era*

After the goal of strengthening the country with a strong transportation network in China was determined, a series of transportation policy upgrades were carried out in China. With the rapid construction of new infrastructure, the transportation industry has accelerated the pace of digital transformation. It is not difficult to see that the new infrastructure for the development of China's transportation industry in recent years is digital infrastructure, whose main role is the digital transformation and network coverage of the transportation system; the purpose of network coverage is to accelerate the operation efficiency of the transportation industry and ensure the high-quality development of China's transportation.

In the face of the current development status of the transportation industry, the traditional security protection system also needs to be upgraded to cope with the emergence of a new competition model in the digital era.

(1) From the basis, we should consider the digital security of the new transportation infrastructure, and build a more comprehensive security plan from the perspective of the overall operation of the integrated three-dimensional transportation network to ensure the overall security of the transportation network.
(2) Enhance the position of information network security in transportation development. In the past, the safety factors of transportation development were mainly the quality of transportation facilities. With the advent of the digital era, information network security has

become an important guarantee to ensure the safe travel of the public, effective coordination of transportation resources, and reasonable emergency response. Information network security also determines the overall operational efficiency of the transportation network.

(3) Upgrade transportation safety solutions from static to dynamic. Upon completion of the digital upgrade of China's transportation network, it will need to face various crises, such as APT attacks and cyber crimes. In the past, it was unable to ensure the safety of the transportation network because the appropriate static security scheme was determined by compliance. It is the focus of the transportation department to create a high-level, detailed and timely dynamic security scheme according to the actual situation of the transportation network in a timely manner.

(4) Abandon the previous "patched" transportation network protection measures, and upgrade the protection measures of collaborative protection and joint protection of the transportation network from the perspective of security top level.

3.1.2 *Strategy first, seven new elements of the safety concept*

Zhou Hongyi, CEO of 360 Group, also said at the Eighth Internet Security Conference, "without security top-level design and methodology, digitalization means 'running barefoot'. The faster you run, the greater the disaster will be". In the digital era, China's transportation development needs to face a new crisis. Only by carrying out the strategy first and formulating a security system that conforms to the characteristics of the digital era can we ensure the construction and development of digital transportation.

After analyzing the characteristics of the digital transportation safety system in the developed countries, it can be seen that the transportation network safety in the digital era mainly includes seven new elements:

(1) **New crisis factors:** Compared with the previous crisis of transportation facilities and individual transportation crisis, the transportation network in the digital era needs to face the new crisis of loopholes and hackers.

(2) **New directions:** In the past, the focus of transportation safety was the rationality of responding to transportation accidents and transportation facilities. In the digital era, transportation safety needs to

respond to human attacks, and the intensity and frequency of attacks will become the focus of response.

(3) **New elements:** The biggest feature of the digital era is the rapid development of technology, and the corresponding transportation safety also needs to be considered at the same time. According to the development speed of digital technology, a reasonable upgrade is matched to ensure that the transportation network security system can resist the attack of new technology.

(4) **New model:** There are various modes of transportation crisis in the digital era. Some safety hazards will not bring harm to the transportation system itself, but can affect the economic development of related industries.

For example, in 2021, the metropolitan transportation agency of New York, the United States, confirmed that the reservation system of the Massachusetts Steamship Bureau of the United States was hacked in April, which resulted in the inability of the customers of the Massachusetts Steamship Bureau to book or change their itinerary through the network or telephone, and then the Massachusetts Steamship Bureau was attacked by blackmail software. Although the two incidents did not affect the normal operation of the New York transportation system, the Massachusetts Steamship Bureau also promptly fixed the loopholes. However, in fact, this incident caused economic losses to the customers of the Massachusetts Steamship Bureau and affected the normal development of the Massachusetts Steamship Bureau.

A similar situation is not uncommon in a digital context. After the arrival of the digital era, public demand industries such as medical, transportation, and food began to suffer from various modes of network attacks. In 2020, United Health Services lost US$67 million in cyber attacks.

(5) **New infrastructure:** New infrastructure is not only the focus of China's transportation development but also the focus of the construction of the transportation safety system. China's overall logistics efficiency and transportation level will continue to improve in future development. The new infrastructure must have the important task of ensuring the safety of transportation, so as to ensure the overall safety of China's transportation network.

(6) **New technology:** While ensuring the network security of the equipment, the transportation equipment production industry also needs to

continuously upgrade its own technology to ensure that the transportation equipment meets the needs of the development of the times in the digital era.

(7) **New system:** The transportation network structure is clearly seen after the birth of the digital twin technology, which undoubtedly increases the security challenges of the transportation network. The transportation department needs to build a new safety system and take the actual operation as the safety standard to ensure the public safety of the transportation network.

3.1.3 *Security core of digital twin technology*

As of 2021, the transportation department of China had constructed a smart mode for the meteorological service to provide support for transportation development and formed an overall security protection mechanism that relies on safety data, digital twin technology and professional skills. Currently, China's transportation network can effectively prevent APT attacks and ensure the efficient operation of the transportation network under the protection of this security system.

By integrating the role of digital twin technology in China's transportation safety network, it can be concluded that the security core of digital twin technology is mainly reflected in four aspects.

3.1.3.1 *Ensure the objectives and results of transportation with the full factor model*

Digital twin technology can build a real simulation model of all elements for the transportation system in China. The more refined and stronger the model, the more predictable the safety hazards and the more comprehensive the protection of transportation.

3.1.3.2 *Real-time monitoring and prevention based on the real environment*

The digital twin security protection system based on the real environment can effectively reduce the security risks brought by the environment or the combination of the environment. For instance, the transportation detection equipment in China will be laid out in a large range and intensively in the key transportation hub areas, with the purpose of ensuring real-time and all-round awareness of the environment and predicting the potential security risks brought by environmental changes.

3.1.3.3 *Multi-technology integration to improve safety level*

Equipped with the characteristics of multi-technology integration, digital twin technology has been effectively combined with AI technology, AR technology, and new generation information technology at present. The innovative new security technology of digital twin has also improved the protection level of China's transportation network security system. In future development, the digital twin will be integrated with more emerging technologies to enhance its own technical value and improve safety protection capabilities.

3.1.3.4 *Real-time data interaction, timely and accurate prediction of safety hazards*

Digital twin, being highly interactive, is able to completely transform physical individuals into real space and virtual space, and infer the possibilities based on the actual situation. At present, this feature has been applied to the field of transportation network security protection. On the basis of the changes in network data, the digital twin can accurately predict the existing management loopholes, predict the security risks in a timely and accurate manner, and protect the safe operation of the transportation network.

Generally speaking, after the advent of the digital era, the security of China's transportation network has been redefined. To develop China into a strong transportation nation, we must strive to build a robust network security system and ensure the safe and high-quality development of China's transportation in the digital era, underpinned by new safety principles and systems.

3.2 In the Digital Twin Era, Transportation Infrastructure Has Entered a New Digital Stage

It is predicted that China will enter a digital and intelligent development period in the next 5~10 years. As stated in the "Outline for the Construction of Strengthening Country with Strong Transportation

Network", "5G, cloud computing, big data, Internet of Things, artificial intelligence and blockchain are the core digital technology to drive China's development". In this change of society, transportation will remain the main development force, and transportation construction will also focus on digitalization and intelligence for upgrading. Compared with the previous vertical development of land, sea, and air, future transportation will focus on the coordinated development of travel transportation, and China's comprehensive three-dimensional transportation network will show a strong leading role.

In fact, the combination of transportation and digital can be regarded as the inevitable law of the development of the times. The digital is the way to connect the virtual space since transportation is the way to connect with the real world, and moreover, the integration of transportation and digital has become inevitable with the combination of the virtual and the real worlds. In order to promote the orderly development of China's digital transportation and intelligent transportation, the new digital infrastructure has established the primary goal of overall planning, industrial layout, and competitiveness improvement of China's transportation departments. The digital upgrading of transportation technology construction is an important guarantee for the development of innovation and integration of transportation.

3.2.1 *Digital transformation of transportation infrastructure*

Digital twin technology plays an important role in the field of transportation in China. In the transportation sector, iTwin Services' digital twin cloud platform has been integrated with another simulation software to suggest an efficient plan for promoting economic recovery. All the achievements of digital twin in the field of transportation cannot be separated from the effective construction of digital infrastructure.

The development of digital infrastructure needs to be analyzed at a macro level. Taking national major transport projects as an example, digital infrastructure is different from traditional transport infrastructure. Digital infrastructure needs to analyze the outstanding performance of project intelligence, consider the coordination of cloud, network, travel, logistics and other aspects, and optimize the construction scheme of digital infrastructure through numerous simulation previews. Generally speaking, the digital transformation of transportation infrastructure is a process to build the cornerstone of transportation safety and support the intelligent development of transportation.

(1) Digital infrastructure needs to be equipped with the safety of traditional transportation facilities, and to further improve the transportation operation efficiency by taking the safety of infrastructure, transportation tools and operating environment into account. For example, analyze the potential risks of accidents, optimize road service facilities, improve safety, and shorten transportation time on the basis of supporting the safe operation of vehicles through digital infrastructure, which is the first step of digital upgrading of transportation facilities.

(2) Digital infrastructure needs to focus on public security construction. Digital twin technology can analyze the important stages of the transportation network, and cooperate with sensor equipment and detection equipment to understand the changes in transportation and people flow. In this process, digital infrastructure should enhance the detection and sensor intensity, complete the scanning detection of people and objects in the form of nonsense, eliminate public security risks, and improve the effective use of transportation resources.

(3) Value of the combination of network application and digital application. In the digital infrastructure, the effective integration of new technologies should be taken into account to ensure that new network technologies and digital technologies can be successfully implemented in the transportation system. Only the construction of long-term, efficient, and emerging technology-integrated transport infrastructure can ensure the overall digital upgrading of the transport network.

(4) Upgrading and innovation of digital technology. The operational effect of digital infrastructure also depends on the level of digital technology. Only by selecting suitable digital applications according to the actual situation of China's transport infrastructure can the value of digital upgrade of transport infrastructure be maximized.

3.2.2 *Digital twin has become a hot spot in transportation construction*

The construction of intelligent transportation is closely related to the digital twin. The new infrastructure of intelligent transportation needs the ability of digital perception, network interconnection, and intelligent disposal. In recent years, the infrastructure construction of intelligent transportation in China has been continuously integrated with digital twin technology, creating a low-cost, low-power, and highly perceived transportation system. In this process, digital twin also gave birth to a lot of traffic construction hot spots.

3.2.2.1 *Construction focus of transportation perception*

The use of vision, radar, and sensors in the process of transportation construction has become more and more frequent in order to improve the perception of China's transportation system. Through the application of these types of equipment, China's transportation network can achieve accurate information collection of transportation elements and flow status, individual accurate identifications, and overall real-time monitoring. These perception devices also support the high-precision modeling of transportation network by digital twin technology.

3.2.2.2 *Construction hotspots of transportation positioning capacity*

The top positioning equipment is also the key point in the process of transportation construction in addition to the perception equipment. The high-precision positioning equipment can accurately perceive the operation state of the transportation elements and build a positioning system in line with the actual situation of transportation in combination with Beidou satellites, in a manner that is convenient for the transportation management department to control and make decisions on the transportation network.

3.2.2.3 *Construction hotspots of transportation network connection capacity*

The digital twin technology has effectively connected with 5G technology to create a reliable and efficient transportation communication system for China's transportation, improving the response speed and disposal speed of the transportation network and playing an effective role in promoting China's intelligent driving technology and unmanned driving technology.

3.2.2.4 *Construction hotspots of transportation digital platform*

The intelligent brain of transportation system is also inseparable from the development of digital infrastructure. Moreover, significant hardware support is required for the information collection, storage, classification, transmission, and the construction of digital twin platforms. The "Outline for the Development Plan of Digital Transportation" issued by the Ministry of Transport of China clearly pointed out that in the future, the transportation department will increase the construction of digital transportation platform, and build a data platform to meet the needs of transportation and capacity based on big data, artificial intelligence technology, digital twin technology, etc., so as to achieve an accurate transportation forecast and assist transportation management to make scientific decisions.

In summary, the new infrastructure of China's future transportation will take digital technology as the core in order to establish more construction hotspots. At the same time, guided by policies and funds, it will continue to strengthen the application depth of digital twin technology; strengthen the integration of 5G, AI, and other new generation information technology; gradually realize the comprehensive digitalization of transportation vehicles, road networks, hubs, and other infrastructure; realize the overall perception of people, vehicles, roads and the environment; and form extensive links, so as to achieve the overall improvement of transportation safety, efficiency, and mass travel experience, promote the transformation of China's transportation industry from high-speed growth to high-quality development, and help China transform from a transportation country to a transportation power.

3.2.3 *Digital twin wins in meeting the needs of transportation in an all-round way*

In the near future, before public travel and transportation, we can find the optimal solution in terms of cost, efficiency, and experience by simply operating the mobile phone. What will China's transportation development present then?

Now, this assumption is no longer a fantasy, but will gradually turn into a reality because digital twin technology can fully help China's transportation to achieve this development goal. The earlier transportation digital models, which could not adequately account for environmental changes or simulate temporal progression, have now been enhanced through digital twin technology.

At present, the digital model of China's transportation system cannot precisely meet the needs of each individual transportation, but it develops at a high speed in this direction. For example, public travel has been efficient and reasonable. On this basis, the transportation department is striving to improve the experience and safety of public travel. It is believed that public travel will be upgraded from efficient to low cost and high experience in the near future.

In other words, the digital twin highlights great value in the process of meeting the needs of mass transportation because it carries out a whole lifecycle preview of the transportation network; it also analyzes the characteristics and value of a single transportation facility. With the continuous improvement of these data, the value of China's transportation system, the rationality of the construction of transportation facilities, and the utilization rate of transportation resources will be greatly enhanced, and a transportation ecosystem with strong vitality will be created.

3.3 Digital Twin Era and Digital Development of Warehousing and Logistics

China's warehousing and logistics industry has also begun to benefit from digital twin with the improvement of the same in the transportation industry; the overall development of the warehousing and logistics industry tends to be digital and intelligent. For example, the warehousing robot created by the digital twin technology can more accurately adjust the angle of a robotic arm, has a higher matching degree with the warehousing environment, and accurately predict possible failure according to the status of the equipment, which greatly improves the overall operation and maintenance efficiency of the warehousing.

As a matter of fact, the digital development of the warehousing and logistics system has become a mainstream trend. With the integration of digital twin technologies, major logistics brands have established intelligent logistics centers and built digital models that can be simulated and tested. With the help of digital model, an intelligent logistics center can detect the operation in various scenarios and make flexible adjustments according to the actual logistics situation to complete self-optimization.

For example, in 2019, DHL, a world-famous express company, released a video about "DHL digital twin intelligent warehouse technology", in which DHL demonstrated the warehouse logistics management and control system built by combining its own supply chain and digital twin technology. In the logistics management and control system of DHL, the digital twin technology can carry out real-time coordination according to the internal situation of DHL and handle various unexpected problems in a timely manner, while improving the storage security.

In general, digital twin brought about six advantages for DHL's intelligent warehouse: (1) optimize warehouse route and improve operation efficiency; (2) build regional early warning system to forecast various conditions in advance; (3) environmental testing to ensure the overall safety of warehousing; (4) visual warehouse management and visual inspection; (5) analyze real-time data and optimize warehousing plan; and (6) enhance the safety of mechanical equipment.

From the logistics and warehousing management and control system of DHL, the advantageous combination of digital twin and the warehousing and logistics industry is mainly reflected in the following three aspects.

3.3.1 *Improve the intelligence of warehousing and logistics industry*

Digital twin has created a virtual twin of the warehousing and logistics system, and obtained the best method of warehousing and logistics operation and maintenance through the means of IOT. At the same time, it has added many functions to the warehousing and logistics system, comprehensively improving the intelligence of the warehousing and logistics industry.

3.3.2 *Lead the warehousing and logistics industry into unmanned mode*

The combination of the new generation of information technology, AI technology, and the warehousing and logistics industry has created a number of unmanned terminals, such as intelligent shelves, intelligent handling robots, intelligent sorting equipment, loading systems, aerial distribution, and robot distribution, but it is digital twin technology that truly creates an unmanned intelligent logistics system.

Digital twin builds a digital model of an intelligent logistics and warehousing system through real logistics data collection, integrates various unmanned terminals in the process, and coordinates the relationship between the unmanned terminals, so as to establish a controllable and verifiable unmanned intelligent logistics and warehousing system.

In addition, digital twin can also carry out the upgrade analysis of unmanned terminals and formulate the overall upgrade scheme of the system, instead of only detecting the real state of logistics and warehousing in real time, which has a profound impact on the development of the logistics and warehousing industry.

3.3.3 *Solve the construction problem of intelligent logistics system*

It is difficult to improve the intelligence of traditional logistics system, and the room for improvement is limited because the intelligence of the logistics system needs to be considered from the overall perspective, rather than improving the intelligence of a certain link and a certain subsystem. After the digital twin technology is integrated into the logistics and warehousing system, the designers of the logistics and warehousing system can effectively combine multiple logistics subsystems, coordinate the cooperation of various links and subsystems, accurately control the upgrading key of the logistics and warehousing system, reduce unnecessary waste of resources, and reduce the probability of system errors.

The system planning of the existing logistics and warehousing system can carry out collaborative planning for each subsystem and each link by using digital twin technology, and build an upgrading framework based on the overall system. In the process, the digital twin can understand the impact of each link on the overall system, after upgrading in real time, and create new processes according to the actual situation, or integrate different links and subsystems, so as to ensure the optimization of the upgrade plan.

In the digital twin era, the digital development of the warehousing and logistics industry is very rapid, which includes not only the digital transformation of the physical logistics and warehousing industry but also the intelligent upgrade of the system. In this process, the digital twin enables stakeholders to identify and resolve potential issues within the logistics system, fostering innovation and new development pathways that enhance the value of the warehousing and logistics industry and accelerate its overall advancement.

3.4 Digital Twin to Create a New Pattern of Intelligent Transportation Development

In October 2020, the National Transportation Communication Information Center held the Directors' Work Exchange Meeting to discuss challenges and strategies in data standardization across provincial transport systems. This meeting, under the theme of "adapting to the new situation, responding to new challenges, applying new technologies, and seeking a new pattern of intelligent transportation development", discussed the development direction of China's intelligent transportation in relation to the "the 14th Five-Year Plan" period, and further discussed hot topics of intelligent transportation such as 5G technology, new infrastructure, big data, and network security.

At the meeting, the head of the Science and Technology Department of the Ministry of Transport of China stated that during the "the 14th Five-Year Plan" period, China's transportation should rely on digital forces to create a new pattern of intelligent transportation development. In particular, after General Secretary Xi gave important instructions such as "accelerating the construction of strengthening country with strong transportation network, accelerating the construction of a new development pattern, and promoting the Construction of new infrastructure and new urbanization and major projects such as transportation and water conservancy", it became imperative for the transportation department to follow such instructions, accelerate the implementation of the specific actions of the "Outline for the Construction of Strengthening Country with Strong Transportation Network", promote the digital and intelligent development of the transportation industry at a new starting point, and better integrate 5G, artificial intelligence, digital technology, and other emerging technologies to better plan the development significance of intelligent transportation.

At present, China's transportation departments have built a big data platform for comprehensive transportation, including a number of intelligent transportation operation supervision and service platforms, and the effect of transportation coordination operation and emergency response has been significantly improved. These achievements are inseparable from the support of digital power, especially the promotion of digital twin technology.

In fact, with the assistance of digital twin, not only has China's transportation department developed a series of technological products, such as transportation data management and analysis platform, transportation simulation and deduction platform, and transportation data storage system, but also the development of the global transportation industry has been changed by digital twining, as digital twin is now a cutting-edge technology, and its deeply integrated industries have undergone tremendous changes.

Case: Digital twin creates new opportunities for aviation industry

Following the outbreak of the COVID-19 pandemic, the global aviation industry entered a period of relative stagnation. However, this phase also presented major airlines with an opportunity to reassess their strategies, explore new development directions, and innovate their business models. During this period, major airlines and aerospace manufacturers worldwide intensified their research and development efforts. For instance, Siemens, Rolls-Royce, and Airbus jointly developed the E-Fan X hybrid-electric aircraft demonstrator, an experimental electric commercial aircraft. Airbus SAS plans to launch the first electric civil aviation aircraft capable of carrying 100 passengers and flying up to 1,000 kilometres by 2030.

From the perspective of energy conservation and environmental protection in the aviation industry, this strategy of Siemens, Rolls-Royce, and Airbus Company is far sighted and meets the needs of the development of the times. However, aircraft R&D is a significant and long-term project. Taking the C919 developed in China as an example, its R&D has been in progress since 2008. By 2021, although the test flight was successfully completed, the C919 has still not been fully put into use. Therefore, it requires airlines to make long-term and large investments to ensure research and development results and further ensure that new models can lead to new breakthroughs.

Facing the technical, strategic, and periodic research and development challenges, the focus of the aviation industry has been how to comprehensively solve these system problems. After the deep application of digital twin technology, major airlines in the world seem to have found new opportunities. For example, Siemens has applied digital twin technology to many aspects in the research and development of its electric civil aviation aircraft:

(1) Siemens has carried out digital simulation of fuel usage and determined the mileage design scheme of the electric civil aviation aircraft according to data changes.
(2) Siemens has also applied digital twin to aircraft power systems. In the process of aircraft research and development, the effects of aerodynamics on the airframe, environmental control, landing gear, and avionics should be considered. Previously, in aircraft development, numerous experiments and modifications were required, as well as large sums of money. However, after the application of digital twin technology, the aerodynamics performance of the aircraft can be simulated and inferred in virtual space. The aerodynamics performance can not only collect comprehensive data but also conduct comprehensive analysis on the design of aerodynamics characteristics, aerodynamics noise, fuselage belly fairing, and wing fairing to optimize the overall design of the aircraft.

At present, major airlines around the world have reached great achievements in research and development by using digital twin technology. For example, GKN Aerofalco Landing Gear Co., Ltd., an enterprise specializing in the production of aircraft landing gears, has upgraded the safety and reliability of its products with the systematic simulation features of digital twin technology, saving 30% of its R&D and production time. GKN Aerofalco Landing Gear Co., Ltd.'s successful model has also been widely applied in the industry.

From the successful experience of Siemens aviation and GKN Aerofalco Landing Gear Co., Ltd., it can be seen that digital twin technology has become a key element in the development of the aviation industry, and the future development of the aviation industry cannot be separated from this technology.

Under normal circumstances, the service life of an aircraft can reach several decades, but in the process, various data of the aircraft changed. Digital twin technology can record and analyze the data of the whole lifecycle of the aircraft, and build the digital model of the aircraft operation to maintain the safety performance of the aircraft to the maximum extent.

Therefore, the integration of modern aviation industry and digital twin technology has become an inevitable trend. Under the support of a large number of digital models, the research and development of aircraft can improve performance, reduce the research and development cycle and costs, and the operation of aircraft can be controlled in a real-time, two-way, transparent, and systematic manner. It can be said that digital twin technology has become the main force to break through the bottleneck of design and development in the aviation industry.

Compared with developed countries in terms of aviation research and development technology and transportation, there is still a significant gap in China's aviation research and development level, but this gap is gradually narrowing with the help of digital twin. The development of China's aviation industry is replicating the innovation path of high-speed rail and the 5G communication industry. We believe that China's aviation industry can seize the opportunity of digital twin to develop into a strong competitor of the international aviation industry.

Case: Tencent autopilot—driving autopilot and intelligent transportation with digital twin technology

Digital twin technology promotes the development of intelligent transportation in an all-round way. It not only creates new opportunities for the aviation industry but also greatly promotes the development of automatic driving technology in China. In September 2020, a Mobile Internet Yuelu Summit was held in Changsha. Su Kuifeng, the general manager of Tencent automatic driving, delivered a speech titled "Digital twin promotes automatic driving and intelligent transportation evolution" at the summit. Su Kuifeng conducted a careful analysis of the fact that the development of digital twin promotes the growth of automatic driving technology and digital twin drives the development of intelligent transportation. Furthermore, Tencent's autonomous driving project (or platform) serves as a case study to demonstrate how digital twin technology creates value and functions within the transportation sector.

Su Kuifeng stated the following at the meeting: "Digital twin technology, supported by real data, realizes the seamless interconnection between virtual space and physical world, provides a free test site for automatic driving and intelligent transportation solutions, and makes people foresee the unlimited possibility of future transportation".

"Tencent has always insisted on consolidating the foundation of industrial development – cloud, simulation and high-precision map, which can not only support the development and operation of automatic driving system, but also help the construction of intelligent transportation through platform services such as data management and vehicle management".

Su Kuifeng then stated the following at the Tenth Annual Meeting of China's Intelligent Transportation Market in March 2021: "The transportation industry is gradually ushering in a change of focusing on vehicles, roads and scenarios, but in essence, it is all about providing services for people. In the future, the development direction of transportation is the comprehensive intelligence centered on people. Based on the dynamic perception and real-time data, Tencent will connect the physical world with the information world by using digital twin technology to forecast, infer and verify the urban transportation system, and at the same time, it will give full play to the connection ability of Tencent C terminal to provide decision-making assistance for the planning and governance of urban transportation".

"On the whole, the future transportation will be based on refined operation and perception, through infrastructure digitalization, cloud data integration, digital twin technology analysis and deduction, combined with transportation indicator system and evaluation tools, to achieve engineering optimization, facility control and C end touch, forming a closed-loop system of sustainable evolution of transportation governance. In this process, Tencent will give full play to its ecological advantages and technological accumulation, help the linkage of automobile, transportation and energy, and contribute more to the construction of strengthening country with strong transportation network and the realization of the goal of reaching carbon peak and carbon neutralization".

As a well-known technology company in China, Tencent Inc. has been playing a positive role in promoting the development of various industries for a long time. The head of Tencent's autonomous driving division indicate that digital twin technology has become a fundamental element of the company's autonomous driving research and development framework. This technology can not only meet the needs of transportation development but can also improve the results of transportation development, and it is of great significance to create a new pattern of intelligent transportation development.

Chapter Summary

An important sign of the development of modern transportation digitalization is the all-round application of digital twin. The safety of China's transportation network will be comprehensively upgraded, digital infrastructure will also be vigorously developed, and the intelligent and digital transformation of the warehousing and logistics industry will also be highlighted in the process of integration of digital twin and transportation. Overall, the integration of digital twin technology marks a new stage in the development of China's transportation sector and enhances its competitiveness within the global transportation system.

Chapter 4

Digital Twin Helps Intelligent Transportation Simulation Upgrade

As early as before the birth of digital twin technology, the transportation industry has carried out the application of various simulation technologies. The simulation of traffic development is carried out via the correlation analysis of traffic individual location data and traffic network passenger flow data. Therefore, after the advent of the digital era, the matching degree and integration degree of the transportation industry and the digital twin are higher. The digital twin realizes the virtual and real synchronization, real-time tracking, and scenario online of the transportation network and comprehensively upgrades the simulation effect of intelligent transportation.

4.1 Simulation Application of Digital Twin in Air Transportation

Beijing Daxing International Airport has always been known as the first of the "seven new world miracles". This large international airport, with an investment of more than RMB 80 billion and a floor area of 1.4 million square meters, is expected to have an annual passenger throughput of 100 million and an aircraft take-off and landing of 800,000, and is at the forefront of the world in terms of scale and performance. More importantly, the completion of Beijing Daxing International Airport has enabled China's transportation to achieve a seamless connection between high-speed rail and aviation, and has brought China's transportation history into a new stage of development.

Since the completion of Beijing Daxing International Airport, the airport has been continuously upgraded in terms of safety, intelligence, and humanized operation. The key force to support Beijing Daxing International Airport in achieving various breakthroughs is technology. After the advent of the digital era, Beijing Daxing International Airport began to integrate with various digital technologies and often carried out digital upgrades, creating a "model project" for the simulation application of digital twin technology in China's air transportation.

The simulation application of digital twin in Beijing Daxing International Airport is mainly shown in the following aspects.

4.1.1 *Digital twin technology realizes all-round simulation of Beijing Daxing International Airport*

Beijing Daxing International Airport has built the world's first "double in and double out" terminal building and adopted a five-finger gallery configuration with central radiation, which makes the functionality of Daxing more prominent. Under the digital twin technology, the three-dimensional model of Beijing Daxing International Airport can be displayed from multiple angles and in all directions, and the perspective can be freely switched, and every architectural detail can be seen in the virtual space.

The airport apron of Daxing International Airport is also the focus of the virtual simulation of digital twin. Under the virtual model established by digital twin, the upper and lower passenger areas, cargo loading areas, refueling areas, parking areas, and specific areas of the aircraft can be clearly displayed. At the same time, the model of virtual space can also predict the departure process of the aircraft, and this process can be detailed to the display of airport runway navigation lights.

In addition, the digital twin of Daxing International Airport also penetrates the simulation effect of the large-scale airport model into the inside of the tower. Through the data collection after the internal scanning of the tower, the digital twin visualizes the inside of the tower, and the working area and working status of the tower can also be clearly demonstrated.

4.1.2 *Digital twin technology realizes route simulation of Beijing Daxing International Airport*

After completing the digital modeling of Beijing Daxing International Airport itself, the digital twin technology has also completed the real simulation of airport routes. We can intuitively see the international routes starting from Beijing Daxing International Airport from the digital model. These routes clearly connect various countries around the world. Moreover, we can see the close relationship between China and the world from this connection network.

4.1.3 *Digital twin technology realizes the operation trend simulation of Beijing Daxing International Airport*

After the completion of the model simulation and route simulation of Beijing Daxing International Airport, the digital twin also brought the flight flow effect and the overall operation of the airport. This simulation technology is a comprehensive management after adding the flight information, real-time flight status, aircraft flight status, airport passenger flow, and cargo flow of the airport to the digital model. The simulation results help airport managers to control the operation status of the airport at all times, optimize the flight arrangement, and improve the operation efficiency of the airport.

In general, the application of digital twin technology in Beijing Daxing International Airport has improved the airport operation efficiency, passenger travel experience, and the safety guarantee level of aircraft operation and has also promoted the development of China's aviation economy. Beijing Daxing International Airport, as a representative transportation hub of digital twin for the digital upgrade of China's air transportation, has a directivity in its upgrade effect and has far-reaching significance for the overall simulation and upgrade of China's intelligent transportation.

In future development, no matter how to upgrade aviation tools, the upgrade effect brought by digital twin in three aspects cannot be ignored:

(1) **Visualization of aviation operation and maintenance management:** The accurate and complete modeling of digital twin technology can monitor all details of the airport in real time, and airport apron monitoring and airport slot analysis can be visualized. This modeling effect helps airport managers to understand the frequent operation situation, ensure the operational efficiency of the airport, and improve the use of aviation resources.
(2) **Visual display of aviation operation:** The visualization of airport routes by digital twin technology can help airport managers to intuitively clarify the flight operation status and facilitate the visual

analysis of operation elements for flight status. For example, after the outbreak of COVID-19, the frequency of temporary flight stoppage continues to increase. Airport managers can intuitively clarify the utilization of airport resources and conduct follow-up analysis and management for the suspended flights with the help of digital twin technology.

(3) **Enhance the airport's emergency response capacity:** The virtual model constructed by the digital twin can analyze the special situations including flight delay, peak passenger flow, large area flight stoppage, and so on by combining the airport flight flow trend and the use of aircraft resources in the rehearsal, so as to facilitate airport management personnel to carry out situation prevention and accident handling in advance. In addition, the virtual model constructed by the digital twin can also assist the managers in analyzing the causes of various emergencies and unexpected situations, and provide real-time and effective emergency decision-making basis for the airport managers.

The completion of Beijing Daxing International Airport represents that China has not only the largest single airport terminal in the world but also the ability to overcome high-end aviation infrastructure problems; the good operation of Beijing Daxing International Airport reflects not only the technical level of China's aviation transportation but also the overall strength of China's transportation development. In the process of construction and operation of Beijing Daxing International Airport, digital twin technology plays a key role. This technical force will continue to improve the intelligence and efficiency of China's aviation and continuously promote the high-quality development of China's aviation.

4.1.4 *Construction foundation of digitalization and informatization of air transportation*

Since the concept of "smart airport" was officially put forward by the civil aviation authorities in 2017, China's aviation development has begun to focus on strengthening digital and information applications to promote smart upgrading, which has laid an important foundation for the sound development of China's air transportation in recent years.

As of 2021, based on the construction idea of "one core, two wings", China's aviation industry has fully established a variety of intelligent platforms such as network security, cloud services, and big data and effectively improved the capacity of aviation transportation resources management, intelligent services, and intelligent operation.

For example, Guangzhou Baiyun International Airport comprehensively enhanced its own intelligence and built an integrated intelligent platform with the characteristics of smart ecology, smart services, smart security, smart commerce, and smart production, which improved the operation efficiency and service level. With the goal of "future airport", Shenzhen Baoan International Airport has carried out comprehensive upgrading of informatization and digitalization, setting a new benchmark and creating a new model for the smart transformation of China's waterway transportation.

In general, there are five bases for the information and digital upgrade of air transportation in China: ① upgrade the personalized and convenient service level of air passengers, ② upgrade the connection efficiency of air ground transportation, ③ upgrade the operation and maintenance management of the logistics warehousing industry, ④ upgrade digital value-added services in the aviation industry, and ⑤ upgrade the intelligence of air transportation infrastructure.

Through digital and information upgrading of these five aspects, we will give full play to the advantages of China's aviation market and aviation resources and create a set of digital twin system that combines big data, cloud computing, artificial intelligence, and other advanced technologies. We will comprehensively improve the development speed and level of China's aviation transportation under the guidance of high-end digital models.

4.1.5 *Air transportation intelligence fusion*

In order to accelerate the process of the integration of air transportation intelligence and ensure the application effect of digital twin technology in the field of air transportation, China's aviation industry has taken a series of digital upgrading and transformation measures in recent years. In general, China's aviation industry has analyzed the sustainable development of digital twin in the aviation field from the top-level perspective, which can be confirmed by the application scope and effect of digital twin technology in major domestic airports. After that, a series of measures for the application of digital twin have been taken by China's aviation industry to ensure the complete landing of the digital twin in the field of air transportation and promote the integration of intelligence and number in air transportation:

(1) Based on the overall needs of air transportation, China's aviation industry has actively integrated advanced technologies such as cloud computing, Internet of Things, and big data and established an intelligent and digital integrated air transportation infrastructure. Based on the digital transportation infrastructure, the aviation industry is able to build a digital twin with comprehensive perception and all elements collection, and take this as the industry's artificial network, cooperate with a variety of emerging technologies to continuously upgrade the intelligence of the aviation industry, accurately associate people, materials, and transportation, and analyze the optimal way of industry development.

Through a comprehensive analysis of the development achievements of China's aviation industry in recent years, it can be found that each major airport has the ability of internal perception, built a data analysis platform, and deepened the application of digital twin, which has significantly improved the two aspects of aviation operation and maintenance and safety assurance. From the perspective of the whole lifecycle of major airports, the business process of China's aviation industry is becoming more and more efficient, the service quality is also constantly improving, and the overall security and stability of the industry are more significant, which is also supported by the simulation and deduction conducted by the digital twin.

Nowadays, China's aviation industry has begun to build a high-precision and multi-coupling digital information model, through which the airport planning, construction, and management are continuously upgraded. From the perspective of model construction effect, the digital twin technology provides airport managers with an integrated, visual, and intelligent airport operation model and provides a basis for comprehensive management and intelligent decision-making. It can be said that various development strategies of modern aviation industry are born in the simulation and deduction process of the digital model.

(2) With the improvement of the application scope and depth of digital twin in the aviation industry, the analysis of airport operation by this technology is more accurate, and the prediction of various development situations is more far-reaching. For example, the law of airport operation becomes more and more obvious with the continuous

accumulation of passenger data, cargo data, operation data, safety data, and aircraft equipment data. With AI technology, various development situations can be predicted by the digital twin, and the development needs of passengers, market, and industry can be met in depth.

(3) The digital twin technology has established a unified platform for the aviation industry and realized the preview management of the whole lifecycle and the whole business cycle. China's aviation industry has integrated an airport intelligent model and built an information management platform integrating the sky and the earth through the visual simulation technology of digital twin. The construction of this platform has overcome various problems of business segmentation management and independent management in the traditional management mode of the aviation industry. On this platform, the onboard single data and business data can be integrated and fused, and the air sky ground business can be integrated and linked. Not only can the airport managers make intelligent decisions on various businesses with one key, but passengers can also enjoy the whole process of one screen-viewing. Both parties can realize real-time and all-round perception of the travel and cargo transportation process.

(4) China's aviation industry has established a complete information security guarantee system according to the characteristics of digital twin applications. The accompanied openness of digital twin has also brought information security problems though the digital twin technology has greatly promoted the development of the aviation industry and promoted the progress of air transportation in China. In recent years, information security attacks have occurred in various fields of transportation in China, and the attack technology and frequency are constantly increasing. Therefore, in the process of using digital twin to promote the development of the aviation industry, the airport information security system must be complete, that is, all links from the real world to the digital space must be secure and free from loopholes and interaction weaknesses. For this situation, China's aviation industry combined with its own characteristics of business flexibility, divided different safety areas in the safety protection system according to the importance of business, and adopted different protection and isolation technologies between different areas to achieve different levels of isolation and access control in different areas. At the same time, in the application process of digital twin technology and the integration process of digital twin and other technologies, the aviation industry first guaranteed the security of its

own digital system, practiced a unified security management platform and operation and maintenance platform, and then added digital twin on this basis, so as to realize the effective control on the user identity and access, and moreover, the requirements of security audit are flexibly enhanced. In general, before the digital twin promoted the integration of China's air transportation intelligence and data, the aviation industry carried out comprehensive safety protection; currently, the security system is upgraded in an all-round way on the basis of the application characteristics of digital twin to ensure the security and stability of digital twin technology.

4.2 Simulation Application of Digital Twin in Ground Transportation

Ground transportation serves as the main mode of transportation operation in China, and the high-quality development of ground transportation is also an important foundation for China to achieve the strategic goal of a strong transportation country. China's ground transportation has also begun to carry out in-depth simulation applications under the development mode of digitalization and informatization and to create a variety of simulation application modes in combination with its own complex environmental characteristics.

At present, the largest role of digital twin in the application of ground transportation lies in the three-dimensional display of transportation construction and transportation facilities operation in various complex environments, as well as the prediction of various accidents and special conditions. It can be said that the application of digital twin leads the ground transportation data to achieve a breakthrough from two-dimensional to three-dimensional and comprehensively combines external data, and what's more, real all-round simulation of the development of ground transportation is carried out.

4.2.1 *Digital twin brings more possibilities for ground transportation*

The application of digital twin in ground transportation mainly represents the replication of virtual space of ground transportation facilities, vehicles, and people flow and then finds the best way for the coordinated development of the three through data analysis. Nowadays, digital twin has created many optimization schemes in the construction and development of ground transportation, which has promoted the high-speed and high-quality development of ground transportation in China.

For example, in the process of expressway or urban transportation development, digital twin can realize the simulation of transportation flow and overall transportation operation, so as to provide a safe, efficient, and economic development plan and continuously optimize various data parameters and implementation strategies.

After summarizing the effects of digital twin in the development of ground transportation, it can be seen that digital twin brings more possibilities for ground transportation and greatly expands the development space of ground transportation.

4.2.1.1 *Possibility of new transportation control strategies*

Digital twin has created a variety of control possibilities for ground transportation, and the actual effect of the new control strategy can be directly detected after digital space simulation. For example, the control focus of ground transportation in China can be upgraded from monitoring flow to real-time dynamic intervention, so as to optimize the use of transportation resources and improve the transportation operation effect.

These new transportation control strategies have many advantages: They can not only reduce the ground transportation congestion but also improve the safety of the transportation network. However, at present, many new control strategies of ground transportation are still in the experimental stage and need to be deeply combined with the actual situation of transportation facilities and ground transportation flow. The landing method and use effect also need to be upgraded. However, these controls may be gradually improving the previous control methods and accelerating the development of ground transportation in China with the promotion of digital twin.

At present, the unlimited possibility brought by digital twin for transportation control has driven the development of the global transportation industry. Taking Aimsun, a newly established software company of Siemens Mobility Group, as an example, it is a service company that models the global transportation industry. This model can simulate not only the normal operation of the transportation network but also the movement of people and goods flow, identify the weak links in the road network system, and optimize the operation mode of traffic flow. Aimsun provides the best way to use resources for the ground transportation of various countries and implements the simulation of the operation of ground transportation to provide a variety of reasonable travel modes. As a result, Aimsun company has developed rapidly in the world.

In the process of understanding the digital modeling products of Aimsun company, the author found that Aimsun observed and analyzed the transportation digital model carefully, observed each transportation area from different perspectives of macro, meso, and micro, simulated each part of the transportation hub carefully, and then carried out correlation analysis of various simulation data, so as to provide customers with more accurate and more practical solutions for transportation problems.

Based on the characteristics of Aimsun's business, at present, the transportation departments of the United States, the United Kingdom, Australia, France, and other developed countries have reached in-depth cooperation with Aimsun. Aimsun conducts real-time simulation of transportation in the important first-tier cities in these countries, forecasts the upcoming transportation disruption in advance, and finds the shortest time to reduce congestion and find solutions for terminal problems, which will make these important cities more intelligent.

4.2.1.2 *Possibility of transportation operation improvement*

At present, many developed countries in the world have formulated transportation development strategies in line with digital and intelligent transformation. For example, Austria regards the cities around the capital Vienna as separate transportation areas, which makes it easier to model these transportation individuals individually using digital twin technology, and then simulate the associated operations of the surrounding areas from a macro level.

Digital twin has improved transportation operation in many aspects and formed various modes at the same time. Digital twin can formulate strategies to improve operation according to the actual situation of different regions, so as to make urban commuting smoother, make urban space continuously enlarged, and improve the utilization rate and rationality of transportation resources.

For instance, after the digital twin modeling, the urban public transport operation can be smoother, and the change time of transportation lights and the allocation of public transport vehicles in key areas can be optimized. At the same time, coordinated management can be carried out on the basis of the development of the city and the change of the environment to ensure that the wisdom of the public transportation system grows simultaneously with the wisdom of the city.

In addition, the simulation brought by digital twin not only involves urban public transport vehicles but also combines the pedestrian flow and bicycle flow in the city. These key traffic flows are critical to the overall impact of urban transportation. With the changes and updates of these data, digital twin can calibrate the accuracy of the virtual model in the shortest time.

4.2.1.3 *Broad prospect of ground transportation development*

With the development of digital twin technology, the geographical and social factors that can be combined with the virtual model become more and more abundant, which means that the virtual model constructed by digital twin is also constantly upgrading, the implementation of various transportation planning is more detailed, and the controllability of transportation development is more comprehensive.

Compared with air transportation and water transportation, the complexity and dynamic of ground transportation is more prominent, and the requirements for digital twin modeling are also higher. In the future, the development of digital twin can give more promotion to ground transportation at the medium and micro levels. This technology will become a significant advantage in the development of ground transportation and open up a broader space for the development of ground transportation.

4.2.2 *How to realize digital twin operation of metro stations with strong passenger flow?*

During the May Day holiday in 2021, the number of tourists in Changsha, Hunan Province, increased dramatically. The passenger flow of Changsha Metro Line ranked first in China for 5 consecutive days. Changsha has also become the first city in Central China with the top passenger flow intensity of the subway. According to the statistics of China's transportation departments, during the May Day holiday of 2021, Changsha Metro transported 15.2113 million passengers and an average daily passenger volume of 2.5352 million. This huge number is a challenge to the operation of urban rail transit in developed countries in the world. However, Changsha perfectly carries this passenger flow operation.

The reason why Changsha Metro can maintain normal operation under such a large passenger flow base is that digital twin has formed the key to security. According to the Changsha Metro network, the Changsha Wuyi Square Metro Hub is the key core of the entire metro network. It connects famous scenic spots and business districts in Changsha, such as the underground mall of International Financial Street, New World Department Store, Pinghetang, IFS International Financial Center, and Huangxing South Road Pedestrian Street. Therefore, this Metro Hub is also the transfer station with the largest passenger flow. During the May Day holiday in 2021, the total number of people entering and leaving the Wuyi Square subway hub station exceeded 1.4 million.

Wuyi Square subway hub is divided into three floors with a total of 8 exits. In terms of construction scale, it is not the world's top subway hub, but why can the Wuyi Square subway hub meet the needs of strong passenger flow in Changsha? It is precisely because before that, the Changsha rail transit management department predicted the rail transit pressure during the May Day holiday period in advance and formulated the corresponding operation strategies.

Changsha rail transit management department has reached a deep cooperation with Beijing 51 WORLD, a famous digital twin company in China. Beijing 51 WORLD and Hunan Infopass jointly built a simulation operation platform for the Changsha rail transit management department for Wuyi Square Station.

As the first test management platform of Changsha smart rail transit, the digital platform has shown strong intelligence, and has realized the dual model drive of "station daily operation management + passenger flow simulation and deduction" of Changsha rail transit, and has also established a good benchmark in the field of national rail transit development.

The Wuyi Square subway hub station can conduct accurate passenger flow simulation with the support of digital twin technology. Based on this actual operation status, Changsha rail transit management department has formulated a good operation strategy in advance. Generally speaking, the simulation operation platform of Changsha Wuyi Square station has two main functions: One is digital operation and maintenance management based on actual operation status, and the other is simulation deduction based on passenger flow simulation and operation simulation.

Rail transit is more intuitive in operation and management than other ground transportation because the core of rail operation and management is passenger flow, and resource allocation is based on passenger flow changes. With the help of digital twin, Changsha Wuyi Square subway hub station can realize the real control of passenger flow, and then match the control strategy of in and out station, boarding and alighting, the carrying capacity and transit capacity of Changsha Wuyi Square subway hub station are effectively improved.

Moreover, in order to ensure that the management department of Changsha Metro can effectively cope with the load pressure caused by high passenger flow, Beijing 51 WORLD, based on the AES full element scenario construction capability, conducted a 1:1 complete restoration of Changsha Wuyi Square Station so that the connection between Changsha Wuyi Square Station and the ground, the status of different periods of time, and the changes in operation status can be perceived in real time, which provides a decision-making basis for site operation management and plan deduction.

Based on the business characteristics of Wuyi Square station, Beijing 51 WORLD has built a set of "5 + 1" business model. The so-called "1" refers to the twin simulation of 1:1 scene of all elements of Changsha Wuyi Square station, which enables Wuyi Square station to realize all-round monitoring of passenger flow, cargo flow, and vehicle transfer. "5" refers to the overall situation awareness, safety protection, real-time response, coordination management, and decision-making assistance of Changsha Metro. Changsha Metro system will not be blocked due to significant changes in passenger flow and will remain in a controllable, visual, and adjustable operation status as a whole through the enhancement of its ability in these five aspects.

The combination of Changsha Metro and digital twin is the guidance of the development direction of China's rail transit. The close combination of Changsha Metro and digital twin technology makes the connection among passenger flow, equipment, and management more closely, which greatly improves the wisdom of rail transit, and also provides safety guarantee for mass transit. It can be seen that digital twin is the main force to help China's intelligent transportation simulation upgrade.

4.3 Simulation Application of Digital Twin in Water Transportation

The development of digital transportation is also the future trend of water transportation. In the process of achieving high-quality development of water transportation, digital and information technology has become a key force. According to the spirit of the "Outline for the Development Plan of Digital Transportation", the development of water transportation needs to explore a new path of integration with digital twin, enhance the core development strength by using the power of digital twin, and promote the wisdom level of construction, operation, and maintenance of water transportation.

2021 is the beginning year of China's "14th Five-Year Plan". In this new stage of development, the consolidation of transportation infrastructure will become the focus of development. Digital technology is no longer a choice but a development model and an upgrading path. Digital twin waterways and digital twin water networks have been gradually implemented in the field of water transportation in China, and the construction of smart water conservancy cannot do without the promotion of digital twin.

Niu Xinqiang, academician of the Chinese Academy of Engineering and president of the Changjiang Institute of Survey, Planning, Design and Research (CISPDR), once said that "Digital China" has now become a national strategy, and "digital economy" has reconstructed the "new prospect" of economic development unprecedentedly. The development of China's water transportation should make use of digital twin technology to build a development model combining the virtual and the real to create a new form of water transportation development. In the "2019 Typical Cases of Shanghai Big Data Collection", the "digital twin water depth real big data platform" created by Tairui Group has become a representative of the integration of water transportation and digital. This digital platform integrated by three-dimensional AR can collect various dynamic data of the whole process of the whole cycle of River Marine Engineering, conduct detailed analysis on the data, and sort out the decision-making of the development of water transportation and water navigation intelligent treatment.

The "digital twin water depth real big data platform" created by Tairui Group is a major breakthrough in the integration of China's water transportation and digital twin technology. It is based on the three-dimensional model of the real scene, simulating the real topography of water and water, and conducting a comprehensive analysis in combination with maritime activities such as terminals and ships. This is very similar to the model of digital twin analysis of environmental factors used by ground transportation in China, but the analysis of water transportation is more three-dimensional and comprehensive. Among the "digital twin water depth real time big data platform", the maritime safety data, ship AIS real-time information, and remote real-time monitoring video can be clearly reflected, and various information data on the Web can be viewed and inquired in real time, and the operation status of channel transportation can also be monitored and controlled in real time. This platform provides multi-dimensional decision-making support for the maritime transportation management department. It ensures the high-quality development of water transportation in Shanghai.

From the role of "digital twin water depth real-time big data platform", it can be seen that the digital twin has mainly contributed to the water transportation in the following aspects:

(1) **Environmental simulation testing:** The digital twin technology can carry out model simulation on the water and underwater environment, enhance the visual authenticity of the water and underwater environment, and improve the visual monitoring effect through layer tree, hidden terrain, shadow effect, and other functions.
(2) **Vessel operation status inquiry:** "Digital twin water depth real-time big data platform" can accurately locate the ships in the AIS real-time information and query the ship navigation track to understand the ship navigation status.
(3) **Remote real-time monitoring:** The "digital twin water depth real time big data platform", together with the eagle eye monitoring of the key section of the ship, can combine the AIS and video identification functions to obtain the dynamic information of the ship in this section in real time, so as to ensure the safety and smoothness of the key section.

At present, digital twin technology has been deeply applied in many maritime and port shipping units in China, and a professional data simulation platform has been established. The safety assurance system of China's water transportation is supported by real and dynamic data, and the regulatory effect of water transportation is also refined and accurate with the help of digital twin.

Compared with most countries, China has a huge water transportation network, and the development of water transportation is also in a healthy state. However, the author found that there are still problems in China's water transportation, such as shortage of shipping crew, insufficient level of intelligent navigation, and limited economic development of water transportation after careful analysis of the development status of China's water transportation. Combined with the matching of digital twin and transportation development, the author believes that the construction of water digital twin system can improve the operational status of China's water transportation and improve the service efficiency and development strength of the water network.

The water network conditions and shipping depth of China's water transportation are among the best in the world, which provides strong support for the development of China's transportation economy. In accordance with the statistics of China's transportation departments, China's waterway transportation accounts for about 10% of all transportation, but the proportion of waterway transportation exceeds 15%, which indicates that China's Waterway resources have greater advantages. However, it is found that the infrastructure construction of waterway transportation in China needs to be improved, especially the intelligence of infrastructure, which is significantly different from that in developed countries, after being compared with the waterway transportation construction in developed countries.

For example, although China's transport vessels are fully equipped with AIS equipment according to the requirements of the government and the AIS system can also clearly understand the actual navigation situation of the vessels, the installation and use effect of AIS has not been effectively controlled, and the role of AIS has not been fully demonstrated, which has led to the deepening of the separation of shipping information, and moreover, the congestion, queuing, as well as low punctuality rate have not been effectively solved.

The author believes that the optimization idea of water transportation development can start from the following three aspects combined with the application characteristics of digital twin technology:

(1) Water transportation needs to be improved in terms of digitalization and informatization from the construction of infrastructure, and a comprehensive digital system needs to be formed for transportation vehicles, loading tools, and transfer facilities.

(2) The ship, port, channel, and other elements should break the information barrier and carry out collaborative operation.
(3) Improve the delicacy of channel operation and maintenance management and control, and enhance the controllability of water transportation management.

It is not difficult to see that the development of water transportation in China can be significantly improved from these three aspects, and the simulation effect of digital twin technology can perfectly solve these problems.

In general, the simulation application of digital twin in China's water transportation mainly aims at solving the existing blocking points of water transportation development, making up for the existing deficiencies of China's water network, making clear the development direction more intuitively under the simulation of array word twin, and learning from the experience of digital transformation of water transportation in developed countries to accelerate the breakthrough of China's water transportation.

Case: The digital twin protects Chang'e 5 to come back home

In December 2020, China's Chang'e 5 lunar probe returned from the moon and landed successfully in the predetermined area of Siziwang Banner, Inner Mongolia. This marks the successful completion of the system engineering with the highest complexity and the largest technology span in China's aerospace field, which is also a milestone breakthrough in China's aerospace field.

Digital twin technology also plays an important role in this major project in the aerospace field. First, the digital twin technology "visual simulation of sampling and packaging" was carried out for the sampling task of Chang'e 5. In the digital model of Chang'e 5, the digital twin made a real reappearance of all elements of the mechanical arm, drilling mechanism, sealing, and packaging device and other complex mechanical structures of Chang'e 5 detector. According to the data from the integrated spacecraft of Chang'e 5, the real-time changes of the joint angles of the robotic arm and the workflow of the robotic arm, the drilling mechanism and the sealing and packaging device are perfectly demonstrated in the virtual space. At the same time, the digital twin technology also carries out real-time visual analysis on the packaging situation according to the end position data of the Chang'e 5 robotic arm, the motor speed of the drilling mechanism, the driving motor current, the force sensor and the temperature, and other data, and provides the operation decision-making basis for each segment of the lunar sampling according to the data changes. Although Chang'e 5 is on the moon which is far away from the earth, every movement and every detail is carried out in front of the eyes of China's space technology team.

In addition, the digital twin technology also enables our space team to have a clear perception of the topography of the moon. The "sampling and packaging visual monitoring system" equipped by Chang'e 5 can capture the imaging data of the spacecraft camera in real time, collect various remote sensing data, provide visual means through digital twin, and conduct multi-dimensional analysis on various data, so as to help the space team accurately master the surface information of the Chang'e 5 working area and make scientific research and judgment decisions.

It can be said that with the support of digital twin, the difficulty of the "moon exploration" mission has been greatly reduced, and the working details of Chang'e 5 have been accurately controlled. Therefore, in the process of collecting lunar regolith, China's aerospace expert team can handle all kinds of situations in an orderly manner, and China's aerospace technology has to reach world-class level.

Chapter Summary

The intelligent upgrade of modern transportation cannot do without digital channels, and the construction of China's intelligent transportation system cannot do without the support of digital twin. Under the application of digital twin simulation, China's air transportation has repeatedly reached new heights, China's ground transportation has become more and more smooth and efficient, China's water transportation has completed a breakthrough and transformation, and China's intelligent transportation has been fully upgraded. It can be predicted that the simulation features of digital twin in future development will play a more effective role, which will further the strategy of a strong transportation country in China.

Chapter 5

Digital Twin Leads to New Breakthroughs in Intelligent Transportation

The intelligent transformation of China's transportation to "people are happy with their travel and materials are excellent with their flow" cannot be separated from the support of digital forces. Under the guidance of the "Outline for the Development Plan of Digital Transportation", China's transportation has made many breakthroughs in the process of digital upgrading. Digital twin, as guidance to promote the intelligent upgrade of transportation, has brought "intelligent transportation, digital travel", promoted the rapid development of automatic driving technology, and drawn a broad blueprint for a strong transportation country.

5.1 Digital Twin is the Key Technology to Promote the Construction of Intelligent Transportation

The development of science and technology has ushered in an era of wisdom, and digital technology is an indispensable key force in the era of wisdom. The digital twin technology has created a connection channel for the transportation field to switch between the virtual and the real simultaneously, and has realized the transportation transformation with the overall improvement of safety, efficiency, and travel experience. This important course of intelligent transportation development is also an inevitable trend of global transportation development from the simulation of a car, a road, and a person to the complete replication of transportation hub and transportation network.

Driven by the digital twin, China's transportation infrastructure has undergone tremendous changes in recent years, and public travel and life have been changed due to the development of this infrastructure.

5.1.1 *Security upgrade*

With the in-depth implementation of the strategy of strengthening the country through transportation, China's transportation industry has entered a stage of rapid development, and transportation has also brought more impetus to China's economic development. The development of the transportation industry is bound to drive the upgrading of the transportation safety system. The digital twin is still the top priority in the process of upgrading the transportation safety system in China.

Through the digital twin simulation model, the bearing capacity of the transportation facilities and the operation status of the transportation network can be accurately predicted. Reasonable construction of safety measures in line with the actual needs of transportation development can maximize the safety of the transportation transport network.

For example, during the construction of the Guizhou digital twin transportation system, the layout of the perception equipment of the transportation hub is the construction optimization obtained through the digital twin modeling and simulation. Under the full element simulation, the layout of the perception equipment can achieve full coverage without dead ends, and the monitoring, operation, and maintenance management of the key transportation hubs by the transportation management department is more effective.

5.1.2 *Upgrading of identification capability*

The identification ability of transportation network also depends on the intelligence of infrastructure. In recent years, China has increased the installation of bayonet, video, and law enforcement forensics equipment and established dozens of illegal analysis models and feature vehicle models in order to improve the identification ability of China's comprehensive three-dimensional transportation network. Through the screening of the track data of transportation flow and pedestrian flow, the transportation management department can timely detect all kinds of illegal acts, and label the illegal objects with "portrait", which is convenient for the public security department to handle in a timely manner.

The improvement of the recognition ability of China's transportation network is also the result of digital twin technology. The digital modeling of illegal conditions and vehicle features enhances the recognition effect of the transportation network and ensures the healthy operation of the transportation network.

5.1.3 *Intelligent upgrade of public travel*

The rational utilization rate of transportation facilities in China is constantly improving, and the intelligence of public travel is also constantly improving with the support of digital twin technology. In the virtual space developed by the digital twin, the transportation department can find the construction plan of seamless switching and seamless connection of transportation tools, which can greatly reduce the travel time of the public, improve the efficiency of transportation services, and make the public easily find the best transportation plan before, during, and after travel. The transportation convenience and living convenience are continuously improved.

5.1.4 *Improvement of transportation data collection capacity*

The "Outline for the Development Plan of Digital Transportation" issued by the Ministry of Transport of China clearly states the following:

Building a digital collection system

(I) Promote all elements and whole cycle digitalization of transportation infrastructure: Promote the digitalization of all elements and whole cycle of transportation infrastructure planning, design, construction, maintenance and operation management. Build a high-precision transportation geographic information platform covering the whole country, improve the information of transportation engineering and other elements, realize the three-dimensional digital presentation of physical facilities, and support professional navigation applications such as automatic driving and large-sized transportation in all-weather complex transportation scenes. For major transportation infrastructure projects, realize the health performance monitoring of infrastructure in the whole life cycle, and popularize and apply the engineering quality control technology based on the Internet of Things.

(II) **Layout of all-round transportation perception network at important nodes:** Promote the coverage of transportation perception network in key sections and navigation sections of railway, highway and waterway fields, as well as important nodes such as tunnels, bridges, interchange hubs and ship locks. Promote the simultaneous planning and construction of transportation perception network and transportation infrastructure, deepen the application of road side intelligent terminals such as highway ETC gantry, establish a cloud connected perception network, and enable the "mute facilities" to have multi-dimensional monitoring, intelligent network connection, accurate control and collaborative service capabilities. Focus on social data integration applications such as crowdsourcing and mobile signaling. Build a basic cloud platform of transportation control network with interconnection of transportation tools, infrastructure and transportation environment. Speed up the application of Beidou navigation in free flow toll collection, automatic driving, vehicle road collaboration, maritime search and rescue, port automatic operation, collection and distribution scheduling and other fields.

(III) **Promote the intelligent transportation tools and operation equipment:** Encourage the application of terminal equipment with multi-dimensional perception, high-precision positioning and intelligent network connection functions, and improve the capability of remote monitoring, fault diagnosis, risk early warning and optimal control of transportation tools. Promote the research and development of collaborative technology of automatic driving and vehicle road, and carry out the construction of special test site. Encourage logistics parks, ports, railways and airport freight stations to widely apply Internet of Things, automatic driving and other technologies, and promote the large-scale application of automated three-dimensional warehouses, automated guided vehicles (AGV), intelligent transportation and sorting, and loading and unloading equipment. Promote the development of self driving vessels, automated terminals and yards, and strengthen information sharing and business collaboration between port and shipping logistics and upstream and downstream enterprises.

As of 2021, in the construction of the past three years, the data collection capacity of China's transportation network infrastructure has been greatly improved. This also supports the full implementation of digital twin technology in the field of transportation in China. Nowadays, most of the transportation infrastructure in third-tier cities and above in China has the ability to collect dynamic data. This ability accelerates the construction of smart cities, enhances the effect of transportation management, improves the collaborative management ability of transportation networks, and makes the operation and maintenance management of intra city and inter city transportation smoother.

5.1.5 *Environmental protection capacity*

In 2018, the Ministry of Transport of China proposed the development goal of "comprehensively promoting the construction of green transportation infrastructure", and clarified the construction focus of green transportation infrastructure.

First, thoroughly implement the idea of ecological civilization. According to the development goals of the two stages of 2020 and 2035, promote the transformation and upgrading of transportation, improve quality and efficiency, and better serve the overall requirements of the construction of strengthening country with strong transportation network and a beautiful China. In 2020, achieve the established goal of "successfully completing the task of transportation pollution prevention and control".

Second, do a good job in the implementation of the tough task of pollution prevention and control. Implement the instructions on "green production and lifestyle", "solving outstanding environmental problems" and "strengthening the Party's leadership over ecological civilization construction", and focus on a number of key tasks.

(IV) Coordinate the spatial layout of transportation infrastructure and comprehensively promote the construction of green transportation infrastructure.

(V) Promote the application of new energy and clean energy such as shore power and LNG in ports.

(VI) Promote the innovation of transportation science and technology, promote the development of intelligent transportation and intensive logistics, and promote the green packaging of express delivery industry.

(IV) Do a good job in adjusting the transportation structure, and carry out special actions such as diesel truck pollution control, ship pollution control, port facilities pollution control, and traffic road environmental pollution control, so as to complete the task of pollution control in the transportation industry.

(V) Strengthen safety supervision and emergency response capacity building.

(VI) Promote international cooperation in green transportation and actively participate in global environmental governance of transportation.

(VII) Carry out green travel, improve rural travel conditions, strengthen the publicity and guidance of green transportation, and promote green transportation national actions such as green organ culture.

(VIII) Deepen the reform of comprehensive transportation system and mechanism, strengthen the construction of laws and regulations, strengthen the guidance of economic policies, strengthen the guidance of evaluation, and improve the ecological civilization governance system.

Third, comprehensively strengthen the leadership of the party on green transportation. It is necessary to strictly implement the tasks of the ecological environment protection responsibility list and specify the timetable, roadmap and responsible person. The main leaders of each department of each unit shall take the overall responsibility for the ecological environment quality of their own department, and the responsible units of each special action shall formulate specific action plans.

During more than three years of development, the green development of China's transportation infrastructure has achieved remarkable results, which also depends on the promotion of digital twin technology. Compared with the previous insufficient combination of China's transportation construction and the environment, and the damage to the environment caused by the construction of transportation facilities, the digital twin technology has found the construction path with the least environmental damage and the highest utilization rate through the combination of environmental and transportation facilities simulation, which not only improves the development effect of China's green transportation but also saves the cost of transportation construction.

In recent years, for instance, Tongzi County, Guizhou Province, has implemented the development model of "avoidance–protection–restoration" of transportation infrastructure construction, enabling the transportation infrastructure to actively connect with natural resources and ecological environment, effectively protecting cultivated land and ecological environment resources, and improving the sustainable development effect of urban economy and transportation.

In future development, digital twin technology will continue to improve the construction effect of transportation infrastructure and highlight the wisdom of transportation construction from automation, networking, greening, sharing, and other aspects. At the same time, combined with the development needs of the strategy of strengthening country with strong transportation network, the construction of transportation infrastructure can effectively drive economic development and accelerate the realization of the goal of strengthening the country through transportation.

5.2 Transportation Technology From Virtual Simulation to Digital Twin

In fact, the transportation departments in China have carried out various attempts of virtual simulation technology before the application of digital twin technology to the field of transportation, which is also the initial prototype of transportation simulation. However, at that time, Internet technology was more integrated to track and describe the time and space of transportation movements. This kind of simulation is not all element digital modeling, so the simulation effect is random, the application field mainly exists in the transportation area simulation and does not provide the decision-making basis for the overall development of transportation.

With the development of transportation technology, digital twin technology has been widely applied in the field of transportation. After the application of the digital twin, the transportation system can not only reproduce the previous operating conditions but also grasp the future operation in advance, thus deepening the understanding and analysis of complex transportation conditions and finding the optimization mode of the transportation system. The technical advantages of digital twin are mainly shown in the following four aspects compared with traditional simulation in the field of transportation.

5.2.1 *The simulation model is more flexible and contains all elements*

The simulation model brought by the digital twin completely copies the change law of all elements in the transportation system and the interaction between elements, so the model construction is more flexible, and the regional model construction or all model construction can be carried out according to the actual situation.

5.2.2 *More accurate and interactive simulation model*

As the simulation model brought by digital twin is the whole element, it can more accurately reflect the change of transportation status or the operation status of a single transportation element. At the same time, the simulation model can also reflect the real-time state of individual transportation, and the virtual and real individuals can complete synchronous interaction.

5.2.3 *More open simulation model*

In the simulation model constructed by digital twin, multi-directional and multi-state simulations can be carried out according to various data. Its openness is far more prominent than traditional simulation technology, and the prediction of transportation operation is also more accurate.

5.2.4 *Strong ability to describe transportation conditions*

The simulation model constructed by digital twin can accurately describe the real-time state of the transportation network and carry out various reproduction and deduction on the basis of the requirements of the transportation management department, thus the experimental cost of the transportation construction is reduced and the intelligence of the transportation system is enhanced.

Although digital twin technology is a comprehensive upgrade of traditional virtual simulation, there is no change in application logic. The simulation of the field of transportation by digital twin also follows five steps: data collection of simulation object, digital modeling, preview programming, simulation preview, and result analysis:

(1) **Simulation object data collection:** The data collection of transportation objects is mainly through the sensor equipment in transportation construction. Digital twin has been able to collect dynamic information, real-time information, and complex information groups of people, vehicles, roads, and environment in real time with the upgrading of sensor technology. The more comprehensive the information collected, the more complete the subsequent modeling.

(2) **Digital modeling:** Digital modeling is a data synthesis based on the collection of information in the new space. Digital modeling can clearly describe the relationship and interaction between various data.

(3) **Preview programming:** Preview programming is the core technology of digital twin. Technicians can use professional programming language to realize a real preview of the simulation model based on the actual situation. Under normal circumstances, digital twin technology is commonly used in two programming languages: One is the general high-level programming language. This programming language is the basis of digital twin modeling programming and is also the most commonly used digital twin programming language. The other is a special simulation programming language, which is usually evolved from a common programming language. It simplifies some common simulation work in the form of functions. There are still many limitations even though it can improve the programming efficiency.

(4) **Simulation preview:** The main way for a digital twin to play a role in the field of transportation is simulation preview. Through the simulation preview, we can learn about the methods of construction and upgrading of transportation facilities, as well as the real state of the transportation network, or predict the transportation pressure to be faced by urban transportation in a certain period of time in the future. As the digital twin is an all factor model, the accuracy and authenticity of the simulation results are higher, which can be used as the main basis for transportation management departments to make decisions.

(5) **Results analysis:** The analysis of various data obtained from the simulation is the embodiment of the role of digital twin in promoting transportation development. For example, the measures taken by Changsha Wuyi Square subway station to cope with the May Day holiday's strong transportation are the result of a simulation rehearsal. From the real cases, we can see the effectiveness of the analysis of the results of digital twin simulation preview.

Compared with traditional virtual simulation, the upgrade of digital twin is mainly reflected in multi-angle simulation applications. That is, digital twin can carry out micro, meso and macro planning and simulation of transportation system, so the promotion of digital twin to China's transportation development is more comprehensive.

Micro simulation: The simulation model established by digital twin is more detailed, so its detailed description of transportation is more accurate. For example, the modeling of regional transportation by digital twin can be detailed to the track of each pedestrian, the identification of each vehicle, the changes of individual regional transportation, and the impact of changes on the overall can be clearly demonstrated.

Meso simulation: Meso simulation refers to the simulation of a transportation unit by digital twin. For example, the transportation management department can understand the operation status of a certain road or the transportation situation of a certain period of time in the simulation model of urban transportation. The transportation management department can reasonably allocate transportation resources or clarify the key points of transportation construction through the display of this transportation information.

Macro simulation: The macro simulation aims at the simulation preview of the overall transportation network, which can be a combination of multiple regional simulation. Moreover, these regional models can be closely connected, and the relationship between them can be intuitively reflected under the operation of digital twin technology.

For example, the simulation of the transportation operation in a certain province can help understand the transportation jams, key nodes, and the operation effect of the transportation network in this province.

At present, the application of digital twin technology in China's transportation field is very comprehensive. Through micro, meso, and macro multi-level simulation preview, the current situation of transportation development and the overall performance of the transportation network are evaluated, and the future development of the transportation network is planned.

It seems that the modeling and simulation of transportation by digital twin is only a simple five steps, but it takes a complex process to really play a role, or the digital twin needs to be modeled based on the transportation development problems or the transportation development needs to play the largest role, which requires the transportation management department to follow up the following points in the process of using the digital twin technology:

(1) **Positioning the core:** The key to the application of digital twin is to position the core problem of the transportation system and analyze and describe the problem by using a simulation model. Only when the core problem is positioned can the digital twin modeling be used for targeted deduction and analysis and, moreover, can the improvement and disposal plans be proposed.
(2) **Determine the deduction method:** Although the digital twin technology can perform all element simulation previews, it needs various experiments and improvements to solve the transportation development problems, so the deduction method needs to be determined in the process of using the digital twin technology. For example, digital programming of different schemes is prepared in advance to facilitate the use of digital twin technology to compare experimental differences.
(3) **Systematic data collection:** The simulation preview of the simulation model is a complex process. Through the perception of transportation facilities, data such as the scale, operation rules, and environmental conditions of transportation facilities are comprehensively collected, and then input according to the requirements of modeling, and then the real results are derived. As a result, the data collection of digital twin needs to be systematic, and full element collection should be carried out according to the actual situation. Only when the data is comprehensive can the modeling and deduction of digital twin be correct and accurate. For example, in a specific environment, for transportation information, it not only needs to collect road information, transportation demand information, and environmental conditions information but also needs to collect noise information, vehicle fuel consumption information, vehicle speed change information, etc.
(4) **Parameter estimation:** There are still random data in addition to fixed data in transportation operations though the perception of modern transportation infrastructure has been comprehensively improved. These random data change with the environment, time, and climate change at any time. In the process of digital modeling, the parameters of these random data are estimated. The so-called parameter estimation refers to the estimation of this value with reference to

the data obtained from previous experience. The more accurate the data estimation, the better the deduction effect of the simulation model will be.

(5) **Results evaluation system:** The results of digital twin simulation are more accurate than traditional virtual simulation but also need to be corrected and compensated. After the simulation model deducts the performance results, transportation managers need to consider whether to adjust some variables or modify some random parameters according to the results, which can effectively improve the simulation accuracy of digital twin.

5.3 Fully Support Intelligent Transportation Construction with Digital Twin Technology

In short, digital twin can be regarded as an analysis tool for transportation development and has penetrated into all aspects of transportation development, fully supporting the intelligent construction of China's transportation. At present, digital twin has made remarkable achievements in several aspects in the development of China's transportation.

5.3.1 *Research on the construction of intelligent transportation*

After entering the digital era, the intelligent development of China's transportation and the construction of digital transportation facilities have started visual analysis at the research stage, which exactly needs the support of digital twin technology. For example, the simulation model was built in advance, and the detection strategy was optimized according to the model simulation in the construction of intelligent transportation in many first-tier cities in China.

Digital twin can realize the comprehensive analysis of various transportation elements, including transportation flow, terrain conditions, road grade, environmental interference, as well as operation status. Under the comprehensive analysis of multiple data, the transportation capacity, flow carrying capacity, and follow-up operation and maintenance management of transportation facilities can be clearly stated. Therefore, digital twin has made great achievements in the construction and research of intelligent transportation in China.

5.3.2 *Actual construction of transportation facilities*

As the all element model constructed by the digital twin is a visual and controllable real model, the construction team can intuitively watch and check the actual construction effect in the actual construction of transportation facilities, timely find out the construction defects and environmental interference, and timely adjust and upgrade. The role of digital twin greatly avoids construction problems and improves the construction efficiency of transportation facilities.

5.3.3 *Design and analysis of intelligent transportation upgrade scheme*

The simulation model constructed by digital twin can fully demonstrate the flexibility of transportation design and intuitively demonstrate the implementation effect of various transportation designs through various simulation deduction. Not only can the best design scheme be selected, but also the design scheme can be upgraded and improved through the comparison of deduction effect. For example, in May 2021, German railway company and German railway freight enterprise Metrans launched the digital twin technology development project of passenger and freight trains. Both parties completed the comprehensive digital transformation of German railway company's fleet through digital twin modeling. Although the transformation project has not been fully implemented, the technical advantages after the transformation have been fully highlighted in the virtual space. Passenger flow, passenger density, and punctuality rate of German railway have been comprehensively improved.

5.3.4 *Transportation safety analysis*

Digital twin can display the operational status of transportation facilities in three dimensions. Transportation designers can directly understand the transportation hazards of transportation hubs and key sections, and then design signal lights and warning lights according to safety issues, so as to reduce the interactive conflict of transportation flow and reduce the incidence of transportation accidents and transportation congestion.

In addition, the digital twin can also carry out transportation accident analysis, reproduce the whole process of the accident through the simulation model, and conduct targeted analysis on the details of the causes leading to the accident, so as to formulate a more comprehensive security strategy.

5.3.5 *Research, development, and testing of new transportation technology*

At present, a number of new technology research and development in the field of transportation in China are related to digital twin. These new technologies are mainly divided into two categories: One is the combination of digital twin and other emerging technologies, and the other is the emerging technology deeply combined with the development of transportation under the deduction of digital twin. For example, the intelligent control system of Beijing Daxing International Airport is a combination of digital twin and other emerging technologies. The data acquisition system in the process of Guizhou intelligent transportation construction is an information system upgrade based on the data twin deduction.

5.3.6 *Application of intelligent transportation system (ITS)*

Digital twin is an important technology in the development of China's intelligent transportation, and its simulation characteristics have played a catalytic role in the upgrading of China's intelligent transportation system:

(1) Through digital modeling, the overall graph of China's transportation network is displayed, and the key hubs of the transportation network are clearly shown.
(2) Digital modeling is used to show the movement track and detailed behavior of individual transportation.
(3) Through digital modeling, the relationship and interaction among transportation individuals are demonstrated.
(4) The operation and maintenance strategy of the transportation network are demonstrated through digital modeling.
(5) The management strategy of transportation network is upgraded through digital modeling.

(6) Simulate the application effect of other technologies in the field of transportation.
(7) The urban transportation construction is optimized through digital modeling.
(8) The state of urban public transport is demonstrated through digital modeling.
(9) Provide decision data support for transportation development.

In general, the value of digital twin in the field of transportation is embodied in the careful perception of every detail of development from the three dimensions of the past, the present, and the future. For the past of transportation development, digital twin can carry out data precipitation, which is convenient to correct the current development mode; at the moment of transportation development, digital twin can realize real-time perception of all elements and adjust the development mode of intelligent scheduling; for the future of transportation development, digital twin can be simulated and inferred to accurately predict the possible future situation, so digital twin will be so popular in the field of transportation.

In the era of rapid development of cloud computing, big data, and AI technology, the application of digital twin in the field of transportation is more convenient. For example, in the aspect of transportation macro management, the overall intelligence of the transportation can be improved by the calculation deduction and prediction ability of digital twin, and moreover, this intelligence can be expanded from details to the overall situation, playing a practical effect.

5.4 Specific Ways for Digital Twin to Fully Support the Construction of Intelligent Transportation

Following the development pace of developed countries in transportation, China has now absorbed the experience of intelligent transportation upgrading from overseas and laid a solid foundation for intelligent transportation construction. Through the analysis of the digital development results of China's transportation network, the author summarizes the specific ways of digital twin to fully support the construction of intelligent transportation.

5.4.1 *Enhance the decision-making of China's intelligent transportation*

On the basis of improving the ability of data collection and integration of transportation facilities in China, digital twin provides solutions for transportation development for transportation managers. With the in-depth application of digital twin technology, China's comprehensive three-dimensional transportation network has built a "smart brain" with complete functions, and various development situations can be responded to in a timely manner. At the same time, this "smart brain" has spread all over the country, and the decision-making ability of intelligent transportation has been greatly improved.

Although there are differences in the levels and functions of "smart brain" in different regions, there is no significant difference in the overall construction. It is usually composed of database, digital model, deduction programming, and decision-making platform. Before the birth of various decision-making, digital twin technology can carry out the process of problem discovery, data collection, model deduction, and analysis and then provide a decision-making basis for managers.

For example, the construction of China's urban rail transit transportation system needs to face problems, such as long operating mileage, large coverage, and long operating time. How to build a stable, efficient, and safe urban rail transit network cannot be fully guaranteed only by relying on the construction experience and knowledge accumulation of the construction team. On the basis of the needs of urban transportation and environmental characteristics, the main framework of the rail transit network can be built independently under the operation of "smart brain", and the experience and knowledge of the construction team can become the main support for the sound and upgrading of the rail transit network, and the construction effect and speed can be greatly improved.

For example, since China's railway bears great transportation pressure, the scheduling difficulty of China's railway transportation network has always been high. During the Spring Festival and Golden Week, the utilization of railway resources, the punctuality rate of trains, and the safety of train operation are all problems that need to be solved by the railway management departments. With the assistance of the "smart brain", the real-time status of trains, tracks, people flow, and logistics can be collected and analyzed, and the intelligent decision-making for efficient and safe operation of the railway is born.

5.4.2 *Enhance the use of big data technology*

As China's transportation develops in a high-speed way, the problems faced by China's transportation industry are also increasing, including but not limited to urban transportation congestion, transportation pollution, as well as environmental damage. The best way to solve the problems associated with these developments is to understand the causes of the problems in a timely manner and optimize transportation development measures.

The integration of digital twin and transportation has improved the perception ability of China's transportation network, the large-scale deployment of monitoring and induction equipment such as sensors and cameras, and the collection of relevant data has become more and more abundant, including climate data, people flow data, logistics data, and transportation condition data, which has also accelerated the application of big data technology in China's transportation field. As these data grow constantly, transportation management departments across the country have built a complete intelligent transportation database, which includes data management system, one key storage system, document classification system, and graphic recognition system, which has also strengthened the basic strength of China's intelligent transportation.

5.4.3 *Accelerate the integration of intelligence and digital in transportation*

Digital twin can conduct comprehensive analysis on the data and information collected from China's comprehensive three-dimensional transportation network, build a collection of unified standards and rules, and integrate various types of data automatically to build a complex digital model.

The simulation model built by digital twin appears to be a pure digital technology but at the same time integrates the intelligence of transportation operation, which is mainly reflected in two aspects:

(1) Digital simulation enhances the value of dynamic data utilization. The intelligence of China's transportation network is mainly reflected in the analysis and deduction of dynamic data. A certain logic deduction can be conducted by the digital twin for the relationship and interaction between data so that the situation deduction and estimation can be proceeded to make the data itself intelligent.

(2) Digital integration improves the precision of intelligent transportation. Digital twin classifies data into three levels in the process of digital modeling, namely basic data, key parameters, as well as decision data. The basic data supports the framework of the simulation model. The key parameters determine the simulation deduction of the model, and the decision data is the analysis result of the first two data. As a result, the transportation development decision made by digital twin technology is based on the real data of the transportation foundation and development law, with higher precision and fineness.

The development of China's transportation industry has experienced many upgrades from virtual simulation to the use of digital twin. Each improvement of digital technology will bring optimization effect to China's transportation construction, management, operation, and service. Digital twin technology will become the toolbox of intelligent transportation, bring more help to China's transportation intelligent development, and create a more convenient and smooth way.

5.5 Digital Twin Drives the Rapid Development of Automatic Driving Technology

In September 2021, the Ministry of Transport of China issued the "Action Plan for New Infrastructure Construction in the Field of Transportation (2021–2025)". The plan clearly states, "With the overall goal of accelerating the construction of a strong transportation country, we adhere to innovation-driven and intelligent development, take digitalization, networking and intellectualization as the main line, organize and promote a number of key projects of new transportation infrastructure, create an influential model of new transportation infrastructure, create an innovative development environment, promote the development of new infrastructure from point to area, improve the efficiency, expand functions increase momentum of transportation, and continuously enhance the sense of gain, sense of happiness and sense of safety of the people". Other policy documents also mentioned that China's transportation construction focuses on the intelligence of roads, waterways, ports, and hubs and deeply integrates technologies such as digital twin, AI, big data, and cloud computing.

It can be seen from the content of the action plan that modern technology is applied to build a transportation network with car road coordination, which helps the transportation industry to achieve digital transformation, so as to obtain broad development space and support from a number of policies. In the industry related to the construction content of the "Action Plan for New Infrastructure Construction in the Field of Transportation (2021–2025)", the field of automatic driving, representing the future technology, is a representative. In recent years, the development of the automatic driving industry cannot be separated from the promotion of digital twin.

In September 2021, Sun Chitian, Tencent's technical director of automatic driving simulation, made a speech on "digital twin simulation accelerating automatic driving" in the "Oriental Wind Leading and Intelligent Driving Leap - China Auto Valley 2021 Intelligent Automobile Industry Innovation and Development Forum". In this speech, Sun Chitian detailed the promotion of digital twin technology to intelligent driving and, moreover, how Tencent Inc. implemented intelligent driving technology under the support of digital twin. Sun Chitian said the following:

> Tencent Automatic Driving Department has been focusing on one thing for the past five years. We have positioned ourselves as a tool box and accelerator for research and development of autopilot. Therefore, our department focuses on four directions: TAD Sim, a virtual simulation system based on digital twin technology and Tencent game technology; the second is TAD Pilot; the third is TAD Cloud; and the fourth is the high precision map service.
>
> At first, Tencent Automatic Driving Department came from Tencent map. We started to make Tencent high-precision map from Tencent map, and then started to do automatic driving. In 2017, we officially established the automatic driving business center, and really started to do work related to automatic driving business. In recent years, we have also built the trinity technology platform, high-precision map platform, cloud development platform and virtual simulation platform.
>
> We have clearly experienced two difficulties in the past few years whether it is our own in-house self driving development or joint communication projects with Tencent partners. The first is the training data set required by the perception algorithm. Each technology company in the world invests a large amount of money in data collection and

labeling every year, with an annual magnitude of USD 1 billion. This cost is relatively large consumption for all technology companies. The second is the test verification. It is to verify the security, time and money, which is familiar for all. I would like to focus on OTA here. The Ministry of Industry and Information Technology of the People's Republic of China issued a request some time ago. In the future, Internet connected vehicles are not allowed to use automatic OTA, but are subject to the verification of all OTA algorithms by the regulatory authorities. Such a request is equivalent to a higher requirement for the automatic industry algorithm verification test platform. We can think of the automatic driving algorithm in the future. If it is upgraded through OTA, it must be verified by the third testing agency.

Of course, the use of simulation to solve the above-mentioned problems must be based on a simulation platform. It took Tencent four and a half years to build such a virtual simulation platform. Our goal is to build an oasis system similar to that in the film "Ready Player One". There was a new film "Free Guy" some time ago, which is also such a concept. What kind of work do we do? We have summarized four core simulation capabilities: ① In terms of the geometric restoration of the scene, the focus is on the three-dimensional restoration of the scene, which is generally based on the three-dimensional reconstruction technology. Of course, each technology path is not the same. After the three-dimensional scene is restored, our purpose is to run sensor simulation, and carry out accurate sensor simulation based on various physical models. The accurate simulation results of the sensor are also based on the accurate restoration of the real world, so that it is possible to obtain more accurate sensor results. ② In terms of the logical restoration of the scene, it involves the element generation of the dynamic scene, how to restore the real world motion logic very accurately. ③ It mainly focuses on the control algorithm and dynamic simulation in terms of the physical reduction of the scene, which involves the calculation of vehicle dynamics model, friction force and tire model, relating to physical properties. ④ The reason for the high concurrent operation in the cloud is clear. The magnitude of the test scenario library that we need to face now is much larger than that of the previous ADAS, which cannot be achieved locally.

It took us four and a half years to build such a set of TAD Sim for automatic driving system. We built a three-dimensional environment based on high-precision map, and mapped the world in from real to virtual. In the meantime, the scene elements and attributes were restored based on the real value system, and were used in the algorithm and training test. Also, we will generate a lot of dynamic elements to help us generate a variety of test scenarios.

When it comes to the details, the first one is to share with you the test work on how to restore a static scene in a test area. Firstly, we make the labeling based on the real scene. We have built a set of tool chain in order to reduce the cost of labeling and greatly enrich the scene of the data set, one is the algorithm of scene generation, the other is the algorithm of scene recognition, and the two algorithms are the iteration method of closed-loop mutual improvement. We converted the data with 100% correct labeling data (image, point cloud) through pure virtual data, based on which we converted the data set with completely correct labeling results to train the perception algorithm. The improvement of perception algorithm will further improve the efficiency of the tool chain for generating three-dimensional scenes in virtual space based on the collected data. As a result, we have been running this tool chain for three and a half years. Both the conversion efficiency and the generation efficiency are relatively high. At present, we also establish static scenes in the direction of automatic driving simulation through this system, and it is also to establish urban level transportation scenarios in the application direction of broader intelligent transportation.

The cost of data storage and calculation is very high when 3D reconstruction of scene cloud is carried out. Is there any way to solve the problem? It is to generate large-scale automatic scenes by combining the multi-source geographic data with the artificial intelligence network.

Cloud support is also a very important tool for cost reduction and efficiency enhancement when we generate urban scenarios. If a city area of about 500 square kilometres is generated, it will take us about 22 hours; if it is put in Tencent cloud, it can be generated within 1 hour. The key is that the basic cost is zero, and the only cost is the electricity fee, which is all, so this method can greatly meet the needs of urban modeling.

Let's share with you the problem of how to restore the dynamic scene, that is, how to accumulate the scene library when we establish the test system. We have developed the TAD Sim stand-alone version and cloud version. These two systems have taken such a long time, but in fact, they also face some challenges. The core issue is authenticity. We have made a lot of explorations in our transportation flow models, including those cooperated with colleges and universities, self-developed transportation flow models, rule-based models, and neuro network models. What's more, we are also trying to explore how to map real world transportation scenarios into the virtual testing environment. At present, we believe that the path is the extraction of transportation flow through real-time data of real transportation and the integration of lightning and photography, which can effectively supplement the real transportation simulation scene library and improve the authenticity of the automatic driving simulation test scene library.

From Sun Chitian's introduction to the research and development of Tencent's automatic driving technology, it can be seen that the digital twin simulation ability has helped Tencent Inc. complete the research and development of the core technology of automatic driving. The digital twin reproduces the real scene of the vehicles and builds an open simulation and testing platform for Tencent Inc. in the virtual space so that the test vehicles can also realize the real environment test in the stagnant state.

This successful case provides a concrete example for the automatic driving industry. Combined with digital twin, auto driving technology can conduct specific tests to accurately analyze the auto driving function of vehicles. In short, in the test platform built by Tencent Inc., the digital model of the vehicle can be driven in the virtual environment of the full element simulation, and the driving data can be sent back to the real-world connected vehicles assisted by driving. All data calculation is completed through the digital platform, and the driver can also feel the real driving experience from this mode and accurately perceive the driving experience of automatic driving and auxiliary driving.

From the introduction of Sun Chitian, it can be seen that at present, Tencent Inc. has not completely solved the problem of "mapping the real world transportation scene to the virtual test environment", which means that the simulation of automatic driving technology needs the supplement of the real transportation scene, and the scalability and flexibility of automatic driving need to be improved. However, digital twin has greatly improved the development speed of automatic driving technology, producing a profound impact on the intelligent development of transportation in China.

Generally speaking, digital twin technology provides the following conveniences for the development of automatic driving:

(1) Digital twin provides a safe and risk-free testing space for automatic driving, and the space can be continuously expanded with the enrichment of data, and can also reflect the parameter variables of vehicles.

(2) Digital twin has created a development platform for the research and development of automatic driving. The limits of automatic driving and Internet connected vehicle technology can be explored by relevant enterprises to ensure the safe landing of various technologies.
(3) Digital twin improves the research and development and testing efficiency of automatic driving technology.
(4) Digital twin technology reduces the research and development cost of automatic driving technology.

Case: Digital twin, builds a digital chemical factory of FAW Hongqi

On 28 May 2021, at the 2021 Alibaba Cloud Summit, Alibaba Cloud and the 9th Institute of Machinery Industry jointly released the "automobile digital factory 1.0" industry solution, in which the digital twin generated the largest highlight.

Huang Zheng, an architect of Alibaba Cloud Automobile Industry, explains that a car generally consists of more than 3,000 categories and 30,000–50,000 types of parts and components. The production process is complicated with machines, including multiple processes such as stamping, welding, coating, and general assembly, and the processing technology alone can reach more than 200 kinds.

After the construction of automobile digital factory by using digital twin technology, the management and production effect of automobile factory will be greatly improved. For example, under normal circumstances, the total floor area of the whole process production workshop of a new energy automobile factory is generally 700,000–800,000 square meters, so it is difficult to achieve meticulous factory management in such a large area. However, in the digital factory, every action of the new energy automobile factory can be reflected in the screen of the office, and the workshop operation and the production situation can also be clearly demonstrated.

Wu Jian, product manager of Alibaba Cloud Industrial Brain digital twin, also said that digital twin reduces the technical requirements of workers. For example, automobile production workers no longer need a rigid CAD foundation, nor professional drawing knowledge, because the simulation model built by digital twin is enough to intuitively show various problems.

Wu Jian also added that a digital twin factory should contain at least three key areas of technology: First, the digital twin project must build a carrier platform of digital twin, i.e. the operation system of digital twin;

second, the digital twin factory must have a three-dimensional technical team to transform the physical world into a virtual space for three-dimensional presentation through data conversion; third, in order to ensure that the digital model can accurately and efficiently collect various types of data, build a comprehensive digital model through digital twin technology, and connect the digital model with the real production data of the factory. With these technologies, the digital twin factory can clearly show the real situation in the virtual space and provide the decision-making basis that can interact in real time.

Alibaba Cloud and the 9th Institute of Machinery Industry jointly released "automobile digital factory 1.0", which was first applied to FAW Hongqi new energy factory. Through the deployment of millions of data acquisition equipment and sensors in the five workshops of stamping, welding, coating, assembly, and battery of FAW Hongqi new energy factory, a complete digital factory model has been established. This is not only a technological breakthrough in the automobile industry but also the beginning of the application of digital twin in the field of new energy vehicles in China. With the empowerment of digital factory, the product research and development efficiency, production efficiency, and technological innovation of FAW Hongqi's new energy factory have been greatly improved, and the brand power and market competitiveness of FAW Hongqi have become more and more powerful.

Chapter Summary

The application of digital twin technology in China's transportation field is deepening from the new infrastructure to transportation technology, and then to the intelligent upgrading of China's transportation system, which also promotes the development of China's intelligent transportation to make many achievements. In a manner of speaking, digital twin is also the main force for the construction and development of smart systems, and the mainstream trend of global transportation development, instead of only being an important foundation for the construction of digital transportation in China. Grasping this tuyere in time can make new breakthroughs in China's transportation and climb to a new height.

Chapter 6

The Application Status of Digital Twin Technology in Western Transportation

The booming development of China's transportation is promoting China to transform from a transportationcountry into a transportation power. The road of intelligent transformation of the transportation developed countries has become the main object of our study in this process. Although China's level of transportation development is close to the international level, many European and American countries started their intelligent transportation earlier, and their transformation experience and upgrading mode are still worthy of our reference and analysis.

6.1 Application Status of Digital Twin Technology in Transportation in Europe and America

Since the digital twin was used by the United States in the field of aerospace and aviation, this technology has taken root and thrived in the field of transportation. In more than ten years of development, digital twin has led a number of European and American countries to take the lead in realizing the upgrade of intelligent transportation, and comprehensively promoted domestic economic development. For example, the 27 countries in Europe jointly proposed the concept of "fifth generation transportation" in 2015, which aimed at the network management of the road network of more than 30,000 kilometers in Europe, and the identification of key sections of the overall network according to the unified rules. With the support of digital twin technology, the "fifth generation transportation" in Europe has developed smoothly. With the improvement of the effectiveness of road network management, congestion, transportation casualties, and carbon emissions in 27 European countries have been significantly reduced.

Currently, the transportation of Germany, France, the United Kingdom, Italy, and other countries has achieved outstanding development results through digital twin technology, and these results provide a reference for China's intelligent transportation upgrade, indicating the direction.

6.1.1 *Application status of digital twin technology in German transportation*

The main field of digital twin technology used in Germany is railway, and the depth of digital twin application in other transportation fields is not prominent. For example, the digital twin cooperation project between the German railway company and the railway freight enterprise Metrans, which we mentioned earlier, has transformed the operation mode of German trains, improved the operation efficiency, and obtained the support of the German government of 3.5 million euros in the process. This shows how much German railway attaches importance to digital twin technology.

At present, the digital twin system developed by the German railway is still second to none in the world. Its digital twin simulation accuracy is extremely meticulous, and it can even be refined to the real-time perception of air conditioning and wheel wear in the vehicle. What's more, the brand new digital twin train set created by the German Railway Department has entered the stage of implementation and is expected to be fully put into use by the end of 2021.

It is worth the attention of relevant personnel in China's railway industry that the simulation of digital twin is not the core of the cooperation between the German railway company and Metrans. The German railway company and Metrans prefer to develop a new measurement system through digital twin technology, in which the situation of German railway freight can be comprehensively improved, and it has also created a "future railway system" with superior intelligence.

6.1.2 *Application status of digital twin technology in French transportation*

The application of digital twin technology in France is more extensive and has achieved remarkable results:

(1) France has applied the digital twin technology to the construction of highway information inquiry system, and in this system, the management department can clearly learn about the real-time operation of each highway. Seasonal variations in the flow of the Paris autobahn network, for example, can be clearly reflected in the system.
(2) France has completed an upgrade of perception of key junction locations on the road network through digital twin. Through digital modeling, the French road management department timely understands the weak links of the road network. After the laying of the targeted sensing equipment, the French Road Department is more effective in the monitoring and operation of roads.
(3) In France, the combination of digital twin technology and a new generation of information technology has enabled the management of city fleets. France has established the management system of the city fleet by installing the positioning perception equipment for the exclusive fleet, and the positioning accuracy of the vehicle can be ensured at about 1 meter during urban driving. At present, this system has been applied to the French urban transportation management and the French police system.

6.1.3 *Application status of digital twin technology in transportation in the United Kingdom*

The United Kingdom has a more practical use of digital twin, which has solved many of its transport problems:

(1) By using digital twin technology, the United Kingdom has effectively alleviated road congestion, reduced transportation pollution, and optimized the coordinated development of people, vehicles, and the environment. The method to achieve this effect is not complicated. The United Kingdom installed a small remote information

processing device on public vehicles to accurately perceive the operation of public vehicles, which not only controlled the resource utilization status of the urban road network but also improved the response and processing speed of traffic accidents and special situations.

(2) The United Kingdom has applied digital twin technology to project management of London's M25 expressway. M25 expressway is an early Expressway built in the United Kingdom in 1995. With the development of the times, the operating pressure of M25 Expressway continues to increase. For the purpose of ensuring the normal use of M25 expressway, the United Kingdom uses digital twin technology to sense the real-time operation status of M25 expressway and adjusts the speed limit of M25 Expressway according to the real-time status, so as to effectively reduce the congestion of M25 Expressway and ensure the safety of M25 high-speed highway.

(3) The United Kingdom employs digital twin technology to improve the real-time transportation and travel information perception system and carries out detailed perception and upgrading of the central area of the British transportation network, highways with high transportation volume, as well as hot tourism cities, ensuring that the transportation in these areas is under control.

6.1.4 *Application status of digital twin technology in transportation in Italy*

The application of digital twin technology in Italy is mainly reflected in the field of public transport. First of all, Italy used digital twin simulation in building automated bus systems. The public transport management department can analyze the optimal solution of the bus driving and parking positions, and optimize the bus entrance and exit areas, parking areas, and fuel filling areas according to the actual situation of urban areas in virtue of the real simulation, thus upgrading the matching between the public transport system and the city.

Moreover, Italy has used digital twin technology to effectively relieve urban transportation pressure. At present, the ground transportation in Italy is facing unreasonable proportion of automobiles, trams, trains, and private cars, and the stability and safety of transportation in various cities have been affected to varying degrees. Italy has developed a 5T transportation management system using digital twin technology in order to solve this problem, which consists of 10 subsystems, 9 of which are ITS systems, and the other is the coordination management system of other subsystems. Turin, Italy, has effectively solved the problem of transportation congestion by using the 5T system. The main way is to improve the priority of public transport vehicles and rescue vehicles according to the distribution of transportation vehicles and the use of roads, thus changing the choice of public travel and accelerating the operation of urban transportation.

6.1.5 *Application status of digital twin technology in transportation in Austria*

The transportation development level of Austria still has a significant gap compared with that of China. However, the methods of Austria to improve the transportation conditions by using the new generation information technology and digital twin technology are worth learning from by the personnel of the relevant industry.

As the economy develops, Austria began to encounter various kinds of urban transportation and inter city transportation jams since 1995, which led to the adverse impact on the economic and social development of Austria. In response to this situation, Austria has designed an intelligent transportation management system for the expressway in Salzburg, a key transportation hub city. Due to the influence of neighboring Germany on the expressway transportation in Salzburg, the transportation pressure is huge during the peak travel season, and the congestion of this section directly affects the normal operation of several expressways in Austria.

As a result, Austria began to significantly improve the perception of the highway in Salzburg. The project has been under construction since 1996 and will continue to improve in subsequent development. After the awareness of the Salzburg expressway has been improved in place, the transportation management department of Austria can accurately sense the number of vehicles on the road, the speed, and the density of the key sections, and then carry out targeted optimization and maintain the transportation order of the key sections. After the application of digital twin technology, the operation and maintenance management of the Salzburg expressway is more reasonable, and the real-time information of the expressway can be timely transmitted to the driving vehicles through information technology so that the transportation management department and the driving vehicles can make subsequent decisions according to the actual situation of the expressway. With the replication of this intelligent model, the high-speed road network in Austria has been improved as a whole, and the economic development has also been promoted in an all-round way.

6.2 Development Status of Western Intelligent Transportation

In comparison with the development of China's intelligent transportation, many European and American countries have obvious advantages in the construction of intelligent transportation system, which is mainly because these countries use advanced technology in the field of transportation earlier. In the case of the United States, the United States can establish the intelligent transportation system to ensure safety, improve the environment, and improve efficiency earlier and more comprehensively since the United States has used the new generation of information technology, digital communication technology, sensor technology, network control technology, and artificial intelligence technology earlier than China.

As of 2021, the construction and development of China's intelligent transportation has shortened the gap with the transportation powers in Europe and the United States and even exceeded in urban transportation intelligent scheduling, highway intelligent operation and maintenance, etc.; however, the overall combination of people, vehicles, roads, and environment still has obvious deficiencies, which is mainly reflected in the maturity of intelligent transportation system (ITS). As a result, China's transportation sector can also find ways to improve the development effect through the analysis of the development of ITS in European and American countries.

6.2.1 *Development of ITS in the United States*

The United States has established its own ITS since the late 1960s, but it was in the 1980s that the ITS was actually applied. In 1987, the United States established the Mobility 2000 organization, which is the prototype of ITS America.

In 1990, the United States established the intelligent vehicle road system organization (IVHSAMERICA), whose main task is to provide the United States Department of Transportation with the requirements, objectives, goals, plans, and progress of the IVHS program. IVHSAMERICA is also the first organization to apply simulation technology to the transportation field through computer simulation; IVHSAMERICA continues to improve the capacity of the U.S. road network.

In 1995, the U.S. Department of Transportation issued the "national intelligent transportation system project plan" for the first time, which explicitly mentioned that the U.S. transportation intelligent system will carry out the development of seven major areas and 29 user service functions. Since then, the U.S. government has stepped up the establishment of ITS and has allocated hundreds of millions of dollars for this project.

Today, ITS in the United States has reached the international first-class level, and its intelligence is very prominent. Overall, ITS in the United States includes seven systems, which promote the development of intelligent transportation in the United States from multiple perspectives:

(1) **Transportation management system:** The transportation management system of the United States mainly includes urban transportation light control, highway transportation monitoring, transportation accident handling, and transportation congestion handling. This system is responsible for not only ensuring the normal operation of transportation in the United States every day but also providing decision-making basis for transportation development research and transportation condition evaluation and presenting the transportation development results in a three-dimensional manner.

(2) **Transportation demand management system:** The United States ITS also collects relevant information on the travel needs and habits of the public, so as to improve the transportation development model. This system ensures that the development direction of intelligent transportation in the United States is deeply consistent with the travel needs of the public and continuously improves the efficiency and satisfaction of public travel.

(3) **Public transport operation management system:** The management of public transport operation by ITS in the United States is very meticulous. It improves not only the stability and safety of the public transport system but also the efficiency of public transport operation in line with urban development.

(4) **State transportation management system:** The state transportation vehicles in the United States are also managed by the ITS system. The ITS system can monitor the operation information and transportation information of the state transportation vehicles in real time, and then make reasonable allocation of the state transportation vehicles.
(5) **Electronic toll collection system:** The electronic toll collection system of the United States was developed earlier and is now a more mature system in the world. The electronic toll collection system of the United States is a sub-system of the ITS system, which can automatically complete all kinds of toll collection for ground transportation, and the rationality, accuracy, and convenience of toll collection are more prominent, which not only improves the transportation capacity and efficiency of the United States but also ensures the economic income of the United States transportation.
(6) **Emergency management system:** ITS in the United States has a high response and reporting capacity to transportation accidents and emergencies and has prepared a full range of emergency resources. For example, the United States has the highest utilization rate of transportation medical helicopters.
(7) **Vehicle control and safety systems:** The U.S. transportation infrastructure has good perception and data transmission capabilities and can provide safety guidance according to transportation conditions and vehicle driving conditions to ensure the safe driving of vehicles.

6.2.2 *Development of ITS in Japan*

Japan is the most prominent country in terms of ITS construction except for the United States. This is not because Japan is good at developing in the field of intelligence but because Japan needs ITS to solve its own transportation development problems. As of 2021, the population of Japan is 126 million, and the number of vehicles in Japan is as high as 74 million. Transportation accidents in Japan will cause about 1 million deaths and injuries every year, in accordance with the statistics of Japan's transportation department, and transportation congestion will cause Japan's economic losses of several trillion yen every year. The main way to solve these problems is the construction and development of ITS.

Currently, Japan's ITS has been integrated with five national economic departments; these five departments are the Ministry of Construction, the

Police Surveillance Agency, the Ministry of International Trade and Industry, the Ministry of Transport, and the Ministry of Posts and Telecommunications. Under this joint development model, Japan's ITS has ranked among the top in the world. Compared to the construction of ITS in the United States, the construction and development of ITS in Japan mainly have the following characteristics.

6.2.2.1 *ITS realizes vehicle road collaboration in Japan*

In the process of ITS research and development and construction, Japan has adopted the joint development model, focusing on the layout of ITS in the field of logistics and transportation and services for vulnerable groups. From 2014 to 2018, the Japanese government promoted the development of "vehicle road collaborative system" by means of cross-sector design in the form of public–private cooperation and carried out various research and development and experiments under the principle of "vehicle road collaborative system and service expansion":

(1) The Japanese government has carried out basic technology research and development and cooperative operation, and developed a vehicle road collaborative system through public–private cooperation. Through various technological research and development, the occurrence of transportation accidents and transportation congestion has been effectively reduced.
(2) The Japanese government has carried out the test and promotion of vehicle road collaborative application. On the one hand, the Japanese government promoted the logistics transportation demonstration of vehicle road collaboration mode, and on the other hand, the Japanese government promoted the demonstration of vehicle road collaboration service for vulnerable groups. The efficiency of logistics transportation and travel experience of vulnerable groups in Japan was improved alternatively under the two measures, and the transportation operation effect was greatly improved.

6.2.2.2 *ITS realized the upgrade of ETC system in Japan*

In the process of ITS system development, Japan has completed the upgrade of ETC system, creating the world's first high-capacity two-way communication vehicle road collaborative system through DSRC. In this regard, Japan's ITS functionality has surpassed that of the United States.

Japan's ETC 2.0 system is able to promote real-time information interaction and plan the best driving route, which is mainly due to the integration of Japan's ETC 2.0 system, vehicle navigation system, and VICS, and the realization of collaborative management of vehicles and roads. As of 2019, Japan's ETC utilization rate has reached 92%, and this data has maintained a continuous growth.

We can feel the intelligent development of Japan's transportation from the ETC 2.0 system in Japan. Compared with ETC in most countries, ETC 2.0 in Japan has the functions of automatic toll collection, real-time interaction of road information, planning the best travel route, etc., which not only effectively solves the problem of transportation congestion in Japan's high-speed, but also improves transportation safety and optimizes transportation charges.

In general, there are two aspects of Japan's ETC 2.0 system that are worth learning from in China's transportation industry:

(1) **With road planning capability:** The Japanese ETC 2.0 system investigates the driving path of vehicles through high-speed payment information, which provides the current status information of the road for the driving drivers. It effectively reduces transportation congestion and transportation accidents, and improves transportation efficiency through real-time information interaction. In 2021, the Japanese ETC 2.0 system was able to provide urban drivers with transportation congestion information within a radius of 1000 kilometres, which has optimized the travel decisions of the public.
(2) **With safety enhancement capability:** Japanese ETC 2.0 system can plan the travel plan for the driver with the shortest travel time, the lowest cost, or the least impact on the environment, and each plan is based on reducing congestion. This intelligent planning capability not only improves the transportation safety guarantee but also promotes the green development of transportation.

In terms of ITS construction and development, the Japanese ITS construction ideas and development results are more worthy of learning by the people in China's transportation industry. The wisdom reflected in this development model is more in line with the needs of China's transportation development and has a certain role in promoting China's strategic goal of becoming a transportation power.

6.2.3 *Development of ITS in Europe*

In general, the ITS development in European countries with developed transportation mainly focuses on several aspects. The transportation development in most European countries has been greatly accelerated through the improvement of intelligence in these transportation areas.

6.2.3.1 *Transportation management*

Developed countries in Europe, such as Italy, France, the United Kingdom, and Germany, have established mature ITS and achieved good results in transportation management. The ITS of these developed countries can effectively allocate the transportation resources of key cities and road sections, thus greatly reducing the occurrence of transportation congestion and transportation accidents. At present, other European countries are also strengthening the construction of ITS. It can be said that transportation management is the main purpose of establishing ITS in many countries.

6.2.3.2 *Travel planning*

The ITS that can achieve the effect of travel planning has a high degree of intelligence, which requires a relatively sound infrastructure of ITS itself. Not many countries in Europe have reached this level, but the overall development is in this direction. Especially after the digital twin technology is applied to the field of transportation, the ITS in most countries can quickly have the ability of travel planning by using digital twin modeling and simulation.

6.2.3.3 *Travel monitoring*

European countries follow the development model of improving transportation perception in the process of establishing ITS under the influence of the transportation development in Germany, France, and other countries. Under this mode, ITS has good travel monitoring capabilities. However, except for Germany, France, and other strong countries, most European countries only have good travel monitoring capabilities for ITS, and the control power is still insufficient.

6.2.3.4 *Vehicle control*

Reasonable scheduling of transportation resources is the original intention of the establishment of ITS. The establishment of ITS in Europe and other countries is to improve the utilization rate of their own transportation resources. In addition to reasonable planning of road resources, the allocation of public transport, transportation, and other vehicles is also a major construction focus.

6.2.3.5 *Automatic toll collection*

Although the automatic toll collection capacity of ITS in Europe and other countries cannot reach the high intelligence of Japan, the automation has been fully highlighted. Taking the United Kingdom as an example, there are various automatic toll collection channels, electronic toll collection channels, credit card toll collection channels, and cash toll collection channels without change in the United Kingdom. Drivers can choose different channels according to their own habits, which effectively relieves the congestion at the entrance of highways, and all kinds of toll collection are under the control of ITS.

6.3 Development Status of Digital Transportation in the United States

Digital transportation is an important foundation for the United States to build a digital city and a digital country. Therefore, the U.S. government provides a lot of support in the process of building digital transportation and provides direction guidance through various policies. Generally speaking, the main strategy of establishing digital transportation in the United States is to realize the digitalization and informatization of urban transportation management first, and then connect and integrate the digital cities. For example, the United States has strengthened the construction of urban transportation intelligent transportation system, fully utilized the new generation of information technology and perception technology to enhance the perception of transportation system, and then combined with digital twin technology to reduce transportation congestion, improve transportation operation capacity, and improve safety conditions. When the urban transportation intelligence highlights, the United States will spread this effect among cities, and in the end, the connection point is formed.

In terms of transportation, the United States has enhanced the coordination among vehicles, roads, and people, continuously improved the intelligence and efficiency of transportation, provided better transportation methods and travel routes for transportation vehicles, and the transportation management department will carry out reasonable counseling, reduce accidents and save transportation time according to the road network transportation conditions, which has contributed to the economic development of the United States.

On the whole, the construction direction of digital transportation in the United States is to ensure that the transportation operation is in the best condition, make use of digital technology and new generation information technology to make up for the existing transportation defects, save transportation resources and transportation energy, and maximize the flexibility, safety, and efficiency of the transportation network. This development direction fully reflects the value of digital transportation and highlights the wisdom and economy of transportation. Therefore, this direction is worth learning from and thinking about in China's transportation industry.

6.3.1 *System composition of digital transportation in the United States*

Today, the digital transportation system in the United States is relatively mature and shows great advantages. It can be concluded through the analysis of the operation mode and main functions of digital transportation in the United States that digital transportation in the United States can be divided into two parts: transportation service system and transportation service system.

6.3.1.1 *Travel service system*

"Travel service system" refers to the service system of digital and intelligent development of transportation in the United States. The system integrates the information platforms of digital railway, digital highway, digital waterway, and digital aviation in the United States and constitutes a public information platform open to the public through information integration and coordination. With the capability of providing integrated services of transportation, including transportation policy information inquiry, travel information inquiry, electronic ticket sales, and automatic toll collection, this platform can realize passenger information collection and freight information tracking, providing accurate and effective data support for the transportation management department, and facilitating the transportation management department to make optimal development decisions. In addition, the "travel service system" of U.S. digital transportation is also the best way to connect transportation and economy, which promotes the development of U.S. economy.

6.3.1.2 *Transportation service system*

"Transportation service system" is the core system of transportation in the United States. It is closely connected with "travel service system even though it is a separate part of digital transportation in the United States". The "transportation service system" closely follows all kinds of information related to transportation in the "travel service system" and accurately controls the passenger flow, cargo flow, passenger and cargo distribution, flow changes, and cargo market demand through this data analysis. After accurate transportation information is collected, the "transportation service system" will use digital twin technology to construct a virtual model, simulate transportation resources and transportation networks, and formulate the best transportation plan and management decisions. In this way, the U.S. Transportation Management Department has improved management efficiency and responsiveness.

In addition, the U.S. "transportation service system" can also coordinate the transportation resources of the transportation network, optimize the allocation and joint operation of passenger and freight transportation, and realize the integration of transportation resources, which also highlights the intelligence of the U.S. digital transportation system.

6.3.2 *Public transportation management technology in the United States*

Urban transportation development in the United States is mainly based on real-time monitoring and intelligent management, which is mainly reflected in the public transport management technology in the United

States. The United States has not only equipped positioning perception equipment for public transport vehicles but also laid monitoring and transmission equipment in the urban road network. In this way, the U.S. public transport management department can realize real-time control of public transport vehicles according to the actual situation of the road, adjust the operating frequency of public transport vehicles, ensure the smoothness of key public transport sections, improve the priority of public transport in key periods, and realize the intelligent management of public transport system.

In addition, on the basis of enhancing the perception and control ability of public transportation system, the United States has established a new digital model of urban public transportation by using digital technology and new generation communication technology, enhanced the functions of information service, electronic toll collection, real-time scheduling of passenger and freight transportation, and intelligent management, and provided passengers with safer, more comfortable and more convenient public transportation services, which has created greater economic and social benefits.

6.3.3 *Intelligent parking guidance system in the United States*

Intelligent parking guidance system is an important transportation management system to optimize urban transportation resources and reduce transportation congestion. This system appears to be a transportation service, but it can optimize the construction scheme of urban transportation facilities, so as to improve the rationality and utilization of transportation resources, and at the same time, it can improve the operating efficiency of smart cities and promote economic development.

The effect of intelligent parking guidance system in the United States is more prominent in comparison with the current situation of China's transportation development. The U.S. transportation management department has fully perceived the urban pedestrian flow, transportation conditions, parking conditions, and transfer conditions by using the digital twin simulation model. According to the characteristics of urban public travel, the public travel mode is optimized by means of information transmission and signboard reminders, and the parking resources near the key transportation sections are reminded in real time to provide the driver with a parking plan reference.

The construction of intelligent parking guidance system in the United States is not complicated, mainly through the parking lot data collection, the data collection of parking spaces on the side of the road, the data collection of key sections and congested sections, and then the solution proposed by the system calculation. The construction of this system mainly lies in the information collection of parking resources, and then the parking information is embedded into the transportation management system, and then it is sent to the driver or it makes the display plate prompt through data transmission in real time. The necessity is very prominent even though the construction of this system is not complicated, which is also the main basis for improving the intelligence of the transportation system.

6.3.4 *Travel information management in the United States*

Today, with the development of digital transportation in the United States, we can see that its greatest feature is comprehensiveness and practicality. It is not a simple upgrade of traditional transportation management system, but a transformation of the transportation system according to the intelligent virtual model. This transformation not only improves the technical capacity of the transportation system in the United States but also solves the problems of transportation planning, scheduling, monitoring, services, and other problems, so as to have effectively promoted the development of transportation.

In the United States digital transportation system, in addition to functional command, scheduling, and guidance, there is an important information management system. The purpose of establishing a travel information management system in the United States is to ensure that transportation development meets the needs of the public and maintains a clear perception of the travel and transportation status of the public. However, with the deepening application of digital technology, the United States has obtained more transportation development optimization decisions from the travel information management system. For example, the United States has improved the community transportation operation according to the public travel information analysis, upgraded the transportation safety and accident prevention system, and accelerated the freight speed. The development of these transportation areas has also improved public travel satisfaction.

In addition, since the outbreak of COVID-19, mass travel information management system can also be used in epidemic prevention work. However, the use of travel information management in the United States is far less than that in China, which is a short board of the development of digital transportation in the United States.

6.4 Digital Twin Application of British Railways

Although the UK is the earliest country in terms of railway development, the development level of British railways has long been caught up with and surpassed by China as the problem of public and private operation has not been completely solved. In recent years, the application of digital twin technology by the British transportation department has been continuously improved. After solving many road transportation development problems, the UK has started to apply the digital twin technology to the railway field and comprehensively and effectively improve the efficiency of British railway transportation, accelerating the development of the railway industry.

For example, under the support of digital twin technology, the efficiency of UK railway operation has been effectively improved. In 2020, the UK's main rail line, Thameslink, reached 24 trains per hour, double the number in previous years.

The British railway operation and management departments have been continuously integrating the railway network and establishing a data analysis center to make the railway operation information more comprehensive and accurate, which also supports the improvement and upgrading of the British railway network.

6.4.1 *Challenges of digital transportation in the UK*

The development of the British railway industry has been greatly impacted after the outbreak of COVID-19, especially in railway construction, and the UK has even entered a period of development stagnation. The British railway still entered a difficult transition period; nevertheless, the UK government announced in September 2020 that it would terminate the 24-year railway franchise model and create a new development model that is easier and more effective.

The British railway industry has entered a new stage of construction and development after the year of 2021, which has also brought many new challenges to the British railway:

(1) The COVID-19 epidemic has led to limited population flow in the United Kingdom. The railway department in the United Kingdom is unable to sense the railway demand for public travel, and the normal work of the railway department has also been affected.
(2) The British railway construction team is not perfect. The UK also needs to integrate and upgrade the railway construction and management team to ensure that the railway development follows the pace of social development after the termination of the railway franchise model.
(3) The British railway development technology needs to be improved, and the problems of railway construction, operation, and maintenance and management have not been fully solved. For example, the United Kingdom encountered many problems in the construction and transformation of London Paddington station in 2020. With the help of digital twin technology, the British railway department finally solved the problem of commuter navigation at Pattington station.

In general, the British railway has shown the problem of insufficient basic strength at the key turning point, and the deepening of the application of digital technology has helped the British railway sector to improve this situation.

6.4.2 *Breakthroughs made by digital transportation in the United Kingdom*

John Nolan, the BIM project manager of the British railway network, said in the construction of the Exeter Central Rail Station in 2020, "the goal of the British railway network is to be at the forefront of digital model development. With the help of collision checking technology, the British railway network company was able to collect the real state of the project and reflect it into the virtual environment. In the end, it created the digital models of Exeter station and Pattington station. These two projects not only meet the initially envisaged project requirements, but also provide valuable digital asset records that can be used in many other projects in the future. Further site visits can be reduced to avoid delays and unnecessary costs".

From John Nolan's discussion, it can be seen that digital twin technology has become the main tool to solve the problem of railway development in the United Kingdom. Through the digital twin modeling and simulation, the British railway development has solved a variety of problems, such as environmental inspection, infrastructure construction, as well as cost saving.

In addition, since September 2020, the British Railway Administration has put forward the development strategy of clean, environmental protection, safety, and stability, and the realization of these development goals also requires the use of digital twin technology. Under the digital twin modeling and simulation, the British railway and the environment can continue to improve the fit, and the construction and upgrading of the railway network can be more secure.

Case: Digital twin upgrades Rolls-Royce jet engine

In 2020, as a defense company, Rolls-Royce and British Aerospace Department jointly deployed a specific plan to monitor the production and use of its jet engine through digital twin. Since then, the flight mode, flight conditions, and use method of each Rolls-Royce jet engine can be clearly demonstrated in the British Aerospace Department.

In fact, this is not the first time that Rolls-Royce has used digital twin technology. Before that, Rolls-Royce had upgraded the R&D and production technology of jet engines through digital twin technology. Stewart Hughes, the chief information and digital officer of Rolls-Royce, also stated that through the application of digital twin technology, the company has improved, upgraded, and optimized the service life of engines and the method, and the company's jet engines can provide more flexible services for the aircraft.

The R&D and manufacturing of jet engine is a very high-cost industry. The application of digital twin technology has greatly reduced the overall R&D cost of this industry. In the production process of Rolls-Royce's jet engine, digital twin mainly obtains the optimized data of system configuration, technology upgrading, and production manufacturing through digital modeling and simulation, so as to accelerate the design process of jet engine.

Nowadays, digital twin has been used throughout the entire process of Rolls-Royce's production of jet engines. Rolls-Royce has also optimized the manufacturing plan of jet engines through the fan blades of digital twin modeling, which has improved the fuel consumption efficiency of jet engines by 25%. At the same time, it also helps Rolls-Royce to extend the maintenance interval of other engines by 50%, which brings Rolls-Royce a huge profit.

As of 2021, Rolls-Royce has applied digital twin technology to various fields, such as research and development, production, and sales, and the value of digital twin is also rising. We believe that digital twin will become the core technology of the company and play a greater role in the future development of Rolls-Royce.

Chapter Summary

Europe and America are the birth areas of digital twin, and their application of digital twin is earlier and deeper; therefore, the application effect of digital twin in Europe and America is more significant. In the process of taking advantage of digital twin to achieve advanced development in China's transportation field, the European and American countries' use of digital twin methods, channels, and skills is worth learning from the people in China's transportation industry. In-depth analysis of these experiences can accelerate and improve the application effect of digital twin.

Chapter 7

Digital Transportation Management in Developed Countries

The advantages shown by the developed countries in the world are not due to their early start nor more comprehensive application of transportation technology. More importantly, these countries have explored transportation development strategies suitable for their national conditions in the field of transportation technology, which has effectively improved the value of transportation construction, transportation development, transportation operation, and maintenance management. This is also the main reason why these countries have developed into a transportation power.

7.1 Development and Experience of Digital Transportation in Developed Countries

In recent years, the construction of China's overall three-dimensional transportation network has achieved remarkable results under the promotion of the strategy of a strong transportation country. According to statistics of the transportation department, China has ranked first in the world in terms of mileage of transportation route construction and transportation technology upgrading level in recent years. It can be seen that the development strategy of "giving priority to transportation" in China is promoting the realization of the goal of a strong transportation country.

There is still a gap compared with developed countries though the overall situation of China's transportation development is very good. This gap exists in not only the development of transportation construction but also the field of transportation to serve the people's livelihood and stimulate the national economy. It is more and more obvious that the transportation developed countries use digital power to drive the overall development of the country, especially after the arrival of the digital era. These development experiences and achievements are worth thinking and learning from the people in the transportation industry in China in exploring the development path suitable for the national conditions.

7.1.1 *Concept of intelligent transportation in developed countries*

Nowadays, the main way of transportation quality change is still to enhance transportation intelligence, which requires the deep combination of the transportation industry and the new generation of information technology, data communication and transmission technology, electronic sensor technology, digital technology, and artificial intelligence technology to build an efficient response and intelligent and independent transportation management system.

From the current situation of intelligent transportation construction in developed countries, we can see that their definition of intelligent transportation is very comprehensive. These countries believe that intelligent transportation is not only the intelligent transportation itself but also the control and support of transportation management, transportation communication, public travel, and transportation construction and development by fully using emerging transportation technologies, making the transportation network an important force for national perception, connection, analysis, control, and driving national development. In the meantime, the new transportation technology also needs to ensure the development of transportation safety, give full play to the efficiency of transportation infrastructure, improve the level of transportation operation and maintenance management, and provide a steady stream of power for the sustainable development of the country.

7.1.2 *Development trend of intelligent transportation in developed countries*

Based on the analysis of the development process of intelligent transportation in developed countries in Europe and the United States, it can be concluded that the future development of intelligent transportation mainly follows the following trends combined with the development of intelligent transportation in China.

7.1.2.1 *Integrated and coordinated development of transportation*

In the development of intelligent transportation in developed countries, the coordinated development of various modes of transportation is the main feature. It is also the main way to improve the effect of intelligent transportation development to open up the blocking points of effective links of various modes of transportation. The transportation departments in developed countries have realized the information interaction among various modes of transportation, established an integrated and interactive digital model, conducted overall scheduling of the transportation network, and improved the interaction among them during this process.

7.1.2.2 *Integration and sharing of intelligent transportation system*

The construction process of intelligent transportation in developed countries mainly includes the construction of urban intelligent transportation system first, and then the connection with intercity intelligent transportation system, finally forming the overall intelligent transportation basic network. After the formation of the intelligent transportation network, the transportation management department will carry out the construction of a unified digital model and jointly share it with the public security management department, urban construction department, public health department, and other departments, so as to break the standoff of transportation resources and realize cross regional and cross industry resource links and to ensure that the intelligent transportation drives the overall development of the country, which is also a major trend in the development of intelligent transportation.

7.1.2.3 *Intelligent guarantee for safe operation of transportation system*

No matter which era, the focus of transportation development cannot be separated from safety guarantee. In the process of intelligent transportation development, safety development has become a more important content. Through the application of digital technology, new generation information technology, artificial intelligence technology, and other emerging technologies, all countries have begun to focus on solving the potential problems in transportation safety. The intelligent performance of the safe operation of the transportation system is also a major trend in the development of intelligent transportation.

7.1.2.4 *Establish a complete technical system and management system*

In order to promote the long-term stable development of the country, intelligent transportation must establish a sound technical system and management system, which is also a common measure for the development of intelligent transportation in developed countries. For example, in the digital development stage, German railway not only relies on digital twin technology to establish a railway construction model but also establishes a management model according to the actual operation of each railway. The establishment of these two models has improved the development speed and results of German railway.

This experience should also be used for reference in the development of China's intelligent transportation. Meanwhile, on the basis of the actual situation of transportation in China, a more complete management model should be established, so as to maximize the value of intelligent transportation development on the premise of meeting the needs of public travel and transportation.

7.2 Demonstration of Intelligent Transportation Achievements in Developed Countries

The development of science and technology has brought social progress and accelerated the flow speed and frequency of public. This development trend has brought a variety of transportation pressures while improving the public life experience. In the face of various transportation development problems, intelligent and information development has become the main way to solve these problems. At present, various developed countries in the world are solving the uncoordinated relationship among people, vehicles, roads, and the environment through various ways and have shown a lot of representative development results.

7.2.1 *Monitoring and coordination management of intelligent transportation in developed countries*

The development results of intelligent transportation in developed countries are first reflected in transportation monitoring and coordination management, which is also a major breakthrough made by them using digital twin technology.

7.2.1.1 *Transportation flow monitoring*

The transportation developed countries have realized real-time monitoring of transportation and people flow with the all-round layout of transportation perception equipment. The transportation management system can clearly show the situation of congestion points and smooth roads and convey this information to drivers and transportation management personnel in real time. For example, the United States has fully integrated transportation intelligence monitoring and management and road system. The average speed of each route will be displayed in real time when using navigation software such as Google Maps during rush hours, and the driver can select the best route according to the real-time situation of the road.

7.2.1.2 *Intelligent signal light*

The intelligent signal light is the most common infrastructure in the intelligent transportation system, which has greatly promoted the national transportation development. The intelligent transportation system can clearly sense the transportation conditions in each direction and each route through the perception of various sensor devices. The intelligent transportation system can adjust the time interval of the signal lights according to these real-time data, so as to achieve the highest efficiency of road operation.

For example, in many cities in the United States, the left turn light of the transportation lights is in red for a long time at night, but as long as a vehicle enters the left turn lane, the next time the transportation lights are switched, the left turn light will be automatically switched to green. In this way, night transportation in the United States reduces a large number of unnecessary left turn light waiting time, ensuring smooth straight transportation.

For example, most cities in China also cancel the waiting time for left turn lights at night but in different ways. After 11 p.m. in most cities in China, left turn lights and direct transportation lights will be combined to turn on simultaneously, which is the most effective optimization measure combined with the actual transportation situation in China.

7.2.1.3 *High-speed limit adjustment*

The biggest difference between the smart high-speed system in developed countries and the high-speed system in China is that the speed limit in most developed countries is not fixed. For example, in many European and American countries, the speed limit will be adjusted in real time in accordance with the actual situation. The intelligent transportation management system adjusts the speed limit of different directions and different sections by perceiving the high-speed transportation flow to ensure the best state of smooth high-speed, and also reduces unnecessary waste of transportation resources. Such transportation management measures seem simple but have far-reaching significance for transportation development. For example, the United States effectively reduced the number of times of braking when driving through such high-speed speed limit adjustment, thus reducing fuel consumption, road wear, and exhaust pollution.

7.2.1.4 *Transportation intelligent management equipment*

Many developed countries have placed intelligent management equipment in important transportation hubs. For example, the United States has established sidewalk smart indicators in many cities in order to enhance road pedestrian safety. These smart devices will automatically adjust the switching of nearby signal lights according to the number of pedestrians and the waiting time, which not only optimizes road management but also ensures the safety of pedestrians at night.

7.2.1.5 *Transportation network optimization*

At present, the transportation network construction plans in developed countries are the final plans after multiple rehearsals and optimization of virtual models. The rationality, efficiency, and intelligence of transportation infrastructure construction have been comprehensively improved with the assistance of digital technology and new generation information technology. For example, the number of motor vehicles in Tokyo of Japan is about 8 million, and the population of Tokyo is over 16 million, which represents that Tokyo has an average of 0.5 vehicles per person, which is a global leader. However, the travel rate of Tokyo's household vehicles is very low, and Tokyo's household vehicles are suspended for about two-thirds of the year. The reason is that the construction of intelligent transportation in Japan has effectively solved the problem of public travel, coupled with scientific management and policy guidance, household vehicles are no longer the first choice for public travel in Tokyo.

In addition, the per capita annual delay rate of Japanese road transportation is only 5 hours/year, which is far below the global average level. The annual death toll of Japanese transportation accidents is only 8000, which is also the lowest level in the world.

7.2.2 *Toll collection and parking management of intelligent transportation in developed countries*

In addition to monitoring and coordinating transportation resources, the results of intelligent transportation construction in developed countries are also reflected in transportation charges and parking management:

(1) The toll collection methods in developed countries are much more than those in China. This is because the intelligent transportation systems in these countries are more carefully designed in terms of toll collection methods and floating range, and their convenience is also more prominent. For example, in many developed countries, the road toll standards are changing every day, and the toll standards in rush hours are significantly higher than those in idle hours. In this way, transportation congestion in these countries has also been eased.

In addition to toll collection methods, many developed countries have also established humanized automatic toll collection systems. The design of these automatic toll collection systems is worth learning in the field of transportation in China. For example, high-speed automatic toll collection devices in the United States can be installed on lease. No installation fee is required, but a monthly rental of US $1 is required. After the installation of such automatic payment equipment, the vehicles can enjoy a 10%–20% discount on high road fees. The lease of this kind of equipment can be terminated at any time, so its practicality is very high. In addition, the United States has also set up a variety of modes according to the habits of the public, and even many states have retained the way of coin payment, which has also enhanced the humanized characteristics of transportation charges.

(2) In developed countries, the management of parking resources is more detailed, which greatly relieves the pressure of road transportation. Similarly, in a case of the United States, parking lots in key cities and parking lots near key transportation hubs in the United States are equipped with perception equipment to learn about the parking resources in these areas in real time. After that, the intelligent transportation management system will sync this information to navigation information and transportation prompt information to help drivers find parking resources in the shortest time.

On this basis, the United States has also set up vehicle positioning services in various large parking lots. With the information prompt in the unfamiliar environment, the driver can find the parking vehicle in the shortest time.

In addition, the United States also integrates the automatic toll collection system with the parking management system, so that drivers can pay various fees together, which improves the payment efficiency and reduces the cost of toll collection. This is also the development result of intelligent transportation in most transportation-developed countries.

7.2.3 *Transportation assistant driving function in transportation developed countries*

With the development of transportation technology, the industry of intelligent driving and automatic driving has been booming. The United States has more experience advantage in this technology field. However, the author believes that Japan and Singapore may be able to become fully popularized countries of automatic driving earlier after a comprehensive comparison of the development level of intelligent transportation in developed countries. The major reason is that Japan and Singapore have not only made more breakthroughs and innovations in intelligent driving and automatic driving but also have more advantages in terms of transportation infrastructure, relevant laws and policies, and public acceptance.

Taking Japan as an example, Japan has made many breakthroughs in the field of automatic driving in recent years. Japan has realized accurate perception of the distance between vehicles and obstacles through a new generation of perception equipment and has set up a safer avoidance procedure through digital twin simulation. In the face of an emergency, Japan's self-driving vehicles have higher safety in braking and avoidance measures.

What's more, Japan has also designed a detailed navigation system on the basis of transportation congestion, which can not only cope with complex road conditions, but also carry out automatic switching of roads to improve the safety and intelligence of vehicles.

Currently, the new road transportation law of Japan has provided legal protection for automatic driving L3 vehicles on the road, and the full implementation of automatic driving is gradually realized.

7.2.4 *Logistics and transportation management in transportation-developed countries*

The intelligent development of transportation has also improved the management effect of logistics transportation, especially in logistics tracking and positioning, and real-time state feedback, and the intelligent system has made a significant breakthrough. Under the linkage of satellite transportation positioning technology, new generation information technology, and digital technology, the perception ability of logistics and transportation conditions in transportation developed countries can be accurate to every minute and kilometre. The driving route, transportation type, departure time, and arrival time of transportation vehicles can be clearly displayed, which not only greatly improves the management effect of transportation but also accelerates the development of logistics transportation industry.

In comparison with the promotion of the development of intelligent transportation to the logistics transportation industry, the developed countries have more advantages than China in the following aspects.

7.2.4.1 *Loose policies to promote development*

In order to promote the development of logistics transportation industry, most of the transportation developed countries have provided support in road, railway, aviation, navigation, and other aspects in the process of intelligent transportation construction. For example, the United States has promulgated and upgraded the joint transportation law several times in order to promote the development of the transportation industry, which has also been implemented in the development process of intelligent transportation and deepened the integration of intelligent transportation and the transportation industry.

7.2.4.2 *Comprehensively promote the intelligent development model of the logistics transportation industry*

Many transportation-developed countries have greatly promoted the intelligent development mode of logistics transportation industry in the construction of intelligent transportation, and China has also followed this development mode. In the process of development, through the combination of transportation network data and logistics transportation, we have built a smart logistics system that is more in line with the actual situation of our country's transportation. In addition, many developed countries also use digital twin technology to improve the intelligence of logistics transportation network and upgrade the logistics transportation services with the virtual models of various transportation networks and smart cities, which has improved the economic benefits of the industry.

7.3 Digital Transportation Construction in Transportation-Developed Countries

In recent years, China's intelligent transportation construction has made remarkable achievements; however, compared with developed countries, there are still lagging areas. From a technical perspective, this gap is mainly reflected in the cutting-edge and overall nature of transportation technology. As a result, it is necessary to conduct in-depth research on the construction mode and development strategy of overseas developed countries, make up for their own deficiencies in technology, and promote the overall economic development in the process of China's intelligent transportation development.

7.3.1 *Advanced technology of intelligent transportation in Japan*

In June 2019, the Japanese government officially issued the "Declaration on Creating the World's Most Advanced Digital Country", and subsequently established the headquarters of IT comprehensive strategy to vigorously promote the digital and intelligent upgrading of land and transportation.

In June 2020, Japan's transportation department issued the "Land and Transportation White Paper 2020", which focuses on the research progress and social pilot situation in the field of intelligent transportation in Japan. From the "Land and Transportation White Paper 2020", we can see the cutting-edge technologies used in the intelligent transportation strategy by developed countries of transportation, as well as the application effect of these technologies.

7.3.1.1 *Create a national one-card model*

As early as 2001, Japan began to use IC cards for transportation settlement. At that time, this settlement mode was mainly to improve the charging and operating efficiency of rail transit. With the development of the times, Japan began to use IC card as the main tool to connect the national rail transit charges.

As of 2019, there were 9 types of mainstream transport IC cards in Japan. The total number of monthly uses of these 9 types of IC cards reached 250 million times, which shows the important role of rail transit in Japan's transport network. In 2020, Japan officially announced in the "Land and Transportation White Paper 2020" that it will promote the national access of IC cards in the future, so as to improve the national travel efficiency again.

7.3.1.2 *Pilot operation of new transportation policies*

Japan is a country with not only the world's leading intelligent transportation development speed but also the world's leading transportation policy update speed. Japan launched a "smart travel challenge" pilot activity for smart travel in many cities in 2019 and tested the changes in people's travel experience under the new development model and new policies through this new travel service model.

In the "Land and Transportation White Paper 2020", Japan has once again upgraded its transportation policies and started various policy pilot projects accordingly. This policy and construction are promoted simultaneously to ensure the development speed of Japan's intelligent transportation.

7.3.1.3 *Create a new mode of intelligent transportation development*

Japan has successively launched a variety of new transportation modes with the aim of accelerating the development of intelligent transportation, which have promoted the overall development of intelligent transportation in the process of continuous testing and upgrading.

"For example, in 2019, Japan carried out 19 pilot tests on the Mobility as a Service (MaaS) model, including major cities, densely populated suburbs, sparsely populated suburbs, as well as tourist attractions. During the test, the Japanese transportation management department also combined the MaaS model with AI technology and array technology to strengthen the switching of various transportation modes in the MaaS model.

During the test and upgrade for more than one year, the Japanese transportation department has collected detailed test data, and also held a special seminar on transportation operation enterprises for the MaaS model in the process. In March 2020, in the end, the joint guide on MaaS data collaboration was formulated to promote the MaaS model nationwide.

The MaaS model did not significantly promote the development of transportation intelligence in Japan, but its flexible floating mechanism of travel expenses and close connection with tourism, retail, education, medical and other industries have driven the overall economic development of Japan.

The Japanese government has also issued Amendments to the "Law on Promoting the Activation and Regeneration of Regional Public Transportation" and the "Plan for the Identification of New Travel Services" after the success of the MaaS model, which has brought policy support to the new travel services industry in Japan".

From the mode of synchronous development of Japan's intelligent transportation and policies, people in China's transportation industry can consider the promotion and implementation of local new travel modes and analyze the benefits of new travel modes to China's intelligent transportation development.

7.3.1.4 *Comprehensive promotion of ETC 2.0 system*

As mentioned previously, ETC 2.0 built by Japan's intelligent transportation has accelerated the overall development of Japan's intelligent transportation. According to the statistics of Japan's transportation department, more than 70 million vehicles in Japan had been installed with ETC, and the utilization rate of Expressway ETC was about 92.9% as of March 2020.

At first, the main reason for Japan to upgrade and promote ETC 2.0 was to solve the high-speed congestion. According to the statistics of Japan's transportation management department, 30% of the high-speed congestion in Japan is due to the queuing of toll station nodes, and ETC 2.0 effectively improves the people's experience of using ETC, and the queuing of toll station nodes is also alleviated.

ETC 2.0 in Japan has realized the two-way information interaction of the whole process of high-speed road vehicles. The driver can understand the real-time operation status of the highway, and the management department can understand the speed, path, and driving status of the vehicle in the process of driving. In the two-way data transmission process, the transportation management of the Japanese highway management department is more timely and effective.

7.3.1.5 *Improve the effect of intelligent transportation safety management*

The Japanese transportation department formulated the research plan for advanced safety vehicles (ASV) as early as 1991 and cooperated with a number of domestic automobile manufacturers to promote the plan, so as to improve the effect of transportation safety management. ASV system has been widely used in Japanese vehicles nowadays through multiple upgrades and iterations. With the assistance of ASV system, Japanese vehicles can realize various functions, such as independent buffer braking and lane maintenance, and the overall safety of the Japanese transportation system has been continuously improved.

7.3.1.6 *Development of automatic driving technology*

In 2020, Japan's transportation management department officially proposed three development goals for automatic driving technology:

(1) Popularize level 3 automatic driving on national highways.
(2) Popularize unmanned driving travel services in limited areas.
(3) Complete the following driving test of the unmanned driving team behind the truck in the highway scene.

At the same time, Japan's transportation management department issued a series of supporting measures for these three development goals.

The original intention of Japan's development of automatic driving technology is not to upgrade transportation technology but to reduce transportation accidents and solve the problem of insufficient drivers brought by the aging society. In the process of actively promoting the development of automatic driving technology, Japan has also carried out the collaborative application of highway data integration nationwide, and conducted many on-site experiments of automatic driving vehicles, and combined with digital twin to upgrade and renew the automatic driving technology. Therefore, the landing speed of Japan's automatic driving technology and the matching degree with the national transportation conditions are far higher than those of other countries.

7.3.2 *Inspiration and suggestions from Japan's intelligent transportation to China's transportation development*

From the development of intelligent transportation in Japan, people in China's transportation industry can feel that emerging technologies have accelerated the integration of transportation industry and people's livelihood industry, changed the construction mode of transportation infrastructure, expedited new travel mode and new management mode, and promoted the subversive development of society as a whole.

In future development, China's intelligent transportation construction can learn from the experience of developed countries in the following aspects to speed up the development process of China's intelligent transportation.

7.3.2.1 *Improve the transportation service quality of new travel modes*

From the process of intelligent development of transportation in the United States and Japan, it can be seen that one of the key modes of transportation development is to improve the public travel experience and bring convenient and economic transportation services to the public. Combined with the development status of intelligent transportation in China, the author believes that this effect can be improved from two aspects:

(1) Innovate the public travel mode and upgrade the public travel habits by pilot promotion. The promotion of the MaaS model in Japan has adopted the pilot application and national promotion after upgrading and improvement. This model is worth learning from in China, and innovation is in line with China's national conditions travel model, improving according to the actual effect of the pilot and then promoting nationwide under the guidance of the government. This way can accelerate the full implementation of new technologies and new modes of intelligent transportation in China.
(2) Upgrade the technicality of transportation services. From the automatic toll collection system of the United States and the ETC 2.0 system of Japan, it can be seen that the emerging transportation technology can promote the development of national economy while improving the transportation service and accelerating the transportation operation. At present, China's intelligent transportation needs to be improved in the field of transportation services, which can be seen from the toll collection model and transportation guidance.

7.3.2.2 *Build a new model of vehicle road collaboration*

In recent years, with the support of digital twin technology, China's vehicle road collaboration has achieved outstanding development achievements. However, from the results of intelligent transportation development in Japan, the United States, and other countries, it can be seen that vehicle road collaboration development can be more detailed and efficient. For example, Japan's ETC 2.0 improved the information interaction ability among vehicles and roads, vehicles and vehicles, and vehicles and management departments, thus completing the unified management of vehicles and roads, which made Japan's high-speed operation efficiency ranked at the forefront of the world.

In the future, China's transportation development can also learn from this method and take the vehicle road collaboration as the leading factor to improve the wisdom of transportation operation and maintenance management.

7.3.2.3 *Build an intelligent transportation industry ecosystem*

In the process of intelligent transportation development, developed countries such as Japan promote not only the development of transportation itself but also the progress of various industries. For example, the ASV system jointly built by Japan and domestic automobile enterprises improves not only the transportation safety of Japan but also the market competitiveness of Japanese automobile enterprises and drives the development of the Japanese automobile economy.

At present, China's intelligent transportation construction is also integrating multiple industries, but there is a gap in the degree and method of integration. In future development, China's transportation departments can lead the development of this model, strengthen cooperation with relevant industries, or promote cross-border integration of the transportation industry, create a healthy and huge intelligent transportation industry ecosystem, and accelerate China's economic development.

7.4 Experience in Intelligent Transportation Construction in Developed Countries

The intelligent development of China's transportation industry has drawn on a lot of overseas experience in recent years and adjusted and integrated with its own actual situation. These advanced experiences have not only translated into the actual development effect of intelligent transportation but also integrated into various policies of China's transportation development.

For example, the China Urban Rail Transit Association has issued the "Outline of Smart Urban Rail Development of China's Urban Rail Transit", in which the connotation of smart rail transit is clearly analyzed in six aspects in the "connotation signs":

(1) The connotation of smart city rail. Through the application of emerging information technologies such as cloud computing, big data, Internet of Things, artificial intelligence, 5G, satellite communication and blockchain, comprehensive perception, deep interconnection and intelligent integration of physical information such as passengers, facilities, equipment and environment, it has independently evolved and innovated service, operation and construction

management models to build a safe, convenient, efficient, green new generation of Chinese smart urban rail transit.

(2) Intelligence and smart. Intelligence: the use of advanced technology for material (equipment) empowerment, in order to achieve automation, unmanned. Intelligence has the characteristics of natural science, such as intelligent equipment and intelligent system. Smart: The advanced technology is used to empower people and materials (equipment) at the same time, so as to achieve independent acceptance, learning and decision-making to achieve higher efficiency. Smart has the dual characteristics of social science and natural science, such as smart city and smart city rail.

(3) Smart system and smart city rail. The development of smart system and the construction of smart city rail are the internal relations between the two, which means the construction of multiple intelligent systems into a system and ultimately constitute the smart city rail. The "Outline" consists of 10 intelligent systems. It is marked by the realization of "convenient intelligent passenger service", "maximum efficiency and efficiency of intelligent transportation", "green intelligent resources and environment", "full automation of intelligent train operation", "localization of intelligent technical equipment", "digitalization of intelligent infrastructure", "perception of intelligent operation and maintenance security", "efficiency of intelligent network management", "intensification of urban rail cloud and big data platform" and "series of smart city rail technology standards".

(4) Smart city rail and city rail informatization. "Promoting urban rail informatization, developing intelligent systems and building intelligent urban rail" is the theme of the "Outline". A new round of scientific and technological revolution is surging and developing rapidly, which gives birth to emerging information technologies such as cloud computing, big data, Internet of Things, artificial intelligence, 5G, satellite communication, as well as blockchain, thus improving the digitalization, networking and intellectualization to a new height. Therefore, essentially speaking, smart city rail is the crystallization of the integration of various systems and services of urban rail transit by using emerging information technology, and is a concentrated reflection of the new stage of information construction in the field of urban rail transit. As a result, information construction is the core and foundation of smart city rail construction. It can be seen that urban rail informatization and smart urban rail are integrated. The "Outline" is

the top-level design for the construction of smart city rail, and is also the top-level design for the information construction of urban rail transit. Under the guidance of the "Outline", we will promote the informatization of urban rail transit, develop intelligent systems and build smart urban rail transit.

(5) Smart city rail and country with strong transportation network. The construction of smart city rail is an important path and strategic breakthrough for the construction of strengthening country with strong transportation network. With a sense of mission of "strengthening country with strong transportation network, urban rail transit bears the responsibility", the association, after two years of in-depth research, summarized and proposed the main work path of "urban rail transit bears the responsibility", which is to focus on intellectualization and autonomy: first, focus on the technology level, the application of new scientific and technological revolution results, and the deep integration of new information technology and urban rail transit to build a smart urban rail; second, at the system and mechanism level, focus on autonomy, break through key and core technologies, and ensure the safety and controllability of technology and industry as well as the dominant position of the market. Therefore, under the premise of autonomy, the date when the smart city rail is completed is the time when the strong China city rail rises.

(6) Smart city rail standards and city rail information standards. Information construction is the foundation of the construction of smart city rail, so China's urban rail information standard system is also the foundation and an important part of China's smart city rail technology standard system. In accordance with the concept and principle of "standard research first, standard guiding construction", the association simultaneously carried out the top-level design of the urban rail information standard system in the preparation stage of the "Outline", and established and gradually released the China urban rail information standard system composed of four levels of "1352". The advance research and implementation of the urban rail transit information standard system will lay a solid foundation for the research and formulation of the smart urban rail transit technical standard system. Centering on the construction system of smart city rail, we have established technical specifications with independent intellectual property rights in China. Firstly, strengthen policy

research, implement top-level planning, build a technical specification system, and form a comprehensive, coordinated, economically reasonable, open and integrated standard system to comprehensively support and lead the construction of smart city rail through demand-oriented; secondly, study and formulate a number of key core technical specifications, and form a series of standards from the top-level management, supervision and evaluation, operation and application, platform construction, data integration to the bottom perception for sharing key areas, so as to guide the orderly and high-quality development of smart city rail construction projects; thirdly, guide the research, improvement and iterative development of each professional information application system of smart city rail; fourthly, actively connect with national competent departments and international standards organizations, participate in international standards research and compilation, and gradually realize the internationalization of smart city rail technical standards.

It can be seen from the six "connotation signs" in the "Outline of Smart Urban Rail Development of China's Urban Rail Transit" that the smart development of China's rail transit has fully drawn on the development experience of developed countries, and various development goals have completely standardized the international first-class level, playing a comprehensive driving role in the travel experience of the public, industrial economy, and the overall development of the society.

In addition, it can be seen from the "Outline of Smart Urban Rail Development of China's Urban Rail Transit" that the advanced experience of overseas intelligent transportation development has brought inspiration to China's transportation development in the process of China's implementation.

First, the development of intelligent transportation needs to be recognized by government departments. Under the support of relevant policies, the development measures of intelligent transportation can be carried out smoothly. This can be seen from the "Outline of Smart Urban Rail Standard and Urban Rail Informatization Standard". The information planning of urban rail transit in China is bound to learn from the advanced experience of developed countries, but it also needs the drive of government departments. Under the guidance of government departments, various regulatory measures can be effectively popularized.

Second, we need to ensure the continuous investment of funds. In order to fully realize the construction contents in the "Outline of Smart Urban Rail Development of China's Urban Rail Transit", the pressure from the government alone is large, and the development speed is bound to be limited. Therefore, the realization of this goal should learn from the experience of the United States, Japan, and other developed countries and improve the development speed by the mode of cooperation between the transportation industry and other industries to ensure the timely realization of various development goals.

Finally, we should pay close attention to the people-oriented development concept. From the development of intelligent transportation in the United States and Japan, it can be seen that people-oriented is the core of transportation development, because transportation is the livelihood industry, only people-oriented can ensure the development effect.

It is also clearly mentioned in the construction focus and layout blueprint of the "Outline of Smart Urban Rail Development of China's Urban Rail Transit" that the smart development of China's urban rail transit will not deviate from the direction of improving the convenience of mass travel and promoting China's economic development. It can be seen that the development concept of intelligent transportation in China is following the right direction, and the development effect is naturally more guaranteed.

In a word, the developed countries of transportation have brought a lot of valuable experiences to China in their development for many years. These advanced experiences are worth the in-depth thinking of the people in the transportation industry of our country, from which we can learn the essence and apply it to transportation planning. The transportation development of our country can achieve the effect of "taking the stones from other mountains to make the jade, and taking the stones from other mountains to build the peak".

Case: Digital twin helps Mawan Port to become a world-class strong port

On 26 August 2020, the western port area of Shenzhen held a demonstration conference of the achievements of Mawan Smart Port. The first box of containers on the site was lifted smoothly and landed on the intelligent driving trailer, representing the successful completion of the transformation of the first smart port in China.

Mawan Smart Port is China's first fully completed intelligent transformation of automated terminals. This transformation project reflects the intelligent and digital development of China's ports and simultaneously improves social and economic benefits.

In fact, before the smart upgrade of Mawan Port was completed, the port was upgraded in virtual space. The engineering department in charge of the upgrading of Mawan Port used digital twin technology to simulate the whole process of the upgrading project and made three breakthroughs in digital space.

First of all, this project has realized the real-time driving of dynamic data in China's port industry under the scenario of digital twin full elements.

Second, for the first time, China's port industry has used digital twin technology to complete the full cycle simulation coverage of cargo arrival, loading and unloading, transfer, storage, and departure.

In the end, China's ports realized the situation awareness of full amount of industrial equipment for the first time with the combination of digital twin technology and 5G technology.

These three technological breakthroughs have enhanced the smartness of Mawan Port and have also improved the logistics and operational efficiency of Mawan Port.

In fact, the intelligent upgrade of Mawan Port by the digital twin is not complicated and only takes three steps, namely static scene restoration, dynamic environment simulation, and the integration of all elements:

(1) **Static scene restore:** The digital twin collected data from the static scene and restored the 1 million square meter Mawan Port area in the virtual space for 1:1, including not only the topography and environmental characteristics but also more than 100,000 containers.
(2) **Dynamic environment simulation:** The digital twin also analyzes the relationship and role between the static environment data, follows the physical attributes and operation rules of the physical entities, realizes the real-time dynamic simulation of the virtual model, and keeps the virtual model highly active.
(3) **All elements are connected and implemented:** Through the monitoring and perception equipment laid by Mawan port, the digital twin can collect comprehensive port information, conduct data analysis and processing after the simulation and preview of the virtual model, and then return all kinds of decision-making basis to the real upgrading

and management of the port, which not only ensures the full element simulation of Mawan port, but also plays the largest role of the virtual model.

Although the smart upgrade measures of Mawan port are simple, the smart achievements are outstanding. The digital twin platform of Mawan port can realize the real-time dynamic perception of the port and complete the whole process accurate tracking of cargo with the combination of 5G technology and Beidou system, thus realizing the rational deployment of intelligent management of unmanned fleet in the port. Nowadays, the transportation, turnover, and allocation efficiency of Mawan port have been among the best in China's port industry. The upgrading and transformation model of this port will be used for reference by more ports, and the number of smart ports and world-class strong ports in China will continue to increase.

Chapter Summary

We are brought with many inspirations by the road of digital upgrade in developed countries, among which the most clear view is that the effective application of digital twin determines whether the upgrade will be successful or will fail. Since digital twin is applied in the smart concept, construction strategy, and existing achievements of developed countries, the value of digital twin has been fully highlighted in the development of these countries.

Chapter 8

Comparison of Domestic and International Intelligent Transportation in the Digital Twin Era

Transportation is the foundation of national economic development. The intelligent development of transportation industry is the main determinant of the development speed of modern countries. China's "Outline for the Construction of Strengthening Country with Strong Transportation Network" clearly pointed out that intelligent development is an important development direction of China's transportation. Shortening the gap between the level of transportation intelligence will help China to achieve the grand goal of "strengthening country with strong transportation network" under the guidance of this direction.

8.1 Comparison of Domestic and International Intelligent Transportation Under the Background of Digital Twin

From the development process of intelligent transportation in developed countries, it can be seen that digital technology has become the key technology of overseas intelligent transportation development, and digital technology has been throughout the development of foreign transportation industry. Through the application of digital technology, not only has the intelligent and automatic transportation management center been established but also the transportation environment has been improved, the feasibility of various travel models has been provided, and the utilization rate of transportation resources and economy has been steadily improved. These development effects can be seen from the intelligent transportation system (ITS) in European and American countries.

In recent years, with the application of digital twin technology, ITS, a powerful foreign transportation development country, has possessed the ability of real-time collection, transmission, and processing of transportation-related information. Through the combination with other emerging technologies, it has realized the unified coordination and processing of transportation resources and established a real-time and efficient integrated transportation management system. The intelligentization of transportation service and management is becoming increasingly prominent, which has a profound impact on the development of the country.

For example, as the world's leading automobile country, Germany has more than 5,000 million vehicles, but its population is only more than 8,000 million, which represents that Germany's per capita automobile ownership ranks first in the world. Although there is no transportation police on German roads, the ultra-low accident rate and ultra-high transportation efficiency in Germany are completely dependent on the intelligent transportation management system.

Under the support of digital twin, Germany has established a comprehensive digital model of the national transportation network and has found out the control schemes of various transportation lights through simulation and preview. Therefore, in Germany, the transportation operation efficiency is extremely high while ensuring a very low accident rate. In addition, German transportation laws and regulations have been integrated into the education system, and, therefore, the awareness of comprehensive compliance and safety is much higher than that of other countries, which is also the main reason for the orderly development of German transportation.

After comparison, we can see the current problems in the construction of intelligent transportation in China.

8.1.1 *The intelligence of urban transportation system needs to be improved*

China's urban population flow frequency and car ownership began to increase significantly, and urban transportation congestion, transportation safety, environmental pollution, and other issues began to be highlighted with the rapid growth of China's economy. In recent years, the transportation management departments have started to use digital twin technology to analyze and optimize urban transportation in China and put forward improvement suggestions for some prominent transportation problems, so as to realize the effective management of urban transportation in China. However, as a whole, there is still a clear gap between the intelligence of China's urban transportation system and that of developed countries. The urban transportation system cannot respond to the instantaneous high-speed changes of transportation and people flow, and cannot achieve sufficient dynamic management. The management speed of urban road network needs to be improved. Solving these problems will become the main task of China's intelligent transportation development.

8.1.2 *The overall intelligence of China's comprehensive three-dimensional transportation network needs to be improved*

Affected by the huge area of China's transportation network, the overall intelligence of China's comprehensive three-dimensional transportation network is also lagging behind that of developed countries. However, this gap is mainly reflected in the short development time of transportation digitalization and informatization, which will be gradually supplemented with subsequent development.

The author believes that the intelligent development of China's integrated three-dimensional transportation network should learn from the development experience of European and American countries, first improve the intelligence of urban transportation network, and then expand the connection point. In the process, we can make full use of digital twin technology to find the connection mode of intelligent transportation network, and also need to solve the problem of the impact of transportation development on the environment. Under the principle of sustainable and green development, we will strive to improve the intelligence of transportation network, so as to comprehensively drive the economic development of China.

8.1.3 *The intelligence of China's transportation resources regulation and guidance ability needs to be improved*

There is still a gap between the wisdom of overall adjustment and guidance of transportation resources and that of developed countries even though China's transportation has made certain achievements under the mode of coordinated development of vehicles and roads. For example, most of the cities in China lack the closeness of road, railway, aviation, and other modes of transportation, and the efficiency of mode switching needs to be improved. While the United States, Germany, Japan, and other major cities in developed countries have achieved seamless switching of transportation modes, the utilization rate of transportation resources is far higher than that of China.

Moreover, the management of parking space resources in developed countries is also more in place, especially in the urban transportation management system of the United States, which has been fully embedded in the management of parking space resources, which greatly saves the time for drivers to find parking space in an unfamiliar environment. At the same time, the United States also carries out reasonable construction of parking lots according to the actual situation of urban transportation, so as to improve the reasonable allocation of urban transportation resources and enhance the public travel experience.

The concept of transportation developed countries is worth thinking about by people in China's transportation industry since the unreasonable allocation of parking spaces and low positioning efficiency have begun to affect the daily operation of China's transportation with the continuous improvement of China's car ownership, and the congestion of key sections in many cities is also related to parking spaces. Therefore, a comprehensive solution to the problem of parking space resources can effectively improve the operation of urban transportation in China.

8.1.4 *The intelligence of urban transportation infrastructure construction needs to be improved*

At present, the construction of urban transport infrastructure in China is mostly to meet the needs of urban development rather than systematic and intelligent planning. For example, instead of the overall improvement of urban transportation network, the expansion and reconstruction of some urban roads are only to alleviate the congestion problem of this road. This construction mode of "treat head for headache and treat foot for foot pain" leads to the quality change of urban transportation network, the unreasonable allocation of transportation resources, and even affects the construction of smart cities.

From the experience of intelligent transportation construction in developed countries, it can be seen that the construction of urban transportation facilities must be a comprehensive and intelligent planning. Through the simulation of digital twin technology, the best way of urban transportation development can be found, and then the systematic construction planning can be carried out.

Although there are differences between China and foreign countries, this construction concept is the main way to solve the rapid growth of urban transportation demand, urban transportation and transportation, and poor operation. It can promote the comprehensive and diversified development of urban transportation in essence and solve various problems of urban transportation development lagging behind and restricting economic development.

8.1.5 *The intelligence of transportation management needs to be improved*

The construction of intelligent transportation in China has been accelerating in recent years, and the construction results are very significant. However, the intelligence of transportation management needs to be improved since intelligent transportation does not give full play to its application value. For example, in 2019, the Kunming Rail Transit Group announced the operation of the Kunming Metro. In 2019, most of the routes of Kunming Metro were in loss operation. This situation is caused by the factors of high construction cost, but the operational condition is also a reason that cannot be ignored. Generally speaking, the intelligence and rationality of Kunming's comprehensive transportation operation need to be improved. Improving the intelligence of transportation management is also the main way to solve this problem.

In addition, the formulation and implementation of China's transportation management policies lack of timeliness, which also has a significant gap with developed countries. For example, Japan will cooperate with the introduction of various policies during the pilot process of the new travel mode, and immediately upgrade and improve the transportation policy after the success of the pilot, which not only ensures the efficiency of transportation development but also improves the quality of transportation management. Therefore, this is also the focus of China's transportation sector that needs to be improved.

8.2 Key Points of Intelligent Transportation Development in Developed Countries

As of 2021, many developed countries in the world have achieved remarkable results in the development of intelligent transportation, which will guide the development of intelligent transportation in other countries. After summarizing the current situation of transportation construction in these developed countries, it is concluded that the development focus of intelligent transportation in the developed countries is mainly reflected in the following aspects.

8.2.1 *New infrastructure*

Any country that wants to realize the comprehensive digitalization and intelligence of the transportation network cannot do without the support of the new infrastructure. Countries with advanced transport systems are more aggressive in building new infrastructure. For example, in 2021, the new Transportation Secretary of the United States, Peter Buttigieg, was promoted from the mayor of South Bend, Indiana to the Minister of Transportation due to the outstanding construction achievements of the new infrastructure. During his term as the mayor, he focused on the new transportation infrastructure of Indiana, striving to improve the development effect of transportation intelligence and thus won the trust of the people of Indiana. In 2020, Peter Buttigieg also participated in the U.S. general election and won in the state of Iowa.

At present, the United States, Japan, Germany, Italy, and other transport-developed countries are strengthening the new infrastructure, and the construction focus of the new infrastructure is not the upgrading of the existing transport network but the infrastructure construction of intelligent transportation. This is because these developed countries have already experienced the driving effect of intelligent transportation development on national development and have seen the substantial results of intelligent transportation development.

8.2.2 *Strengthen the allocation of transportation resources*

The allocation of transportation resources is an important foundation to improve the intelligence of transportation and city. Through the construction of intelligent transportation system, developed countries have been continuously enhancing the allocation effect of transportation resources and revising the transportation development strategies. For example, in order to reduce domestic fuel consumption and vehicle pollution, the United States has vigorously promoted the smart travel mode of public transportation in recent years and increased the sales price of vehicles to reduce the sales volume of vehicles. The United States has ensured the sustainable development of the fuel field through such methods.

Except for the United States, most of the developed countries are using intelligent transportation systems to enhance the connection of various modes of transportation. For instance, the construction scheme of urban high-speed is planned synchronously in the process of railway construction in Germany, so as to construct the switching time between road and railway.

8.2.3 *Strengthen intelligent transportation management*

At present, most of the developed countries have formulated the development strategy of intelligent, unmanned, and digital transportation operation and maintenance management. The implementation plan of transportation self-management is obtained through digital twin modeling and simulation, and then the number of road management personnel is reduced through the intelligent control of signal lights, gradually realizing the unmanned transportation management.

As of 2021, Germany is still the country with the most outstanding achievements in unmanned transportation management. The German intelligent transportation system not only fully understands the operation of the transportation network but also combines the travel needs of the public, national transportation development, and other data to create an intelligent transportation management system that meets the travel needs of the public and improves the efficiency of transportation.

8.2.4 *Strengthen the formulation of transportation development policies*

With the intelligent development of transportation in various countries, the traditional transportation management policies need to be upgraded and improved. The speed of policy formulation, revision, and upgrading in developed countries is outstanding. All transportation development strategies will be released or changed before, during, and after the release. These policies and strategies are carried out simultaneously to improve the development efficiency of the developed countries and, at the same time, improve the speed of public acceptance of new travel modes and new transportation development modes.

8.2.5 *Enhancing the coordination between urban transportation management and national transportation management*

The development of urban transportation is relatively independent. With the purpose of eliminating the differences in urban transportation development, many ways such as government coordination have enhanced the coordination between urban transportation management and national transportation management, and thus promoted the transportation development of other cities.

For example, the U.S. Department of Transportation has introduced a number of policies to encourage coordinated development of interstate transportation. In this way, the U.S. Department of Transportation can connect the interstate transportation management network more quickly and realize the unified management of national transportation.

However, the author believes that this kind of development concept of Western countries is not suitable for China's national conditions because the construction and management of China's urban transportation generally takes the overall situation as the starting point and pays attention to the intercity transportation interconnection, so the connection between China's urban transportation management networks can be smoother, and the construction of China's comprehensive three-dimensional transportation network can be smoother.

8.3 Key Points of Intelligent Transportation Development at the Current Stage in China

In recent years, a number of documents based on the actual situation of transportation development have been issued by China's transportation management departments, in which the key tasks of China's intelligent transportation development have been mentioned. Among them, the

"Outline for the Construction of Strengthening Country with Strong Transportation Network" issued by the State Council in 2019 takes the lead in providing guidance for the development of intelligent transportation, and the "Outline for the Construction of Strengthening Country with Strong Transportation Network" also specifies the key points for the development of intelligent transportation in China.

Scientific and technological innovation is full of vitality and wisdom:

(1) Strengthen the research and development of cutting-edge key technologies. Aim at the world's scientific and technological frontiers such as the new generation of information technology, artificial intelligence, intelligent manufacturing, new materials and new energy, and strengthen research on forward-looking and disruptive technologies that may lead to changes in the transportation industry. Strengthen the research and development of power transmission system of automobiles, civil aircraft, ships and other equipment, and break through the key technologies of high efficiency, high thrust/ high power engine equipment. Strengthen the research and development of coordinated operation and service technology of regional comprehensive transportation network, coordinated management and control technology of urban comprehensive transportation, safety management and emergency search and rescue technology of inland river shipping based on ship-shore cooperation, etc. Reasonably make overall arrangements for the research and development of technical reserves such as 600 km/h high-speed magnetic levitation system, 400 km/h high-speed wheel-rail (including variable gauge) passenger train system, and low vacuum tube (tunnel) high-speed train.

(2) Vigorously develop intelligent transportation. Promote the deep integration of new technologies such as big data, Internet, artificial intelligence, blockchain and super computing with the transportation industry. Promote the development of data resource enabled transportation, accelerate the integrated development of transportation infrastructure network, transportation service network, energy network and information network, and build extensive and advanced transportation information infrastructure. Establish a comprehensive transportation big data center system and deepen the development of public transportation services and e-government. Promote the application of Beidou satellite navigation system.

(3) Improve the mechanism for scientific and technological innovation. Establish a technological innovation mechanism with enterprises as the main body and deep integration of production, learning, research and application, encourage all kinds of innovation subjects in the transportation industry to establish innovation alliances, and establish a key core technology tackling mechanism. Build a number of innovation platforms with international influence, such as laboratories, test bases and technology innovation centers, strengthen the opening and sharing of resources, and optimize the investment mechanism of scientific research funds. Establish a standard system that adapts to the high-quality development of transportation and strengthen the effective supply of standards in key areas".

After the release of the "Outline for the Construction of Strengthening Country with Strong Transportation Network", 32 academicians of the Chinese Academy of Engineering, 12 units including a number of well-known scientific research institutes, universities, and enterprises, and a total of more than 100 researchers formed thc "strategy of strengthening country with strong transportation network" project team, of which Lu Huapu, the head of the Transportation Research Institute of Tsinghua University, served as the leader of the intelligent transportation subproject of the project team. The research project team of the project also put forward six strategic development focuses on the development strategy of China's "intelligent transportation".

Strategy I: Big data sharing platform and application of transportation cloud technology

Three levels of big data sharing cloud platforms at national, provincial, and municipal levels have been established, and the data has been aggregated from the bottom to the top, realizing cross level, cross region, cross system, cross department, and cross business data sharing, collaborative management, and integrated services; establish big data sharing standards and security management mechanism; establish open sharing model and mechanism of big data between government and enterprises; construction mode: government led, enterprise construction, and operation and maintenance; analyze the transportation demand, optimize the infrastructure and operation management, excavate the potential value of transportation big data, and establish and improve the big data assisted scientific decision-making mechanism; realize intelligent organization and management, such as transportation safety management, congestion management, and sharing management.

Strategy II: Improve the management level of urban intelligent transportation

Taking intelligent transportation as a means, innovate data driven smart and fine management of urban transportation; establish an intelligent transportation management system based on the integration of transportation control, management, decision-making, and service, which is supported by big data; make use of big data, Internet +, artificial intelligence, and other technologies to innovate new modes of transportation management services; promote some areas of intelligent transportation such as precision law enforcement based on big data and convenient Internet services to lead the world's intelligent transportation.

Strategy III: Achieve efficient and convenient one-stop intelligent passenger service, and achieve door-to-door one-stop intelligent freight service

By making use of advanced technologies such as Internet, big data, and electronic payment, the one-stop efficient and convenient door-to-door service is realized through the intellectualization of various links such as driving, parking, hub transfer, terminal travel, and responsive customized services; implement personalized services, diversified services, full service, appointment service, and other intelligent service methods; build a freight cloud platform with information sharing, whole process visibility and intelligent control to realize the integrated information service and transportation service of the whole chain of freight logistics; promote the use of electronic freight waybills; establish multi-modal transport and joint distribution of goods; strengthen the research, development, and application of advanced freight technology; and promote the research and application of unmanned technology in freight vehicles.

Strategy IV: Intelligent improvement of active transportation safety

Construction of transportation safety intelligent analysis, research, and judgment system.

Intelligent improvement of transportation safety facilities.

Construction of intelligent and safe passage.

Intelligent transportation safety prevention and control system of the whole society.

Improve the level of vehicle safety and intelligence, and improve the technical condition of vehicles.

Intelligent means regulate transportation behavior and promote the formation of transportation safety culture.

Strategy V: Coordinated and integrated development of vehicles and roads

Improve transportation efficiency, improve transportation safety, and promote energy conservation and environmental protection.

Priority will be given to promoting unmanned driving operation in long-distance freight and buses.

Strategy VI: Achieve breakthroughs in key technologies of intelligent integrated transportation

Based on the transportation big data sharing platform, a national comprehensive transportation intelligent monitoring and intelligent decision-making platform covering the whole transportation mode is established, and the docking with the urban intelligent platform is realized.

Build a new generation of intelligent transportation systems based on Beidou navigation system. Around the continuous navigation, location service, emergency rescue, and other fields, the Beidou system is applied in the transportation field on a large scale, and the Beidou infrastructure integration, application demonstration integration, and operation service integration are implemented.

Promote the application of Beidou navigation system in intelligent transportation, and build an emergency rescue system based on Beidou navigation system for transportation monitoring, management, highway toll collection, urban public transportation, parking, and transportation accidents.

China's "Outline for the Development Plan of Digital Transportation" also clearly pointed out the following:

Development objectives

By 2025, a new step will be made in the digital upgrading of all elements and the full cycle of transportation infrastructure and transportation equipment, and the digital acquisition system and network transmission system will be basically formed. Transportation has become the main civil industry of Beidou navigation. The public network such as the fifth generation of mobile communication (5G) and the new generation of satellite communication system has been initially applied in the industry. The application level of transportation big data has been greatly improved, travel information services have been fully covered, the logistics service platform and integration have entered a new stage, and the industry governance and public service capabilities have been significantly improved. The transportation and automobile, electronics, software, communication, Internet services and other industries are deeply integrated, and the new business types and new technology application level remain advanced in the world.

The transportation infrastructure will be digitalized in all elements and cycles, and the transportation control network integrating heaven and earth will be basically formed by 2035. The on-demand instant travel service will be widely used. Moreover, China will become a major international standard setter or participant in the field of digital transportation, and the overall competitiveness of the digital transportation industry will lead the world.

Building a digital collection system

(1) **Promote all elements and the whole cycle digitalization of transportation infrastructure:** Promote the digitalization of all elements and the whole cycle of transportation infrastructure planning, design, construction, maintenance, and operation management. Build a high-precision transportation geographic information platform covering the whole country, improve the information of transportation engineering and other elements, realize the three-dimensional digital presentation of physical facilities, and support professional navigation applications such as automatic driving and large-sized transportation in all-weather complex transportation scenes. For major transportation infrastructure projects, realize the health performance monitoring of infrastructure in the whole life cycle, and popularize and apply the engineering quality control technology based on the Internet of Things.

(2) **Layout of all-round transportation perception network at important nodes:** Promote the coverage of transportation perception network in key sections and navigation sections of railway, highway,

and waterway fields, as well as important nodes such as tunnels, bridges, interchange hubs, and ship locks. Promote the simultaneous planning and construction of transportation perception network and transportation infrastructure, deepen the application of roadside intelligent terminals such as highway ETC gantry, establish a cloud-connected perception network, and enable the "mute facilities" to have multi-dimensional monitoring, intelligent network connection, accurate control, and collaborative service capabilities. Focus on social data integration applications such as crowdsourcing and mobile signaling. Build a basic cloud platform of transportation control network with interconnection of transportation tools, infrastructure, and transportation environment. Speed up the application of Beidou navigation in free flow toll collection, automatic driving, vehicle road collaboration, maritime search and rescue, port automatic operation, collection and distribution scheduling and other fields.

(3) **Promote the intelligent transportation tools and operation equipment:** Encourage the application of terminal equipment with multi-dimensional perception, high-precision positioning, and intelligent network connection functions and improve the capability of remote monitoring, fault diagnosis, risk early warning, and optimal control of transportation tools. Promote the research and development of collaborative technology for automatic driving and vehicle road, and carry out the construction of special test site. Encourage logistics parks, ports, railways, and airport freight stations to widely apply Internet of Things, automatic driving, and other technologies, and promote the large-scale application of automated three-dimensional warehouses, automated guided vehicles (AGV), intelligent transportation and sorting, and loading and unloading equipment. Promote the development of self-driving vessels, automated terminals and yards, and strengthen information sharing and business collaboration between port and shipping logistics and upstream and downstream enterprises.

Build a network transmission system

Promote the integrated construction of transportation infrastructure and information infrastructure, promote the deep integration of special transportation network and Skynet and public network, promote the deployment and application of Internet of vehicles, 5G and satellite communication information network, improve the national highway communication information network, form a multi network integrated transportation information communication network, and provide wide coverage, low delay, high reliability, and large bandwidth network communication services.

Build an intelligent application system

(1) **Create digital travel assistant:** Promote the full open sharing and integrated development of various information, such as transportation and tourism. Platform enterprises are encouraged to deepen the integration of multi-source data and integrate online and offline resources, encourage full open access of various transportation ticket systems, create digital travel assistants, and provide passengers with "door-to-door" whole journey customized services. Advocate the concept of "travel as a service" and link travel demand and service resources with data to make travel an on-demand and real-time service and make travel easier. Create a "smart mobile space" where passengers travel and business, shopping and consumption, and leisure and entertainment are mutually penetrated, bringing a new travel experience. Promote the development of "Internet +" convenient transportation, and encourage and standardize the development of new urban travel services, such as customized public transportation, intelligent parking, intelligent public transportation, automobile maintenance, online booking of taxis, Internet rental of bicycles, and time sharing of minibuses.

(2) **Promote the digitalization of the whole process of logistics:** Vigorously develop "Internet +" new efficient logistics models and new business types, accelerate the digitalization of the whole process of logistics activities, promote the electronization and sharing and mutual recognition of freight documents such as railways, highways, and waterways, and provide "one-stop" logistics services that can be monitored and traced throughout the whole process. Various enterprises are encouraged to accelerate the differentiated development of logistics information platform, promote the information sharing of the whole chain of urban logistics distribution, and improve the information network of rural logistics terminals. Strengthen mutual recognition of logistics-related management information of various departments by relying on various information platforms, and build

an open and sharing mechanism for comprehensive transportation and logistics data resources.

(3) **Promote the modernization of industry governance:** Complete the national comprehensive transportation information platform, improve the use of big data in areas such as decision-making support, safety emergency, command and scheduling, regulatory law enforcement, government services, energy conservation, and environmental protection, and achieve accurate analysis, accurate control, fine management, and meticulous services. Complete the resource catalog and information resource management system, realize the convergence and integration of industry information resources, and improve the ability of information resource sharing and exchange and open services. Establish a decision-making and planning system supported by big data, promote the integration of multi-source data among departments, governments, and enterprises, and improve the level of decision-making and analysis of transportation. By adopting the data and panoramic display mode, the supporting capacity of comprehensive transportation operation monitoring and early warning, public opinion monitoring, safety risk analysis and judgment, dispatching command, energy conservation and environmental protection online monitoring, etc. has been improved. Further promote the "Internet + government services" in the field of transportation to achieve the same issue, the same standard, and the same code of government services. Extend the online service chain and promote the extension and expansion of government services to "microblog, WeChat, and news client". Speed up the improvement of government information systems such as transportation administration, road administration, and maritime affairs, promote the construction of systems such as comprehensive law enforcement of transportation and overload control and networking, improve the intelligence level of law enforcement equipment, and promote online identification and off-site law enforcement.

In 2020, the white paper "Sustainable Development of China's Transportation" issued by the Information Office of the State Council once again mentioned the development focus of China's intelligent transportation and demonstrated the achievements of China's intelligent transportation in recent years.

Parallel development of transportation technology from "following first" to "following, parallel running, and leading"

Through unremitting efforts, the innovation capacity of transportation science and technology has risen significantly, the core technology has gradually become independent and controllable, the infrastructure and transportation equipment have achieved landmark major scientific and technological innovation results, and the sustainable development capacity has been significantly improved. China's transportation technology has entered a new stage of parallel development of "following, parallel running and leading" from the world's first-class level of "following".

The transportation super project has attracted worldwide attention: The technology of high-speed railway, cold railway, plateau railway, and heavy load railway has reached the world-leading level, and the construction technology of special geological roads such as plateau frozen soil, expansive soil, and desert has overcome world-class problems. The key technologies of offshore deep water port construction, the regulation technology of huge river mouth channel, the system regulation technology of long river channel, and the construction technology of large airports are leading in the world. The world's single Beijing–Guangzhou high-speed railway with the longest operating mileage has been fully connected. The Lanzhou–Urumqi High-speed Railway with the longest operating mileage has been completed at one time. The world's first high-speed railway in cold regions, Harbin–Dalian High-speed Railway, has been put into operation. The annual transportation volume of Harbin–Dalian High-speed Railway is the first in the world. The Xueshan No. 1 Tunnel in Guoluo Tibetan Autonomous Prefecture, Qinghai Province, the highest in the world, has been put into operation. The construction of Ya'an–Linzhi section of Sichuan–Tibet Railway commenced. A series of major projects, such as the Hong Kong–Zhuhai–Macao Bridge, Qinling Mountains tunnel group of Xi'an–Chengdu High-speed Railway, Yangshan Port Container Terminal, Qingdao Port Fully Automated Container Terminal, and Yangtze River Estuary Deep Water Channel Management, have attracted the attention of the world. The total scale of highway bridges and tunnels under construction and in service in China is the first in the world. The cable-stayed bridge, suspension bridge, and sea crossing bridge ranked the top 10 in the world in terms of main span. There are 7 cable-stayed bridges, 6 suspension bridges, and 6 sea crossing bridges in China, and 8 of the 10 highest bridges in the world are in China.

Major breakthroughs have been made in transportation equipment technology: Aiming at the world's cutting-edge scientific and technological development "the treasure of the nation", the level of independent research and development of transportation key equipment technology has been greatly improved. The "Fuxing" China standard multiple units with full independent intellectual property rights achieved the world's first speed of 420 km/h intersection and reconnected operation, the world's highest speed of 350 km/h continuous commercial operation in Beijing–Shanghai high-speed railway, Beijing–Tianjin intercity railway, and Beijing–Zhangjiakou high-speed railway, and the intelligent multiple units achieved the automatic driving function of 350 km/h for the first time; the high-speed maglev test sample vehicle with a speed of 600 km/h and the high-speed multiple units with variable rail gauge with a speed of 400 km/h with cross-border interconnection capability are offline. Great breakthroughs have been made in the research and development of special construction machinery such as shield machine. The largest diameter earth pressure balance shield machine, the largest diameter hard rock shield machine, and the largest diameter slurry balance shield machine have been successively developed. The industry of energy saving and new energy vehicles has been booming, keeping pace with the international advanced level. The manufacturing technology of marine machinery special vessels and large-scale automatic specialized complete sets of container equipment is leading the world, and the 300 m saturated diving has made innovative breakthroughs. C919 large passenger aircraft successfully made its first flight. Commercial operation of the regional jet ARJ21 began. The rapid development of express sorting technology. Ocean-going vessels, high-speed multiple units, high-power railway locomotive, marine engineering machinery, and other equipment and technologies are leading the world. Large aircraft, new generation intelligent networking vehicles, and other equipment and technologies are in the ascendant, becoming the "gold card" for China's manufacturing industry to go global.

The pace of intelligent transportation development is accelerated: promote the development of "Internet +" transportation, promote the comprehensive integration of modern information technology and transportation management and services, and improve the level of transportation services. Make full use of emerging technologies such as 5G communication, big data, and artificial intelligence and make continuous breakthroughs in intelligence in transportation infrastructure and equipment. With the increasing popularity of railway, highway, water transport, civil aviation passenger electronic tickets, and online ticket sales, the level of informatization of transportation production scheduling and command has been significantly improved. By the end of 2019, 229 airports and major airlines have realized "paperless" travel. The national highway provincial toll stations have been completely canceled, and the application of new technologies such as highway electronic toll collection system (ETC) has achieved remarkable results. As of the end of 2019, the number of ETC customers nationwide has exceeded 200 million, and the monitoring and information dissemination capabilities of the whole road network, all time and all weather, have been continuously enhanced. The Beidou system is widely used in all fields of transportation. There are 7.6 million road operating vehicles, 33,300 express mail trunk vehicles, 1,369 department system business ships, 10,863 navigation aids on water, 109 coastal foundation strengthening stations, 352 general aircraft applied to the Beidou system, and 3 transport aircraft applied to the Beidou system, Beijing–Zhangjiakou high-speed railway has become the first intelligent high speed railway in the world that adopts Beidou satellite navigation system and realizes automatic driving and other functions. The application of smart roads has gradually deepened, and technologies such as smart ports and intelligent shipping have been widely used. Intelligent delivery facilities are all over major cities in China, and automatic sorting covers backbone distribution centers of major express delivery enterprises. "Introduce the management standard of automatic driving road test and the construction guide of closed test site, promulgate the standard of intelligent ship, establish the sea test site of unmanned ship, and promote the demonstration application of unmanned aerial vehicle in express delivery and other fields".

From the research results of China's various transportation development policies and the project group of "strategic research on strengthening country with strong transportation network", it can be seen that the focus of China's current intelligent transportation development is mainly reflected in the upgrading, application, and innovation of transportation technology. As the main force of global transportation development, digital twin can effectively promote the development of China's transportation and support China to achieve the strategic goal of a strong country with a strong transportation network as soon as possible.

Chapter Summary

In terms of the current situation of digital and intelligent upgrading in the field of transportation in China, the overall level of intelligence is still lagging behind that of developed countries. However, after comparing the development mode, development speed, and development achievements, the path of digital and intelligent upgrading in China is clearer, and the application mode and application field of digital twin are more obvious. It is believed that under the guidance of a number of policies such as the "Outline for the Construction of Strengthening Country with Strong Transportation Network", the "Outline for the Development Plan of Digital Transportation", and the "Sustainable Development of Transportation in China" white papers, the gap between China's intelligent transportation and the international first-class level will continue to narrow and eventually achieve breakthroughs and surpluses.

Chapter 9

Impact of Digital Twin on Global Transportation Development

The transportation industry in various countries has achieved qualitative change and development with the application of digital twin in the global scope, and the global intelligent transportation upgrade has become a major trend. Nowadays, through the deep combination of digital twin and transportation, China has gradually built a new extensive and advanced transportation infrastructure, realizing the real-time perception of transportation. The intelligent upgrade of this development model has had a positive and far-reaching impact on China's transportation industry, reducing the substantial gap between China and the transportation-developed countries.

9.1 How to Realize the Comprehensive Digitalization of China's Transportation Infrastructure

In 2021, after the full launch of China's "14th Five-Year Plan", the State Council issued the "Outline for the National Comprehensive Three-dimensional Transportation Network", which is another important document after the "Outline for the Construction of Strengthening Country with Strong Transportation Network". The document emphasizes the importance of digital and intelligent development of transportation and sets specific development goals, and moreover, it points out the important reference object for China to build a strong transportation country in the scheduled time.

The "Outline for the National Comprehensive Three-dimensional Transportation Network" clearly states the following:

> **Improve the level of smart development:** Speed up the improvement of the technological innovation capacity of transportation and promote the digitalization and networking of transportation infrastructure. Promote the application of satellite communication technology, new generation communication technology, high-resolution remote sensing satellite, artificial intelligence, and other industries, build a comprehensive, replaceable, and safe industrial Beidou high-precision basic service network, and promote the large-scale application of industrial Beidou terminals. Build a high-precision transportation geographic information platform and accelerate the independent innovation and application of building information model technology in various fields. Lay out a comprehensive transportation perception system, plan and construct simultaneously with the transportation infrastructure, deploy active early warning facilities for key parts, and improve multi-dimensional monitoring, accurate control, and collaborative service capabilities. Strengthen the research and development of intelligent transportation tools and key special equipment, and promote the application of intelligent Internet-connected vehicles (intelligent vehicles, automatic driving, vehicle road collaboration) and intelligent general aircraft. Encourage logistics parks, ports, airports, and freight terminals to widely apply Internet of Things, automation, and other technologies, and promote the application of automated three-dimensional warehouses, guided transport vehicles, intelligent transportation and sorting, and loading and unloading equipment. Establish a comprehensive transportation big data center system and improve the comprehensive transportation information platform. Improve the open and sharing mechanism of scientific and technological resources and build a number of innovation platforms with international influence.
>
> **Accelerate the intellectualization of existing facilities:** New technologies will be used to enable the development of transportation infrastructure, strengthen the upgrading of existing transportation infrastructure, and improve the utilization efficiency and service level of the facilities. Modern control technology is used to improve the intelligence level of train dispatching command and transportation management of the whole railway network. Promote the intellectualization of road network management and travel information services, and improve

road transportation monitoring equipment and supporting networks. Strengthen online monitoring of the operation status of high-grade inland waterways, and promote the development of ship shore coordination, automatic terminals, and yards. Develop a new generation of air transportation control systems, promote the intellectualization of air transportation services, flow management, and airspace management, and promote information sharing among all parties. Drive the coordinated development of smart vehicles and smart cities, construct a perception system integrating urban roads, buildings, and public facilities, and build a smart travel platform based on urban information model platform and integrating urban dynamic and static data.

If we say that the "Outline for the Construction of Strengthening Country with Strong Transportation Network" specifies the important tasks for the development of intelligent transportation in the next 10 years, then the "Outline for the National Comprehensive Three-dimensional Transportation Network" points out the specific content of the development of intelligent transportation in the next ten years. From the content of "promoting the level of intelligent development" in the "Outline for the National Comprehensive Three-dimensional Transportation Network", it can be seen that the focus of China's transportation intelligent development is the digital construction level of transportation technology facilities, and comprehensive digital infrastructure is an important foundation to support the digital upgrading of China's transportation. In the "Outline for the National Comprehensive Three-dimensional Transportation Network", it also clearly states that China will complete the extensive advanced transportation information infrastructure, realize the comprehensive perception of the transportation network, and build an intelligent, efficient, and real-time transportation operation and maintenance management system by 2035. This is not only the development goal of the transportation industry but also the main way to accelerate the development of transportation in China.

As a matter of fact, it has been proposed in the "Outline for the Development Plan of Digital Transportation" released in 2020 that China's transportation development should "build a digital collection system" in the future, which includes three key points: ① Promote the digitalization of all elements and cycles of transportation infrastructure. ② Lay out all-round transportation perception network of important nodes. ③ Promote the intellectualization of transportation tools and operation equipment.

Under this direction, China's transportation sector will create a high-precision transportation service network, build multiple information interaction platforms, and use digital twin technology to establish a virtual model of transportation operation and maintenance management by adopting the way of integrating the new generation of communication technology, artificial intelligence technology, and digital technology. These measures not only enhance the digital nature of China's transport infrastructure but also bring substantial development results.

Based on the direction guidance and construction contents of several documents such as the "Outline for the Construction of Strengthening Country with Strong Transportation Network", the "Outline for the Development Plan of Digital Transportation" and the "Outline for the National Comprehensive Three-dimensional Transportation Network", it can be concluded that the comprehensive digitalization of China's transportation infrastructure in the future is mainly through the following ways.

9.1.1 *Digitalization of whole lifecycle of transportation infrastructure*

To achieve it, China is required to strengthen the digital infrastructure in the field of transportation and complete the digital upgrading of the existing transportation infrastructure, so as to build a comprehensive digital network of transportation and strengthen the digital management of the whole lifecycle of transportation infrastructure.

In addition, in the process of construction, transformation, and upgrading of transportation infrastructure, the construction goal is to support the real-time transmission of dynamic data such as BIM and GIS, so as to ensure the digital construction results of transportation infrastructure.

9.1.2 *Enhance the perception of transportation network*

The monitoring and perception ability of China's terminals, airports, key hubs, rail, and other transportation resources will be upgraded in a digital manner. Based on the perception ability and real-time data transmission ability, the monitoring, operation, and maintenance ability of China's transportation network will be enhanced.

9.1.3 *Enhance the intelligence and advancement of transportation technology*

With the deep integration of digital twin technology, new generation information technology, and other cutting-edge technologies, it will promote the development of China's unmanned driving technology and unmanned transportation management mode, enhance the integration of China's transportation industry and other industries, and strengthen the overall drive of the transportation industry to China's economy.

As of 2021, the comprehensive digital development of China's transport infrastructure has achieved certain results, but these development results are mainly reflected in the second point. The transportation operation and maintenance management and intelligent development have been effectively promoted with the enhancement of the perception ability of China's transportation network, but the digital upgrading of transportation infrastructure and the intelligent upgrading of transportation technology need to be improved. It is believed that China's transportation departments will increase investment in these aspects to ensure that the digital level of China's transportation can keep pace with that of developed countries in the future.

9.2 New Momentum for Transportation Development in the Digital Era

Nowadays, digital twin technology has penetrated into all walks of life in China and also triggered a data wave in the commercial market, driving the development of sensor and big data industries. Generally speaking, a new development momentum has been given by the digital twin technology to this era, allowing the traditional transportation industry in China to enter an integrated, digital, and intelligent development era, and making diversified, heterogeneous, and massive data become an important support for transportation development.

Guided by the "Outline for the Construction of Strengthening Country with Strong Transportation Network", China's transportation has made major breakthroughs in the fields of highway construction mileage, transportation equipment level, and transportation technology upgrading and has formed a new development trend of the transportation industry.

At the intelligent transportation technology development forum held in Shenzhen in 2020, Luo Ruifa, President of Shenzhen Intelligent Transportation Industry Promotion Association, delivered a speech named "new momentum of digital transportation in the context of new infrastructure", in which Luo Ruifa made a clear analysis of the time development promoted by digital power in four aspects, and clarified the overall changes brought by digital power to China's market.

Luo Ruifa said the following:

New infrastructure

The new infrastructure is a new pattern and development direction under the great changes. At present, a new pattern of development with domestic major cycle as the main body and international and domestic dual cycle promoting each other has been formed. In the stage of economic globalization, China actually needs to do three things: ① Build a new base point relying on new infrastructure. ② By virtue of localization of connection breakpoints. ③ Make use of new consumption to break through the barriers.

We will seize the strategic opportunity of major changes and lead the fourth industrial revolution with new infrastructure. Looking back to the whole industrial revolution, the first industrial revolution centered on steam engine, the second on electrification, and the third on electronic information. A new round of scientific and technological revolution is actually transitioning to artificial intelligence and big data.

The new infrastructure is at the right time to build a new pattern. According to the definition of the National Development and Reform Commission, new infrastructure is divided into three categories: ① information infrastructure, ② Integration of infrastructure, and ③ innovative infrastructure.

The essence of new infrastructure is digitalization, uncertain changes, and determined transformation and upgrading trend of digitalization. Entering the stage of transformation and upgrading of digitalization, the whole society will be reconstructed around digitalization to build a new era featuring more open, sharing, scientific, precise governance, intensive, and efficient development.

The new infrastructure will restructure the production and living mode of an intelligent society through digitalization. From the traditional physical society to the new era of information and data centered on digitalization and data, cloud-tube end covers a new era of digitalization, virtualization, and ultimately digital twin.

New transportation

Transportation has entered a new stage of development, and digital transportation is the trend and key to future transportation development. Why is it that the more infrastructure is needed in the downward economic cycle, the more strategic and leading investment is needed? As mentioned in the "Outline for the Construction of Strengthening Country with Strong Transportation Network", new transportation is related to the production, circulation, consumption, and people's life of the whole society, as well as the improvement of the comprehensive national strength and industrial transformation and upgrading of the whole country.

New transportation and new infrastructure emerged at a historic moment, enabling new transportation with new infrastructure. The new infrastructure provides a low-level digital base for digital transportation, and the new infrastructure provides a new momentum foundation for new transportation, becoming an important bearing and support for the smart society.

"Driven by factors", digital transportation will usher in a new era of digital transformation of transportation. Traditional transportation is a physical world. After digitalization, a virtual space is created. The physical space is mapped by data, and the virtual space enables the physical space. In this way, the future of digital transportation consists of a data-based main line and three systems. The three systems are digital acquisition system, network transmission system, and intelligent application system.

New vision

At the macro level, the new transportation helped realize the upgrading of both industry and consumption. New infrastructure is at the national level, including the main core industries clearly defined by the National Development and Reform Commission, including 5G, artificial intelligence, big data center, and satellite. In fact, the base of the new infrastructure will provide upgrading support and foundation for the supply chain, production and manufacturing, circulation, and consumption.

As for the industrial level of intelligent transportation, the ultimate goal will be to form a digital transportation control network integrating heaven and earth. Constitute the digitalization of all elements and the whole cycle, and realize the implementation of the new vision of MaaS service platform.

At the industrial level, from ETC to V2X, we will build a new ecology of intelligent network connection. The current ETC has several main features and attributes: ① accurate identification, ② BYO safety payment, ③ a large group base of 220 million users, and ④ vehicle, pedestrian, and road connection channel.

Starting from ETC and from the two dimensions of capability line and scenario line, it will make a comprehensive layout of ETC+, ETC-X, and 5G- V2X, build a smart vehicle + smart road, realize “make transportation smarter”, expand multiple application scenarios in vehicle, road, field, and metro areas, realize “make life easier”, and help the ecological construction of intelligent Internet infrastructure, intelligent Internet connected vehicles, and MaaS platform services.

The traditional road has become an intelligent network infrastructure. Vehicles and roads together form a road transportation system to build a new intelligent network road transportation system to serve the MaaS under the 5G environment.

The horizontal axis is the scene expansion, known as “ETC+”. ETC+ road area, field area, vehicle area, and metro area, from the back to the front, from the front to the front of intelligent network connection. Road area, from traditional ETC to free flow ETC, free flow ETC to smart road. ETC can go down the highway for gas filling, parking, charging, and station management. In metro area, ETC can be used for transportation management and urban management.

ETC+ makes life easier. ETC is integrated with Beidou/GPS, laser/MMW radar, V-box, and LTE/5G to create a more intelligent car. Road end technology integration refers to the integration with new technologies, such as edge computing, cloud computing, video, artificial intelligence, big data, blockchain and digital twin, and the way to wisdom, which is the future smart infrastructure.

New energy

New infrastructure is the means and new kinetic energy is the goal. How to leap from the means of new infrastructure to the goal of new kinetic energy requires new exploration and new ideas.

Give full play to the technological advantages of the new infrastructure, strengthen the new technology to empower transportation, promote the application of new technology in the transportation field, and provide the underlying support for transportation digitalization. New technologies such as 5G, big data, artificial intelligence, blockchain, and digital twin have laid a solid foundation for the new infrastructure, which is finally driven by application.

Digital transportation deeply integrates the new generation of information technology, new infrastructure, and transportation, aiming at the personalized transportation needs of people, vehicles, and objects, and builds a "whole object, whole process, and full coverage" new sustainable future transportation scene. The travel of people and the flow of goods, from warehouse to warehouse, will generate integrated and uninterrupted travel, including timely distribution, UAV, unmanned vehicle, and other new logistics methods, including shared travel, intelligent parking, sharing, and green parking.

A number of new services have emerged, including MaaS travel, trunk line connection, urban and rural connection, warehouse and distribution integrated logistics services, integrated government services, and new transportation management. In the future, travel will be more flexible, more nimble, more user-friendly, and more scientific and technological in the organization mode of flexible transportation based on demand.

At the same time, new capital is also essential. The old roads cannot lead to new places. In terms of the allocation of capital environment, the construction of the capital system and capital mode, we need to inject new elements in an all-round way to open up new models.

In terms of the new model, the entire business model, production and operation, market form, value form, organization method, innovation system, etc. have been changed. In particular, the generation of the middle platform is actually based on the middle platform in the platform, which then empowers and drives the front desk, including data middle platform, business middle platform, and algorithm middle platform. The middle platform has brought about many new changes in the mode of production, process, and consumption ports.

The integrated development of digital transportation and all walks of life has enabled it to break new boundaries from production and consumption, offline to online, and spawn the continuous penetration, integration, and emergence of new retail, new finance, new technology, and

> other new types of business. The traditional industrial chain is chain type, and the new format driven by digital transportation will become network and platform type, and become a closely connected symbiotic network supported and empowered by data.
>
> In a word, intelligent transportation is expected to reach 3.0 from 2.0. Using the law of Amara: People tend to overestimate the near term but underestimate the long term. Driven by this round of new technology revolution and the empowerment of new infrastructure, it is believed that it will take 3 years for new infrastructure to generate new kinetic energy. However, if we look at the 10-year cycle and walk for 10 years, we will look back and see that in the future, the industrial upgrading and consumption upgrading of the whole country will surmount the road of a thousand miles and arrive at a new "poem and distance".

It is not difficult to see from this speech that the promotion of digital power to all walks of life in China is a bottom-up improvement. Digital power improves the development of China's new infrastructure so that China's production and operation, market form, and scientific and technological innovation have solid foundation strength. At the same time, it also enables China's transportation to achieve dual upgrading of industry and consumption.

Compared with the development level of developed countries, upgrading both industry and consumption is the main goal of transportation development. On the one hand, the intelligent upgrade of the transportation industry improves the public travel and transportation experience and, on the other hand, promotes economic and social development, both of which are indispensable. Every improvement of transportation made by digital power centers on these two directions.

In recent years, for instance, China has continued to strengthen the construction of digital infrastructure for transportation, making the transportation infrastructure have scientific and technological advantages, and strengthening the empowerment of digital technology and information technology on transportation. The development mode achieves the improvement of public travel efficiency and travel safety guarantee, the intelligent upgrade of the city, and the linkage and efficient development of the economy.

Moreover, the overall needs of people, vehicles, roads, and the environment are connected by digital forces, and visual, controllable, and predictable transportation scenarios are built. Numerous transportation innovations have emerged, and the urban density and population density of China have been greatly improved. All these achievements are the result of the combination of digital technology and transportation.

In addition, it can be seen from Luo Ruifa's speech on "new momentum of digital transportation in the context of new infrastructure" that digital transportation will remain dominant for a long time in the future, transportation development will be continuously quantified, and transportation data will be more important. In the future, the flexibility, humanization, technology, and other characteristics of China's transportation will be born through data processing, which will be the biggest development feature of digital transportation.

9.2.1 *Transportation big data collection technology*

In general, the perception ability of China's transportation network has been greatly developed driven by the digital twin. The scope and depth of transportation data collection of China's railway, highway, aviation, and waterway have been comprehensively strengthened, as all kinds of sensor equipment and intelligent terminals have begun to be laid substantially, which provides basic support for digital twin simulation modeling.

In the process of application of digital twin technology, the transportation big data collection technology has also been upgraded many times. Including the upgrading of mobile terminal collection method, the upgrading of real-time transportation data transmission method, and the updating of transportation infrastructure data, the upgrading of these data collection technologies has laid a foundation for the overall digital development of China's transportation and also supported the establishment of a more comprehensive and detailed transportation digital model in China.

Compared with the traditional transportation development model, the development of transportation big data collection technology is more like a transportation construction, but this construction is the infrastructure construction of virtual space. Through this kind of data, more new digital technologies can be built to replace traditional transportation technology. For example, big data modeling brings full element simulation and prediction, improves the accuracy of prediction, and improves transportation safety protection measures, but the upgrading of traditional transportation safety protection measures mainly relies on experience summary and accident analysis. Therefore, the transportation big data collection technology can be regarded as the basic development momentum given to the field of transportation in the digital era.

9.2.2 *Transportation big data mining technology*

Many people believe that the largest role of digital twin technology is data analysis and application. In fact, the largest role of digital twin technology is to excavate data value and promote the development of the industry. For example, as a result of joint analysis, optimization, and simulation based on various transportation data, most of the intelligent upgrading strategies in China's transportation industry are promoted by digital twin. It not only provides more possibilities for the transportation data to play a greater value but also fully demonstrates the linkage between transportation data.

At present, the digital twin technology has reflected the data mining ability in the field of transportation services. For example, the upgrade of navigation technology is based on the analysis and optimization of transportation data. The intelligence and convenience of navigation technology can be continuously improved under the support of comprehensive transportation data. It is believed that in the near future, China's navigation service level can also reach the level of developed countries, and the average speed of road vehicles, nearby parking information, optimal travel time, and other information can be fully displayed.

9.2.3 *Transportation big data integration technology*

The integration of China's transportation industry and other industries has begun to improve after the advent of the digital era, which is also the result of the digital twin in-depth mining of the value of transportation data. At present, with the purpose of obtaining more development results, the transportation industry in China has started to integrate with other industries. However, digital technology needs to solve an important problem in this process, which is the effective integration of transportation data and other industry data.

As early as 2015, the United States, Japan, Germany, and other developed countries began to develop this field of transportation. For example, the integrated development of the transportation and communication services industry in the United States not only improves the level of transportation services but also expands the market scope of the communication industry; the integration of Japan's transportation and automobile production industry has not only improved transportation safety protection but also enhanced the market competitiveness of the automobile industry. These high-quality integration cases bring more enterprises into China's transportation industry and also promote the development of China's transportation big data integration technology.

The sharing and integration of transportation data in other industries needs to ensure the safety of China's transportation development. Under the premise of ensuring the safety of digital transportation system, the transportation industry can develop cross-industry multi-source and different data processing technology, build a multi-source and different data sharing platform, and enhance the transmission and interaction between data.

In addition, the sharing and integration of transportation data with other industries also needs to bring a two-way promotion effect. For example, the development of China's navigation technology is based on the analysis and optimization of transportation service data, and the navigation technology improves the efficiency of China's transportation operation. This two-way promotion is the effect of data integration.

In 2020, the "Outline for the Development Plan of Digital Transportation" issued by the Ministry of Transport of China also clearly pointed out that China will build a digital collection system and focus on promoting all elements and full cycle digitalization of transportation infrastructure, layout of all-round transportation perception network of important nodes, and promotion of intelligent transportation tools and operation equipment; build an intelligent application system, focusing on building digital travel assistants, promoting the digitalization of the whole process of logistics and promoting the modernization of industry governance.

The realization of these development goals also requires the strong integration of China's transportation data, so as to promote the economic integration of the transportation industry and other industries, which is not only the development trend of the transportation industry but also an important way for China's economic development.

In future development, digital power will give more momentum to China's transportation development, and the value of digital technology and transportation data will also be unlimited. The collision between transportation data and the collision between transportation data and other industry data will bring more help to China's goal of becoming a strong country.

9.3 Transportation Digitalization and Globalization

Throughout the development of the global transportation industry, most of them are marked by transportation capacity. The growth of transportation capacity represents the progress of the transportation industry. After the advent of the digital era, this development model has not changed. Digital technology promotes the intelligent development of China's transportation and also improves China's transportation capacity. With the development of transportation capacity, it brings China's technological innovation and social productivity improvement, and keeps China's economic, political, and cultural development at an efficient pace.

The overall improvement of transport capacity of various countries makes the economic trade, personnel, and goods between countries in the world more closely in the process of promoting global economic development by digital power, and their mutual dependence continues to deepen, which means that China's transportation must keep up with the pace of world economic development, so as to maintain an effective link of international trade. Looking at the development mode of transportation globalization in the developed countries of the world, combined with the current situation of China's transportation intelligent development, it can be concluded that there are mainly two digital paths for China's transportation globalization.

First, we will eliminate the unbalanced development of the transportation industry between China and other countries through digital forces, and enhance the interaction between China's transportation and other countries. At present, the gap between countries in the world in transportation capacity is relatively obvious, and the economic operation capacity of developed countries in transportation is more prominent. As of 2021, the transportation efficiency still needs to be improved even though China's transportation capacity has gradually upgraded to the forefront of the world, resulting in the process of China's economic globalization being limited. From the process of transportation globalization in developed countries, it can be seen that the premise of transportation globalization is the smooth and three-dimensional domestic transportation, improving the switching efficiency of different modes of transportation, and extending the coverage of transportation network to remote areas, which can effectively drive the national economic development. In other words, only by breaking the imbalance of domestic transportation can the balance of international transportation be achieved. Therefore, we should focus on building an efficient and smooth transportation network in the process of promoting the development of China's transportation industry by digital forces, so as to improve the development speed of China's transportation globalization.

Second, develop the transportation industry with national interests as the leading factor. The overall layout of China's transportation should be guided by national interests and aimed at promoting China's economic development. This requires China's transportation industry to follow the following development models:

(1) Digital technology and new generation information technology are widely used in the transportation industry to strengthen the informatization and intelligent transformation of the original transportation system, improve the modernization level of transportation equipment, and make scientific decisions on transportation management methods.
(2) The transportation efficiency of various regions in China should be improved to ensure that China's comprehensive three-dimensional transportation network can undertake high-speed and professional transportation tasks. At present, the vigorous development of high-speed rail and large-scale transportation equipment in China is an important performance in improving transportation efficiency in various regions.
(3) Strengthen the division of labor and coordination among various transportation departments, integrate and restructure various transportation resources by using digital forces, and guide the development of China's transportation system toward diversification and integration. The division of labor and cooperation, integration, and reorganization of the transportation department are different from the overall upgrading of the transportation system. The division of labor and cooperation of the transportation department focuses on the organic combination of the transportation mode, the reasonable layout of the transportation mode, and the smooth connection of the transportation mode. The cooperation of the transportation department through these modes can show the economy and enhance its value.

China's transportation efficiency and the overall economic benefits of the society will be improved through the coordination and synergy of transportation departments, which is also the original intention of many important transportation engineering constructions in China.

9.3.1 *Development features of transportation digitalization*

The digital development of China's transportation is the main way to solve various problems of transportation globalization. China's transportation capacity is also improving with the development of transportation digitization. However, simply integrating the transportation industry and digital technology cannot be called digital development. Only when digital technology is fully and deeply applied to the transportation field can the transportation industry enter the digital development stage. As mentioned in the "Outline for the Development Plan of Digital Transportation", digital transportation should "build an intelligent application system", and the complete construction of these application systems is the main feature of the development of digital transportation.

9.3.1.1 *Establishment of digital management warehousing system*

In the process of digital technology promoting the development of transportation, the digital transformation of the transportation industry should start from the digital upgrading of the storage management system. Digital technology is used to optimize warehousing, in-stock and stock removal links, and realize automatic identification and classification of goods information, intelligent sorting and allocation, mechanized warehousing and handling, intelligent monitoring, and intelligent operation and maintenance management of warehousing management. The warehousing industry has entered the digital development stage with these digital characteristics.

9.3.1.2 *Establishment of digital transportation management system*

Digital transportation management not only represents the digital management of transportation roads but also includes intelligent route planning, intelligent vehicle matching, and real-time perception of transportation status in transit. The transportation capacity of China's transportation industry will be gradually improved under the combination of digital technology, Internet technology, and artificial intelligence technology, and the intelligence of transportation goods in matching, transportation, testing, and warehousing will be enhanced, which is the digital development of transportation management.

9.3.1.3 *Establishment of digital capacity management system*

With the promotion of digital technology to China's transportation industry, China's transportation capacity is also increasing. Digital management of transportation capacity is also an important feature of the digital development of transportation industry. For example, the establishment of a resource management database for transportation vehicles and transportation staff, the standardization of the transportation process through online operation, the improvement of transportation efficiency, and the enhancement of the accuracy of vehicle allocation and personnel allocation in the transportation process are the main features of digital management of transportation capacity.

9.3.1.4 *Establishment of digital transportation service system*

Transportation services mainly include the construction of the supply chain, the establishment of the transaction settlement system, and transportation market services. Building a digital transportation service system refers to standardizing the management and control of market service standards, service quality, and service process through the use of digital technology, while establishing a digital supply chain, improving the convenience and efficiency of transaction settlement, and guiding the overall digital transformation of transportation services.

The digital development of transportation has strengthened China's basic strength. Relevant people in China's transportation industry should consider the new ideas for the development of the transportation industry under the digital environment of transportation, and combine the characteristics of the broad and huge transportation market in China to find a breakthrough development mode, so as to drive the overall growth of China's economy.

9.3.2 *Future trend of world transportation integration development*

Currently, China's transportation improves its own attributes of integration in the process of digital and intelligent development; the local transportation began to connect effectively under the guidance of digital model, and integration development has become a major trend in China's transportation industry.

For example, in 2016, the Ministry of Transport of China and the National Development and Reform Commission jointly prepared and issued the "Integrated Transportation Planning for Coordinated Development of Beijing–Tianjin–Hebei". In 2020, the National Development and Reform Commission issued the "Integrated Development Plan for Higher Quality Transportation in the Yangtze River Delta region". In 2021, Henan Province subsequently issued the "Planning for Integrated Development of Transportation in Zhengzhou Metropolitan Area (2020–2035)". All kinds of signs indicate that China's transportation has gradually completed the regional integration, and the overall development has shown the characteristics of comprehensive and integration.

In fact, the integrated development of transportation is not only the main mode of China's transportation development but also the integrated development of global transportation under the promotion of digital forces. For example, the European multi-national transportation network has developed from the road interconnection to the all-round three-dimensional interconnection of roads, railways, aviation, and inland rivers with the application of digital technology. Through this kind of transportation connection, the tightness of European countries has been continuously improved, and the European digital communication network has become more and more perfect. At the same time, various supporting policies in response to the trend of transportation integration development have been issued by European countries to ensure the effective link of economic nodes, so as to continuously improve the sustainable development strength of European countries.

The integrated development of world transportation mainly aims to promote each other's economic development and enhance the effective allocation of global resources. After the transportation of various countries has entered the digital development period, this direction has not been affected at all. China's transportation industry will also value the development of this direction and enhance the integration of the transportation industry from three aspects in future development.

9.3.2.1 *Improve transportation efficiency*

The important premise for the integration of China's transportation industry and global transportation and the good development is to have super transportation efficiency, so as to ensure that China's transportation maintains a good leading position in the process of integrating into the world's transportation and shows high economic value.

9.3.2.2 *Undertake responsibility for the overall development of international transportation*

On the basis of the standards of international transportation, it makes appropriate adjustments to the development of transportation, enhances the effectiveness of integration, and takes the initiative to undertake the responsibility for the overall development of international transportation. For example, with the purpose of achieving a good cooperative relationship, it needs to improve the economic and trade efficiency of the cooperative countries, adjust the international air transport plan, or give more policy support to ensure the efficiency of international transportation connection.

9.3.2.3 *Promote the overall development of China's economy*

Explore more development space in the international market and promote the overall development of China's economy by making use of the overall strength of China's transportation industry. For example, through the advantages of China's transportation capacity, increase the competitiveness of international trade, strengthen the integration of China's economy and the world economy, and promote the international development of more industries with the advantages of transportation.

Chapter Summary

Digital twin technology not only leads the global transportation industry into a rapid development stage but also drives the social and economic effects of transportation in various fields and levels. In general, digital twin has ushered in a new era for modern transportation, in which digital represents high-end, economic, and powerful.

Chapter 10

Digital Twin Improves the Intelligent Development of Transportation Infrastructure

The achievements of vehicle road collaboration, smart high-speed, and full network perception are highlighted in all directions looking at the transportation development planning and construction based on digital twin in China. Digital twin has made breakthrough progress in the process of improving the intelligent development of transportation. However, all of them are to enhance the intelligent attribute of transportation infrastructure when looking at the implementation of digital twin in various fields of China's transportation. Digital twin realizes the real-time interaction between China's transportation in digital space and the real world through digital improvement of transportation infrastructure.

10.1 Application of Digital Twin in Transportation Infrastructure

After China successively issued the "Outline for the Construction of Strengthening Country with Strong Transportation Network", the "Outline for the National Comprehensive Three-dimensional Transportation Network", the "Guidance of the Ministry of Transport on Promoting the Construction of New Infrastructure in the Field of Transport", and the relevant requirements for the development of digital transportation during the "14th Five-Year Plan" period, China's transportation infrastructure will embark on the road of digital transformation and intelligent upgrading. In this process, digital twin technology gives full play to its own value and promotes the high-quality development of China's transportation infrastructure.

According to the "Action Plan for New Infrastructure Construction in the Field of Transportation (2021–2025)" issued by the State Council in September 2021, the main contents of new infrastructure construction in the field of transportation in China can be divided into seven points: ① construction of smart road, ② construction of smart channel, ③ construction of smart port, ④ construction of smart hub, ⑤ construction of transportation information infrastructure, ⑥ construction of transportation innovation infrastructure, and ⑦ completion of standards and specifications.

Among the seven types of infrastructure construction in the field of transportation in China, digital twin technology reflects its own value and improves the construction results.

10.1.1 *Smart road construction*

The operation and maintenance management levels of smart roads in China are directly upgraded by the digital twin simulation features, and moreover, the transportation departments are guided to lay perception equipment over a large range, enhance the real-time monitoring and early warning capabilities of highway gallop facilities, and thus realize the synchronous intelligent management of highway perception and highway infrastructure maintenance.

For example, in recent years, in China's cross sea bridges, super long tunnels, and other transportation projects, transportation infrastructure includes a large number of perception equipment, cross sea bridges and super long tunnels have good perception ability, and the monitoring and early warning of structural disasters, mechanical and electrical failures, transportation accidents and other conditions are more timely, and when this situation occurs, it can be intelligently controlled and disposed.

10.1.2 *Smart channel construction*

Digital twin technology has improved the guarantee ability of channel operation in the construction of a smart channel. The data of waterway, maritime, and water transportation are comprehensively collected and processed in the process of digital infrastructure construction, so as to improve the intelligent level of waterway detection capacity and maintenance.

For example, the digital twin technology is equipped with the capability to detect the navigation status of ships in real time and understand the actions of ships during navigation, including the receipt and management of pollutants from ships, ensuring the sustainable development of China's navigation channels.

In addition, digital twin technology also improves the comprehensive service capacity of China's waterways. Under the integration of digital twin and passive navigation, China has built a number of shore-based facilities that adapt to the docking and transfer of intelligent ships, enhanced the comprehensive service capacity of intelligent ships, promoted the overall intelligent scheduling and automatic control of the waterway, and upgraded the intelligence of China's waterway.

10.1.3 *Smart port construction*

Digital twin is mainly used to improve the automation level of the port in the process of smart port construction. China's smart ports speed up the construction and transformation efficiency of automated container terminals and strengthen the intelligence of large cargo shipping equipment with the application of digital twin technology. China's smart ports have also realized the whole lifecycle digital management of construction and maintenance and the intelligent construction of infrastructure supported by the virtual model.

Digital twin technology also promotes the development of smart logistics industry in China's smart ports. Through the integration of transportation data and logistics transportation industry data, digital twin technology realizes the coordination of port logistics transportation and transportation resources, improves the convenience of logistics transportation industry, and realizes that the whole process of logistics transportation can be detected and controlled.

10.1.4 *Smart hub construction*

Digital twin enhances the intelligent characteristics of hub infrastructure during the process of infrastructure construction of the transportation hub. First, based on the passenger demand, the digital twin technology improves the digital upgrade of the transportation hub, promotes the application of intelligent security inspection equipment, electronic sales system, epidemic prevention, and control system on a large scale, and strengthens the capacity of automatic passenger flow identification and environmental monitoring of the transportation hub.

In addition, through the construction of the digital model of the transportation hub, the digital twin technology promotes the efficient interconnection of various transportation modes, such as urban public transport, urban rail transit, railway, and aviation, and creates a comprehensive three-dimensional urban transportation network so that the transportation hub plays a greater role. The digital twin also enhances the freight capacity of transportation hubs. Through intelligent security inspection and loading and unloading equipment, the efficiency of logistics transportation is improved, and the intelligence of logistics park, logistics station, and logistics platform is upgraded to achieve intelligent matching of various modes of transportation, and the whole process monitoring service of logistics information is provided.

10.1.5 *Construction of transportation information infrastructure*

Digital twin promotes China's big data industry to develop vigorously and combines big data technology and artificial intelligence technology to create a "smart brain" for China's transportation management departments. Therefore, the intelligence of China's integrated transportation information platform is becoming more and more prominent due to the digital twin technology. The reason for this development is that the digital twin technology has enhanced the connection of transportation platforms at all levels by deeply mining the value of transportation data through data modeling and simulation deduction so that the connectivity of transport platforms at all levels has been strengthened.

Digital twin technology has optimized the laying scheme of data transmission equipment in key transportation sections, important transportation hubs, important ports, and other areas in the process of the combination of China's transportation platform and 5G technology, and promoted the pan perception ability of China's Internet of vehicles and ships.

Moreover, digital twin technology also extends the development field of China's passive navigation technology. On the basis of high-precision navigation and positioning of transportation, it provides intelligent services and improves the ability of intelligent search and rescue and intelligent handling of dangerous situations.

10.1.6 *Transportation innovation infrastructure construction*

Digital twin technology has played a significant role in China's important innovative transportation infrastructure construction projects. The application effect of various innovative technologies in transportation infrastructure is mostly demonstrated through digital twin modeling and simulation. This way not only improves the application effect of innovative technologies but also greatly saves the experimental cost of China's transportation innovation infrastructure. Digital twin technology is deeply integrated in the development of transportation technology in China at present, such as automatic driving, intelligent shipping, and intelligent transportation management, and the application effect is very prominent.

In addition, the integration of digital twin technology and other emerging technologies has also enhanced the innovation capacity of China's transportation infrastructure, promoted the healthy development of China's smart logistics and smart cities, and fully demonstrated the application value of digital power.

10.1.7 *Improvement of standards and specifications*

Most of China's major transportation engineering and key transportation facilities construction standards are related to digital twin. Under the digital twin simulation modeling, the construction effect, construction measures, and implementation plans of various transportation engineering can be directly demonstrated. The engineering construction department can formulate various construction specifications and standards in accordance with the visual effect of the simulation model. For example, the smart warehouse created by HUAWEI is built with digital twin technology as the core. The construction standard, operation standard, and management standard of the smart warehouse are also formulated according to the digital model.

It is believed that the construction guidelines and specifications of China's transportation technology projects such as smart roads, smart ports, smart hubs, and intelligent transportation management will be born in the digital twin model in future development, and the construction speed and achievements of China's transportation technology projects will be improved accordingly.

10.2 Digital Twin Helps the Comprehensive Development of New Infrastructure

The State Council of China issued the "Action Plan for New Infrastructure Construction in the Field of Transportation (2021–2025)", which clearly stated that "based on the new development stage, implement the new development concept, build a new development pattern, take the promotion of high-quality transportation development as the theme, speed up the construction of a transportation power as the general goal, adhere to innovation driven and intelligent development, take digitalization, networking and intellectualization as the main line, organize and promote a number of new key transportation infrastructure projects, create an influential new transportation infrastructure model, and create innovative development environment, promote the development of new infrastructure from point to area, improve the efficiency, expand the functions and increase the momentum of transportation, and continuously enhance the sense of gain, happiness and security of the people". It can be seen that in the future, China's new transportation infrastructure is a comprehensive and intelligent livelihood project, with a broad application prospect, and moreover, it produces a far-reaching impact on China's goal of becoming a strong country.

After 2020, digital twin has begun to show off in many intelligent fields in China, and this digital technology has been recognized and favored by more and more industries. DTT (Deloitte Touche Tohmatsu), a leading professional service institution in the world, has also rated digital twin as one of the "five key emerging trends that may trigger disruptive changes in the short term" in the "2020 DTT Technology Trend Report". This technology has rapidly triggered a wave of digital change in various industries.

Looking at the current situation of intelligent development in all walks of life in China, digital twin has been deeply integrated into China's new infrastructure. Digital twin can be found everywhere in major projects, such as smart city, intelligent transportation, and smart industry. Based on this situation, the author clearly sorted out the main ways of digital twin in helping the comprehensive development of new infrastructure in China.

10.2.1 *Smart system construction*

The operating systems and information platforms in various fields in China have begun to show intelligence under the promotion of digital twin technology, which is mainly reflected in the simulation application of digital twin and effective integration with other emerging technologies.

10.2.1.1 *Digital twin can build a digital model of smart city*

After the integration of digital twin technology, 5G technology, and big data technology, through the data collection of urban physical entities, the simulation modeling can be completed in the virtual space, and the virtual model can be simulated in detail. After that, the optimization data can be collected from the simulation and fed back to the hypothetical process of the smart city.

This kind of smart city digital modeling has a variety of promoting effects on new infrastructure:

(1) **Optimization of new infrastructure site selection:** The virtual model of smart city can adjust the location of various infrastructure without cost, and then carry out simulation test to ensure that the location of new infrastructure meets the needs of smart city development. For example, the layout of the 5G base station in the city can be optimized through the virtual model, and the optimal solution can be obtained from both functional and economic aspects.

(2) **The construction efficiency of new infrastructure of smart city can be maximized:** Smart city infrastructure projects can carry out a variety of dynamic simulation in the virtual model, with the assistance of Internet of Things equipment, simulation can be achieved one by one. In view of the simulation effect, the continuous optimization of the new infrastructure is carried out, and the optimization scheme is finally fed back to the real world, and the resource allocation optimization of the new infrastructure is carried out. This way not only improves the construction effect of the new infrastructure, but also saves the experimental cost.

(3) **Maximize the effect of operation management:** Rather than only being used in the construction of new infrastructure in smart cities, digital twin technology also plays an important role in subsequent

operation and management. Digital twin has the functions of virtual model description and diagnosis, and can also predict the development according to the model and provide a decision-making basis. These data can be fed back to the real world to improve the operation and management of new infrastructure. For example, the maintenance of smart high-speed infrastructure in China is a systematic arrangement under the analysis of digital twin technology. In addition, the digital twin simulation also has the characteristics of data visualization, which makes the deployment and management of smart city more effective in front of the intuitive virtual model.

10.2.1.2 *Digital twin can also be integrated with artificial intelligence*

The digital platform constructed by digital twin and artificial intelligence can not only simulate the real situation of physical entities in digital space in real time, but also enable the virtual model to have the ability to self-learn and predict the upcoming situation.

At present, the combination of digital twin and artificial intelligence is used in the research, development, and construction of new infrastructure in many fields in China, and the algorithm model of artificial intelligence is constantly improved, so that the virtual model built by digital twin is more intelligent and has a broader application range.

10.2.1.3 *Digital twin technology can also be integrated with industrial Internet technology*

In the framework of Industry 4.0 in Germany, it is clearly proposed that digital twin is the main way to implement the industrial Internet and can make the industrial Internet play a greater role. From the principal analysis, industrial Internet collects all kinds of industrial data through the network connection of industrial equipment, conducts in-depth analysis on industrial development through a fixed algorithm model, and finally obtains the actual situation of industrial operation and the optimization plan for industrial development.

Digital twin technology can continuously improve the application effect of industrial Internet in this process:

(1) **Industrial management:** Digital twin technology can conduct real modeling and simulation of various situations such as industrial production environment, operators, equipment, and vehicles through industrial data, and conduct simulation preview according to the relationship and dynamic of various industrial elements, so as to finally obtain a more reasonable management mode.
(2) **Production management:** Digital twin technology can be used in all aspects of industrial research and development, design, production, and manufacturing. Combined with big data and artificial intelligence technology, data collection and screening are carried out according to the market demand. Data modeling is carried out according to the collection and screening results, and then the performance of the virtual model is evaluated and optimized, so that products more suitable for the market demand can be designed and developed. The full closed-loop R&D of this virtual space requires a large amount of data collection and screening in the early stage, but the time cost and funds spent are far lower than those traditional design and R&D methods.

Digital twin can simulate and model industrial production lines. It can clearly perceive the actual situation of production equipment, improve the safety and reliability of production lines, and improve production efficiency through the collection of industrial Internet technology. On this basis, digital twin can also be combined with artificial intelligence technology to improve the self-learning and self-upgrading ability of industrial production lines.

Generally speaking, digital twin plays an important role in the construction of intelligent systems and information platforms in various fields, which not only accelerates the construction of infrastructure in various fields, but also improves the construction effect, maximizes the efficiency of various operation systems and information platforms, and completes the digital and intelligent upgrade.

10.2.2 *Intelligent construction in the energy field*

Digital twin technology has also played an important role in the infrastructure construction of China's energy sector. It not only improves the security of energy infrastructure construction but also strengthens the reasonable layout of strong energy construction.

For example, digital twin constructs virtual power system in UHV construction, carries out simulation on transmission lines and substations, and carries out simulation test on the maximum transmission efficiency of each line and station in digital space, from which the best scheme for UHV construction is analyzed.

For example, new energy charging station has become a key project of new infrastructure in smart cities with the rapid development of new energy vehicles in China. Through urban modeling and simulation, the digital twin can obtain the best layout of new energy charging stations in the city and ensure the good development of new energy undertakings in the construction of smart cities, combined with new energy vehicle operation, ownership data, and transportation network data.

10.2.3 *Intelligent construction in the field of transportation*

Digital twin is most widely used in the new infrastructure in the field of transportation and runs through the whole lifecycle of the design, construction, maintenance, and management of the new infrastructure of transportation.

For example, in the construction of China's transportation bridges, tunnels, tracks, hubs, and other projects, the sensor equipment, monitoring equipment, and infrastructure have been laid simultaneously. The purpose is to make the new transportation infrastructure have strong perception. The follow-up work of the new transportation infrastructure can be intelligently optimized under the digital twin simulation, and the transportation management can be more effective, transportation routes can be smoother.

Generally speaking, digital twin technology can optimize the construction effect of the new infrastructure through modeling and simulation from the geological exploration stage of the new infrastructure of transportation, and the characteristics of digital twin visual simulation enhance the possibility of intelligence for the new infrastructure.

10.3 Digital Twin has Gradually Become the Main Technical Support for the New Infrastructure of Transportation

In the past few years, the rapid development of China's transportation industry mainly benefits from the effective use of transportation technology, which ensures the continuous improvement of people's living standards.

With the continuous growth of transportation efficiency, China's transportation capacity fully meets the needs of social development.

However, the requirements of matching degree and high-end transportation technology begin to improve rapidly as China's transportation industry gradually changes from high-speed development to high-quality development; China's transportation industry must position itself as a strong and key scientific and technological capability in order to achieve a new leap from infrastructure.

In combination with the achievements of China's transportation digitalization and intelligent development, through the analysis of the development model of developed countries, it is not difficult to see that digital twin will become the main construction model of China's new transportation infrastructure, and gradually lead China to realize the dream of a strong country and a dream of rejuvenation.

10.3.1 *Digital twin promotes the development of intelligent transportation in China*

Although digital twin technology was first born in the space industry, it has now developed into a core force in various transportation fields. The ability of digital twin in omni-directional perception, calculation, and modeling has realized the description, prediction, and deduction of transportation engineering in virtual space, and has also completed the real-time interaction between virtual space decision-making and the development of the real world, which has optimized the whole lifecycle of China's transportation industry.

Li Yi, the founder of Beijing 51 WORLD, a famous digital twin company in China, once said that the value of digital twin can be reflected in three dimensions: the past, the present, and the future. In the past, digital twin was a kind of precipitation data and the basis of various data analysis; for now, digital twin is the global perception of various scenarios, the intelligent operation and control, and real-time scheduling of various systems; and for the future, digital twin is real simulation deduction, simulation prediction, and auxiliary decision. This view fully demonstrates the great value of digital twin in the development of transportation industry.

In the new era of technology upgrading and innovation, digital twin technology has been fully penetrated into all areas of China's transportation development. From a macro perspective, digital twin is optimizing the management effect of China's transportation, accelerating the connection between software and hardware of China's transportation industry, expanding the application scenarios of new infrastructure, and playing a practical effect in intelligent transportation. It can be seen that the digital twin is promoting the comprehensive development of intelligent transportation in China.

10.3.2 *Digital twin accelerates China's new transportation infrastructure*

The digital era features that everything can be digitalized. Digital twin technology is the channel to digitalize physical entities and then restore virtual space. Digital twin embedded the new infrastructure into the virtual model of China's transportation network in the construction of new infrastructure of China's transportation, simulating the construction effect of new infrastructure from the macro, medium, and micro perspectives so that the new infrastructure can promote, supplement, improve, and upgrade the development of transportation.

At present, the effect of digital twin technology accelerating the new infrastructure construction of China's transportation is mainly reflected in the following aspects.

10.3.2.1 *Transportation display equipment*

Since the simulation platform constructed by digital twin can be regarded as a new type of transportation display equipment, digital twin technology itself can also be regarded as a new infrastructure in the field of transportation. The digital twin technology constructs the virtual model of real transportation scene with super fine and high-performance three-dimensional rendering technology on the basis of the collection of real transportation data. This model can also be regarded as the real, complete, and detailed restoration of transportation scenarios, since this model includes transportation infrastructure information, spatial environment information, urban construction information, and geographical location information.

Moreover, the digital twin can also simulate the changes in the transportation environment, such as light, rain, and fog, and other environmental changes can be reflected in the virtual space. The virtual preview of this transportation scene provides strong support for various transportation management decisions.

10.3.2.2 *Transportation monitoring equipment*

Digital twin technology speeds up the development of transportation monitoring equipment in China and improves the perception of transportation network.

Data collection of digital twin is mainly obtained through perception devices. The new infrastructure construction of China has increased the laying of sensor equipment, monitoring equipment, and data transmission equipment driven by the digital twin technology, and moreover, the use effect of this equipment is improved in combination with artificial intelligence technology and new generation information technology.

The vigorous deployment of transportation perception equipment has enhanced the basic management and control ability of China's transportation, improved the operation and maintenance effect, and also accelerated the digital development of China's transportation.

10.3.2.3 *Unmanned intelligent equipment*

In recent years, China has also increased the investment in unmanned intelligent equipment and unmanned aerial vehicles with the purpose of ensuring the improvement of the perception ability of China's transportation network, and moreover, intelligent signal lights, AI identification, automatic collection equipment, and other equipment began to be fully embedded into the transportation data collection network. Through the increase in the use of this equipment, China's transport management departments have a more comprehensive grasp of the data of all elements of personnel, vehicles, environment, and roads, and a clearer picture of the transportation network and the city.

In addition, digital twin has strengthened the visual analysis and decision-making of China's transportation situation, and fully realized the real-time perception, intelligent additional measurement, early warning, and timely response of the transportation network through the data collection of these unmanned intelligent devices.

10.3.2.4 *Vehicle supervision equipment*

The new infrastructure of China's transportation sector is also reflected in the supervision equipment of special vehicles. By positioning special vehicles such as ambulances and fire trucks and installing monitoring equipment,

China's transportation management departments, medical departments and police departments have effectively controlled these vehicles.

The combination of these vehicles and transportation resources is more effective with the help of digital twin technology. For instance, the transportation management department can provide the best travel plan according to the transportation operation conditions in the process of executing tasks of police vehicles, and adjust the signal lights to improve the travel speed of police vehicles.

10.3.2.5 *Intelligent early warning equipment*

The application of digital twin technology has effectively improved the intelligence of transportation warning equipment in China. The digital twin technology can provide the best laying scheme of early warning equipment according to the environmental characteristics from the macro perspective, while from the micro perspective, the digital twin can enhance the protection effect of the early warning scheme and upgrade the intelligence of the separate early warning equipment through simulation and preview.

10.3.2.6 *Data analysis and control platform*

The transportation data analysis and control platform is the main tool for the development of China's transportation intelligence. The transportation data analysis and control platform can realize multi-dimensional dynamic demonstration of transportation models, demonstration of geographic information combination, and real-time interaction of transportation operation status by virtue of digital twin. What's more, it can also realize cross-domain data integration, deeply excavate digital value, and assist transportation management departments to make scientific, reasonable, and intelligent decisions.

10.3.3 *Guarantee of digital twin for breakthrough development of new transportation infrastructure*

Digital twin technology has brought breakthrough development to the new infrastructure of transportation in China. In order to ensure the development effect of digital twin technology in the field of transportation, in September 2021, China's Ministry of Transport issued the "Action Plan for New Infrastructure Construction in the Field of Transportation (2021–2025)", which explicitly referred to the "safeguard measures" for new infrastructure construction in the field of transportation.

(1) **Strengthen the organization and implementation:** The Ministry of Transport coordinates the new transport infrastructure actions, improves the working mechanism, strengthens the guidance on the implementation of various actions by the provincial transportation authorities, conducts periodic assessment once a year, and makes timely and dynamic adjustment in accordance with the planning policy requirements. All provincial competent departments of transportation shall attach great importance to the implementation of the main responsibility, strengthen the organization and coordination, and speed up the construction of key projects by combining with the pilot program of transportation power. The progress of work and the implementation of key projects shall be reported to the department before 10 December every year. The project undertaking units shall implement specific responsibilities to ensure the smooth implementation of key projects and give full play to their actual effects.

(2) **Promote diversified investment:** The Ministry offers overall support to projects in line with the investment policy during the "14th Five-Year Plan" period and gives full play to the leading role of government investment. All provincial departments in charge of transportation shall actively seek for priority support from local finance for new infrastructure projects. Diversify investment and financing channels by making full use of market mechanism, attract social capital to actively participate in investment, construction, and operation services, give better play to the market's supporting role in the development of new infrastructure industry, and strengthen the construction of risk prevention and control mechanism.

(3) **Encourage innovation and promotion:** Encourage the development of new business modes and new modes based on the new infrastructure of transportation in accordance with the principle of inclusiveness and prudence, encourage collaborative research and integration of upstream and downstream of the industrial chain, and form a collaborative innovation pattern of production, learning, and research. The Ministry played a typical leading role and summarized and promoted important achievements and advanced experience through on-site meetings, experience exchange, scientific and technological awards, and other forms.

The State Railway Administration, the Civil Aviation Administration of China and the State Post Bureau, in accordance with their duties and in accordance with the "Guidelines of the Ministry of Transport on Promoting the Construction of New Infrastructure in the Field of Transport" (JGHF [2020] No. 75) and the relevant requirements of this action plan, independently studied and organized the development of related work in the fields of intelligent railway, intelligent civil aviation, and intelligent post.

In February 2022, China's transportation clearly proposed the strategy of expanding effective investment and development, the main content of which is for new transportation infrastructure such as smart roads, smart ports, and smart hubs. Therefore, the Ministry of Transport of China has issued the "Notice on Actively Expanding Effective Investment in Transport" (hereinafter referred to as the Notice).

The "Notice" pointed out that all localities and units should adhere to the general tone of making progress while maintaining stability, fully, accurately, and comprehensively implement the new development concept, closely focus on the "14th Five-Year Plan" series of planning objectives, tasks, and key projects, moderately carry out infrastructure investment in advance, and solidly do a good job in expanding the effective investment in transportation. It needs to vigorously promote the implementation of the "14th Five-Year Plan" projects, speed up the construction of 102 major national engineering transportation projects and the "6 axis, 7 corridors and 8 channels" strategic backbone channel projects, and accurately fill in the short board of the national comprehensive three-dimensional transportation network; it is necessary to strengthen the organization and implementation, further promote the pilot task of the construction of strengthening country with strong transportation network, focus on summarizing experience, and form replicable and popularized systems, methods and results; it is necessary to improve the quality and depth of preliminary work, promote the early work of projects within the planning as soon as possible, deepen the demonstration of main construction contents such as route location, construction scheme and technical standards, and reasonably determine the investment scale; it is necessary to strengthen project reserve, establish project list, and form a virtuous cycle of reserve, commencement, construction, and completion; it is necessary to strengthen supervision and management, implement the main responsibility and regulatory responsibility, and ensure the smooth implementation of the project.

The "Notice" also requires all localities and units to do a good job in ensuring policy elements such as capital and land, give full play to the leading role of central funds, innovate the investment and financing system and mechanism, strengthen communication and coordination with relevant departments, promote the establishment of coordination and promotion mechanism for major inter ministerial, inter provincial and inter provincial projects, and form a joint effort to promote the construction of major engineering projects. It is necessary to strengthen risk prevention and control, strengthen the coordination mechanism of transportation development planning and fund guarantee, and ensure the timely payment of migrant workers' wages and project funds; it needs to firmly establish the concept of safe development, strictly implement the responsibility system for safe production, establish a dual prevention mechanism of hierarchical risk management and control and hidden danger investigation and control, and promote the construction of boutique products and fine management under the guidance of the safety century-long quality project, so as to ensure the safety production and project quality.

Chapter Summary

Digital infrastructure and digital twin are complementary and mutually reinforcing. Digital twin can improve the digital upgrading effect of transportation infrastructure, and digital infrastructure can ensure the maximum use of digital twin. It can be seen that the beginning of digital and intelligent development of transportation in China is the digital construction of transportation infrastructure. Only by completing the quality improvement and development of digital infrastructure can the intelligent transformation of comprehensive three-dimensional transportation network be realized.

Chapter 11

Look into the Intelligent Transportation Texture and Clarify the Context of Intelligent Transportation

The development of human digital technology has brought new opportunities and challenges to intelligent transportation. Digital twin technology has opened up a new perspective window for the cause of intelligent transportation. By combining the texture of the application of this technology in transportation management, the whole society will have a clearer and more comprehensive view of its future development.

11.1 Digital Twin: New Direction of Future Transportation

The most important feature of digital twin technology is the dynamic simulation of physical objects. The people have higher and higher requirements for the safety, stability, and convenience of the transportation system operation with the increasing complexity of various scenarios in the domestic transportation field, and the trend of digital twin enabling the future transportation development has become a historical necessity.

11.1.1 *Digital clonal plan for future transportation*

The hand-in-hand of transportation and digital twin is more like a grand "clonal" activity aiming at transportation in the real world. In the "clonal" activity, digital twin technology will make full use of data information such as physical model, sensor update, and operation history to conduct multi-dimensional simulation of transportation phenomena described by different physical quantities, probabilities, and scale standards from different disciplines. Not only that but also this technology will completely map it to the virtual digital space, reflecting the whole cycle process of the transportation system. In simple terms, the major mode of future transportation development is that, on the basis of advanced equipment or systems, a digital version of the transportation system is created.

Transportation system is created in reality, and its digital twin is created on the information platform. It is because of its "clonal" nature of transportation reality that it can complete a dynamic simulation of the realization of the transportation system and become a real "dynamic" information system.

The "movement" of the digital twin comes from the components of the main model of the transportation system, instead of occurring autonomously, which mainly includes the data fed back by the sensors on various ontologies in the transportation system, as well as the historical data of the operation of the ontologies. This means that the real-time status of each vehicle and road in the transportation system, as well as the changes in the relevant external environmental conditions, will be restored to its digital "clonal body". Especially when the physical transportation system needs to be changed in design, or the decision-maker needs to learn about and evaluate the specific changes of the transportation system in the special environment, the data engineer can carry out experiments on its digital clonal to avoid the impact on the physical transportation system and improve the decision-making efficiency and reduce the decision-making cost.

At present, the technology community actively constructs and simulates the urban transportation conditions and optimizes the actual transportation control strategy through the evaluation and deduction of the digital clonal of urban transportation by using digital twin technology in the field of transportation control. By using the transportation digital clonal system, technicians can find and deal with key problems in a timely manner, which is an important scenario for the application of the transportation digital clonal plan.

For example, part of the highway in a certain area is subject to frequent fog in autumn and winter. This kind of weather phenomenon is difficult to accurately and timely forecast, with strong suddenness, which can rapidly reduce the visibility in a short period of time and easily lead to transportation accidents. When the transportation conditions of these road sections rely on digital twin technology and the corresponding digital clones are established, the technicians can monitor and predict the transportation conditions in real time on the virtual platform by using dynamic perception data, so as to detect the possible high-risk situations caused by fog in advance.

The ability of digital twin simulation behavior is constantly improved, and the accuracy of the prediction algorithm is also constantly improved with the development of technology. The transportation digital clones will have more and more wonderful performance. Professionals can even calculate multiple actual situations in the future once and for all by using transportation clones to form a prediction of "parallel transportation world". Of course, the requirements for the digital twin system will become higher and higher due to the increasingly complex transportation scenarios, and the digital twin technology will develop rapidly under the higher requirements.

11.1.2 *Insight into future transportation dynamics from a global perspective*

The combination of digital twin and transportation system reflects the technical characteristics of global, real time, and two way. From the date of its birth, digital twin technology has continuously strengthened these characteristics, and people are able to have a deep insight into the future transportation development trend.

11.1.2.1 *Global characteristics of digital twin transportation*

Overall refers that the digital twin technology can run through the entire cycle of many industries, including design, development, manufacturing, service, as well as maintenance, and so on. This feature of digital twin technology not only enables it to complete the auxiliary work of management but also helps users in the industry to use products better.

The comprehensive use of various technologies is demanded by intelligent transportation to ensure the comprehensive perception, calculation, and prediction of different static and dynamic nodes in the transportation system, so as to carry out the overall management of the transportation conditions of the whole city and even the region. This process requires the collection and use of a large number of data for analysis, comprehensive analysis of the analysis results, and then transmission to the management server to assist decision-making. It is difficult to ensure the accuracy, real time, and comprehensiveness of a large number of data collection if traditional technical means are adopted, and loopholes are inevitable in the process. The intelligent transportation system integrated with digital twin technology, like a high-definition projector, projects every detail to the information platform without any details, which can fully adapt to the operation mode of "one brain plus multiple synergetic brains" of intelligent transportation, give full play to resource advantages, and avoid waste in the decision-making process.

11.1.2.2 *Real-time characteristics of digital twin transportation*

Real time is an important feature of transportation system operation, and digital twin technology also has targeted service advantages. The real-time characteristics of digital twin technology can be divided into two forms: one is real-time active service, and the other is prediction service. The former is mainly reflected in the active reporting and analysis of the existing problems, including the damage of transportation facilities needs to be reported in time for maintenance and replacement, and so on. The latter is mainly represented by the prediction of possible situations in advance according to the existing data, which can more significantly reflect the difference between intelligent transportation and traditional transportation. That is, the possible situations in the future can be predicted in advance according to the complex and diversified digital model, so as to change the passive response attitude of humans facing transportation for thousands of years into the active attitude of early management and control.

The real-time feature of digital twin technology also reflects the comprehensive and real-time connection between the virtual digital body and its noumenon. The relationship between the noumenon and the twin is not independent but has a certain real-time connection after using this technology to project the real transportation system into a virtual transportation system, that is, the changes of the real transportation will continue to be reflected in the virtual transportation system in real time.

11.1.2.3 *The two-way characteristics of digital twin transportation*

Two-way refers to the data flow between the noumenon and the twin is not a single dimension but flows with each other. Digital twin transportation is not only a projection from reality to virtual reality nor is it just an output of data from noumenon to a twin system. Virtual twin transportation system can also feed back information to noumenon through technical means. The relevant regulators and participants of the transportation system can pre-judge and automatically check the problems through the information fed back by the data twin and take pre-control actions to avoid the problems.

11.1.3 *5G technology fully enables digital twin intelligent transportation system*

5G technology plays an important role in the construction and application of digital twin intelligent transportation system.

There is real-time massive data interaction between physical entities and their digital twin in the intelligent transportation system, just like the interaction between a healthy human body and the external environment, where various kinds of perception and control information are transmitted through the nervous network in this process to maintain the normal operation of the human body. Similarly, communication network plays an irreplaceable similar role in the digital twin of an intelligent transportation system.

Generally speaking, the faster the speed of nerve transmission, the more sensitive the response of nerve endings, and the wider the distribution of nerves, the higher the flexibility and agility of the human body. The same is true of the digital twin of intelligent transportation. A large number of transportation application scenarios test the twin “artificial network”. The number of distributed nodes, mobility, transmission bandwidth, transmission delay, and loss in the process of data transmission are all testing the quality of “artificial network”, and complying with the development trend of social transportation, resulting in higher and higher requirements.

As a matter of fact, there is a close relationship between the application level of the intelligent transportation digital twin and the support level of the communication network. For instance, it puts forward high requirements for transmission delay, bandwidth, and other elements of the communication network since the intelligent transportation system is inseparable from the production, equipment, and real-time control of unmanned vehicles. Moreover, the entire intelligent transportation digital twin is also an important scenario for 5G communication network applications.

11.1.3.1 *3 Application scenarios of 5G technology*

5G is the fifth-generation communication technology. In September 2015, the International Telecommunications Union (ITU) officially confirmed the three major application scenarios of this technology, namely eMMB, uRLLC, and mMTC.

Enhance Mobile Broadband (eMMB), the upgraded version of mobile broadband currently used, mainly serves the needs of mobile Internet and emphasizes the bandwidth and speed of the network.

Ultra Reliable and Low Latency Communication (uRLLC) mainly serves Internet of Things scenarios. In addition to the industrial Internet, this scenario is the most deeply integrated with the intelligent transportation system, such as the Internet of vehicles and UAVs. Once the delay is long, the network fails to respond to the data in a very short period of time, and it is likely that hazardous accidents will occur.

Massive Machine Type Communication (mMTC) is mainly used in a large number of terminals within a unit area and needs network support, such as intelligent street lights, transportation signals, transportation monitoring, and transportation sensors.

11.1.3.2 *The road to full empowerment of 5G*

Among the three major application scenarios of 5G, two are mainly Internet of Things services. Therefore, 5G technology has the important value of enabling digital twin intelligent transportation.

The construction foundation of digital twin intelligent transportation system is two-way communication, data collection, and operation control of massive Internet of Things terminals, which can be satisfied by 5G mMTC application. The important role of the digital twin intelligent transportation system is to feed back and transmit the results and instructions from the simulation of the virtual transportation system to the real transportation system. The timeliness and reliability guarantee provided by the uRLLC application are also indispensable. In addition, users also need to carry out human–computer interaction through VR technology, image recognition technology, remote control, and other methods in order to effectively control and use the digital twin virtual transportation system, which is inseparable from the support of large bandwidth network transmission in eMBB applications.

In summary, the three major application scenarios of 5G are not separately enabled in the digital twin intelligent transportation system. As a truly integrated communication network, 5G, an advanced network technology, will actively carry out dynamic switching and adjustment under different branch application scenarios of the digital twin intelligent transportation system to achieve continuous empowerment.

11.2 Deep Integration of Digital Twin and 5G Technology

There is an unwritten "law" in the field of communication, that is, a new generation of wireless mobile communication technology will be born every 10 years. However, each time a new generation of "G" appears, it will take 3–5 years to complete the industrial chain and the stage of industrial integration and development, and then officially enter the stage of innovation and application that the whole society can perceive. For instance, 2G brings SMS services, 3G brings smartphone applications, and 4G brings new mobile Internet formats, such as live broadcasting, takeaway, sharing economy, new retail, and online education, which are the products of deep integration of traditional formats and mobile communication technology. Similarly, a new generation of intelligent transportation system will be developed by the integration of 5G technology, digital twin, and transportation field.

11.2.1 *5G technology: a strong outlet for digital twin ground transportation*

5G technology has the characteristics of high bandwidth, low delay, and wide connection and is an important driver for high-quality transformation and upgrading of the transportation industry. The implementation of 5G technology to facilitate the digital transformation of the transportation industry requires the integration and application of infrastructure, technology, and data, as well as innovation and collaboration among industries.

Digital twin is equipped with the technical advantages of digital identification, synchronous visualization, virtual reality interaction, intelligent control, etc., providing new solutions and new paths for businesses, such as transportation perception, road warning, emergency rescue, and intelligent driving, and creating more efficient, convenient, and safe management and service methods. 5G technology plays an important role in information transmission and provides strong support.

Of course, 5G technology helps the transportation industry to achieve digital transformation, and the overall mechanism must be innovated. Mobile communication is closely related to transportation itself. The new mobile communication technology can also change the operation mode and mechanism of the entire transportation industry, instead of only improving the operation efficiency of the transportation industry. The service model of communication operators must be changed again when the era of digital twin landing in the transportation field comes. Communication operators can no longer set a fixed network capacity according to the usage peak value every day as in the past but need to realize intelligent operation together with application enterprises.

Under the support of policies and guidance of various parties, digital upgrading and transformation have penetrated into the whole transportation industry, and the application of digital twin-related technologies has become more and more diversified, and transportation development has entered a new stage of digital transportation.

Under the traditional transportation business, a lot of data is in the form of drawings. In the data sharing and application, construction and operation are difficult to get through. Intelligent transportation is to add a layer of digital infrastructure on the basis of the original civil engineering facilities. Therefore, the digital transformation process of the transportation industry requires the use of more universal communication equipment and the help of more development achievements of communication technology, and 5G is the best choice.

11.2.2 *How to efficiently interact between physical transportation and digital transportation under 5G technology*

The development of communication and transportation has always been closely related. The development of digital transportation is neither a castle in the air nor a simple transfer of traditional information work to the Internet. It is under the integration of 5G technology and implementation of digital twin in the transportation field that new business forms such as vehicle road collaboration, automatic driving, and smart high speed derive a new development trend of efficient interaction between physical and digital transportation.

Digital twin is not only an important application scenario of 5G technology but also an important direction and inevitable trend of the development of the transportation field. In July 2019, the Ministry of Transport issued the "Outline for the Development Plan of Digital Transportation", which clearly proposed the direction of the construction of digital transportation. It is necessary to build a "modern transportation system that takes data as the key element and core driver, and promotes the continuous integration and interaction of physical and virtual space transportation activities". "Digital twin" is highly consistent with the concept of digital transportation system in the outline, which marks that China's intelligent transportation development has entered the digital twin era. The efficient interaction between physical transportation and digital transportation cannot be separated from the important role of 5G technology.

From the physical level to the digital level, the empowerment of 5G on the intelligent transportation interaction process is mainly reflected in the following aspects.

11.2.2.1 *Communication system construction*

Digital collection system, network transmission system, and intelligent application system are indispensable in the field of intelligent transportation. Among them, the application of Internet of Vehicles and satellite communication information network supported by 5G technology will more effectively improve the communication information network and form an intelligent transportation information communication network that integrates physical and digital aspects.

11.2.2.2 *Smart mobile space*

The reason why intelligent transportation needs to be connected from physical to digital level is largely because in order to realize the concept of "MaaS", connect the actual travel transportation space with data level resources, and under the support of 5G technology, it can create a "Internet +" convenient transportation system and develop new types of transportation services, such as customized public transportation, intelligent parking, intelligent bus, network booking taxi, and Internet bicycle rental.

11.2.2.3 *Application scenarios*

In various scenarios of intelligent transportation, it includes the use of different types of systems, such as safety system, transportation management system, vehicle control system, travel demand system, electronic toll collection system, operation management system, operation vehicle operation system, as well as public transportation operation system. The collection and processing, release and exchange, and analysis and application of data information are the key to the interconnection of physical and digital levels in the intelligent transportation system, regardless of the specific application scenarios. This requires the full use of 5G technology to promote the integrated application in vehicle information services, transportation dynamic simulation, transportation flow scheduling, and other aspects and promote the large-scale application of various scenarios of the intelligent transportation system.

11.2.3 *The combination of double effect and 5G technology supports the multi-dimensional dynamic mapping of digital transportation*

The integration of communication, navigation, and time and space is an important starting point for the future construction of a digital transportation system. The technology integration of 5G and intelligent transportation realizes the interconnection and precise coordination of all things in the transportation system, so as to provide highly reliable dual effect intelligent transportation services to society.

With the in-depth application of 5G technology in intelligent transportation, its advantages in supporting digital transportation are revealed. First of all, the technology can fully integrate virtual reality, artificial intelligence, and other technologies into the operation of digital transportation, making the digital twin intelligent transportation a qualitative leap. Second, before the integration of 5G technology into the operation of intelligent transportation, intelligent transportation has made some progress in relying on cloud computing and big data analysis, but the technical requirements put forward by the digital twin construction goal are getting higher and higher. Only 5G can replace some of the original backward information technology and accelerate the digital and diversified dynamic mapping.

5G technology supports multi-dimensional dynamic mapping of digital transportation, mainly focusing on the following points.

11.2.3.1 *Improve the efficiency of transportation data projection*

In the construction of intelligent transportation digital twin, the collection and analysis of data information is extremely important. The higher the collection efficiency, the higher the projection efficiency. 5G communication technology can collect data by using different high-tech terminal equipment, effectively improve the efficiency of transportation data projection, and accurately construct and manage the digital twin of intelligent transportation.

For example, 5G technology can effectively manage the vehicle road collaboration system. Vehicle road collaboration system requires information technology to process information within 100 ms, while 5G technology, with its own communication speed and derivative cloud computing functions, is sufficient to complete this task, providing sufficient conditions for the realization of vehicle road collaborative work. For example, 5G technology can improve the efficiency of transportation management, rely on a variety of advanced sensors, detect different driving behaviors in order, and collect data for the construction of digital twin.

11.2.3.2 *Improve the identification ability*

In the construction of digital transportation management system, the ability to identify transportation conditions is crucial, which requires the use of powerful analysis capabilities of information technology. 5G technology can predict and analyze the road transportation conditions in advance and in real time through various means to ensure the timely identification of accidents. In addition, 5G technology can also accurately judge the abnormal conditions in the transportation process and timely detect the transportation unsafe factors through the super Internet of Things, intelligent video, and other ways.

11.2.3.3 *Speed up the development of unmanned driving technology*

In the process of intelligent transportation system construction, only through the participation of 5G technology can the unmanned driving technology based on the Internet of vehicles be truly realized. With its super efficient data collection capability, 5G technology transfers massive data to the cloud, which solves the problem of data dynamic mapping of unmanned driving and accelerates the development of unmanned driving technology. Similarly, 5G technology also helps achieve accurate management of transportation facilities.

11.2.4 *Strong coupling, 5G technology accelerates the formation and application of digital twin*

By giving full play to its advantages, 5G technology realizes the collection and supervision of transportation big data, solves many problems in the development of intelligent transportation, speeds up the formation and application of digital twin to a large extent, and also promotes the development of China's intelligent transportation in the digital twin intelligent transportation system.

11.2.4.1 *Popularization and safety of optimized application*

The intelligent transportation operation should value more about the information interaction and sharing between the transportation facilities and vehicles, between the transportation facilities, and between the transportation facilities and people after the intervention of 5G technology. It is necessary to actively optimize various applications and improve the participation of all parties in the digital twin intelligent transportation system; thus, the promotion of comprehensive development of the digital twin intelligent transportation system can be achieved. In addition, the digital twin intelligent transportation system will also collect information of various transportation participants in the era of information sharing. The system should also emphasize ensuring the data security of the participants and doing a good job in related confidentiality in the process of collecting and using information, which is also inseparable from the advantages of 5G technology.

11.2.4.2 *Improve the transportation network coordination system*

The combination of digital twin of intelligent transportation and 5G technology will further create a perfect collaborative system with the in-depth integration of 5G technology and intelligent transportation system. Intelligent transportation can be further optimized under the support of 5G technology through continuously improving the digital twin and improving the service quality. In addition, the digital twin of intelligent transportation needs to be maintained with higher performance due to the complexity of intelligent transportation in terms of technology and facilities. The smooth operation of intelligent transportation can be guaranteed to a greater extent with the service of 5G technology.

11.2.4.3 *Train professional talents and strengthen infrastructure construction*

In addition to the need for information technology to support the construction of digital twin, professional management and research team support is also essential during the development of an intelligent transportation system. Efforts should be made to meet the needs of the information age, so as to implement professional construction. Moreover, relevant professionals should be trained to improve the professional ability of intelligent transportation operation and management personnel, adapt to the application of digital twin system, and promote the development of intelligent transportation under the support of 5G technology. In the meantime, the application of 5G technology can in turn promote the construction of intelligent transportation-related infrastructure and promote the efficient operation of information computing facilities and equipment in the system.

With the advantages of high information capacity, large bandwidth, low transmission delay, and high efficiency, 5G technology has the ability to fully couple the construction of intelligent transportation, including improving the efficiency of data collection, promoting the development of unmanned driving technology, and assisting the operation of intelligent transportation system. It is bound to play a greater role in intelligent transportation with the development of 5G communication technology.

11.3 Visual Management of Intelligent Transportation

The existing transportation management system has adopted various means, such as the construction of signal control, bayonet monitoring, transportation guidance, and other business systems with the aim of alleviating various problems of the traditional transportation system, and to a certain extent, the problems are alleviated, but it is still unable to comprehensively control the overall situation of the transportation system, and it is difficult to fundamentally achieve intelligent management of urban transportation. Therefore, the visual management of intelligent transportation has been put on the agenda.

11.3.1 *Visual empowerment and digital twin construction of intelligent transportation ecological network*

Visualization means the deep understanding and insight of the data, the in-depth analysis of the calculation results, and the transformation of digital information into intuitive images or graphic information presented to professionals, so as to facilitate their observation of the calculation and simulation process. With the aid of visualization, the digital twin system will deliver the transportation phenomenon that cannot be directly observed in the traditional way and provide reliable management and service means.

Visualization technology is playing an increasingly important role in the construction of digital twin intelligent transportation ecological network. The coordination, layout, and monitoring capacity of transportation participants and management entities can be comprehensively improved by applying visualization technology.

For instance, in order to use the system to carry out daily transportation operation monitoring and coordination management from multiple dimensions, the transportation management department needs to highly integrate the various business systems of the transportation after the construction of the digital twin system is completed. This integration should make full use of the digital twin system to achieve multi-departmental data collaborative management and improve transportation management and service level. All of these need to be achieved with visual integrated management platform.

The digital twin technology is used to establish a big data visualization platform for intelligent transportation, which can effectively support various transportation management business systems and provide functions such as video monitoring, intelligent checkpoint analysis, transportation operation monitoring, and transportation signal monitoring. On the daily level, the platform can effectively monitor the road transportation conditions in real time, carry out visual management of daily transportation, and help the management team to master the dynamics of transportation supervision service personnel in real time, so as to facilitate timely dispatching. From the overall perspective, it can also help the decision-making team to learn about the law of changes in road network operation conditions and provide data support for the implementation of scientific planning and design.

The digital twin transportation system can realize the analysis, research, and judgment of transportation situation and achieve the purpose of scientific and detailed management when visualization technology fully participates in the information management of transportation infrastructure. Moreover, visualization technology can also support digital twin system and build intelligent transportation ecological network in more aspects.

The data of digital twin intelligent transportation system has the characteristics of massive, diversity, fast, and high value of big data. It is necessary to convert these data into highly accurate and reliable visual information of human–computer interaction through visual empowerment with the purpose of making full use of these characteristics, so as to show its internal laws.

There are mainly three ways for visualization technology to empower digital twin intelligent transportation system.

11.3.1.1 *Combination of static and dynamic interaction*

In this way, user-centered, the ever-changing transportation data flow is edited into static charts and dynamic interactive content, and visual analysis is carried out to help users make decisions related to transportation business. Users can have multiple synchronized views and directly select specific data points for visual analysis under the visual effect of static and dynamic interaction.

11.3.1.2 *Visualization of massive transportation data*

By visual analysis of massive transportation data, the results of transportation data can be presented, the relationship between the data at the global and partial levels can be reflected, the importance of data at different levels can be shown, and moreover, it takes into account the dynamic display of interactive hierarchy and time dimension, and excavates the internal rules of massive data.

11.3.1.3 *Visualization of multi-dimensional transportation data*

In the massive data environment, the level, complexity, and depth of multi-dimensional transportation data flow also increase. For this reason, relevant institutions need to study the visualization technology to clearly identify the characteristics of data, show the distribution of data, help users choose the correct parameters from multi-dimensional, and reduce the misleading of decision results.

The adoption of efficient visualization technology can improve the user experience of digital twin intelligent transportation and improve the efficiency and accuracy of data flow analysis.

11.3.2 *Promote new infrastructure with simultaneity and build three-dimensional transportation map*

Today, the ever-changing transportation changes and the ever-changing industry business are recorded by the surging data and constantly flow as the new infrastructure has become the tide of the times, forming a digital twin system and reacting to the real society. Among them, the three-dimensional transportation map with strong visual simulation and massive spatial basic data is an important factor for the sustainable development of the new infrastructure.

Undoubtedly, digital transportation is an important part of China's new infrastructure, and it also needs its own "infrastructure" and the support of the bottom business. Digital twin transportation system puts forward higher requirements for map application in the digital transformation of intelligent transportation construction. The important role of map application lies in the typical solution of the integration of map visualization services and digital twin transportation. In this area, Singapore has provided us with a good demonstration.

"Virtual Singapore" is a dynamic three-dimensional city model and collaborative platform, which includes a three-dimensional map of Singapore. Unlike traditional 2D maps, this map not only has bottom attributes such as water, vegetation, and transportation infrastructure but also shows terrain, curbs, stairs, or slopes. As a result, the "virtual Singapore" map can be used to calculate and show accessible routes for minorities who will be able to easily understand how to get to bus stations or subway stations. In addition, the map can also provide data not included in traditional maps, such as building height, roof surface, and sunlight, which helps relevant planners adjust according to seasonality to obtain more accurate forecasts.

How about the 3D map? What's the difference with mobile navigation map in the 4G era? How will it take advantage of its visualization features to help achieve digital transportation? From the geographical perspective, the three-dimensional map is not a single concept but contains the multi-dimensional nature of time and space. From the perspective of data, it also has the diversity of different application levels.

11.3.2.1 *Presentation of dynamic data*

On the three-dimensional map, the dynamic data of objects or people in the transportation system are presented by visual elements. For instance, the map adopts thermal analysis, trajectory analysis, and other methods to carry out interactive graphic display, which helps the transportation management department to better understand and solve the transportation planning problems, and forms an intelligent and automatic transportation management mode.

11.3.2.2 *Vector map*

The three-dimensional map adopts the structure of vector composition. The points, lines, and surfaces of the map can be edited and updated, and interactive analysis with vehicles, Internet of Things, and transportation data is carried out by using the vector composition elements to promote the realization of intelligent Internet connection, automatic driving, and unmanned driving. In addition, the map-based vector structure can also be used for multi-dimensional semantic and temporal analysis.

11.3.2.3 *Multi-level layer*

The "three-dimensional" aspect of the three-dimensional map is most intuitively reflected in the multi-level layer. From the perspective of visual interaction effect, the map used for digital twin transportation is not only a plane picture but also a multi-layer map that can be classified and compared according to different transportation business areas. From the river trend in nature, to the road network of infrastructure construction, to the factory and shopping malls from the commercial perspective, or to the transportation management level of congestion, different sub-items can be used as the specific map layer to achieve visual classification after data integration.

11.3.3 *Build a "cloud road" for transportation to realize the dynamic perception of human, material, and road data*

Under the development of intelligent transportation, not only data perception and analysis need to be performed from a separate road section but also the overall expansion of human, material, road, and other factors to the network level for state analysis needs to be conducted to build a new transportation "cloud road".

Transportation data perception technology is the application foundation of all intelligent transportation systems. In order to realize intelligent transportation, it is necessary to achieve scientific collection and efficient processing of transportation flow data. Under the technical environment of a combination of visualization and digital twin, the transportation signal can be more evenly distributed on the time-space platform simulated by the digital twin by improving the time-space precision of transportation data collection, and the effective dynamic perception of people, objects, and roads can be realized.

Before the arrival of digital twin technology, there had been many technological changes in the field of transportation information perception. Hardware detection technologies such as induction coil, camera, infrared, piezoelectricity tube, and GPS have been widely applied. During the transformation, the transportation information perception technology has also changed from the original manual, single-point, and static collection to the automatic, data fusion, and dynamic collection, and the collection methods show the characteristics of diversity. As a matter of fact, no matter what kind of transportation data, it has the dual attributes of time and space. The sequence of state data of people, objects, and roads reflects the timeliness of transportation data, while the source location reflects its spatial nature. The "cloud road" of transportation has been formed on the basis of the digital twin of intelligent transportation, focusing on the dynamic perception and visual presentation of the data of people, objects, and roads, and resulting in different applications.

For example, congestion is a chronic disease of modern urban transportation. By using the transportation "cloud road" to retrieve the urban transportation travel data under the fine granularity visual state and present it in the thermal dynamic state, the transportation management resources can be dynamically allocated according to different road sections and time. On the basis of the same transportation thermal analysis model, digital twin technology can also be used to analyze key hub sections or intersections, so as to scientifically and dynamically adjust the interval frequency of transportation lights and quickly resolve road congestion. In addition, the analysis of the historical accumulation of massive transportation tracks and cross-border time and space tracks can also help the public security and transportation management departments to build an information model of transportation violations and prevent transportation congestion caused by violations at the source. The massive data of "cloud road" can also present the characteristics of urban boundaries and development, and support the scientific planning of cities.

When the "cloud road" based on the digital twin enables real-time aggregation of data of people, things, and roads and the completion of quantitative calculation, the transportation managers and participants can efficiently implement their respective transportation operations on a digital twin "map", and achieve a significant reduction in the cost of transportation activities by more advanced means. For example, by applying the dynamic data provided by "cloud road", a new transportation control mode based on the electronic fence technology is established, and the intelligent transformation of the control mode is completed, so as to automatically guide vehicles to bypass the illegal areas. For another example, we can observe the transportation operation mode of a city and a highway from the perspective of "god" of hologram time and space and explore the source of solving transportation problems. These changes are an important foundation for building a smart city and a smart society.

11.4 Intelligent Application of Digital Twin Technology

Digital twin is the key direction and application scenario for the development of the transportation industry, and also the key means to complete the visual presentation of intelligent transportation. Digital twin technology will realize dynamic virtual scenes at multiple levels through various intelligent applications, which can not only synchronize the real transportation but also make full use of the constructed information model to complete scientific research, analysis, and prediction of the real transportation.

11.4.1 *Create three digital levels of transportation digital clones*

From the technical level, the digital twin technology used to create transportation digital clonal can be divided into three digital levels, namely data access level, computing simulation level, and output application level.

11.4.1.1 *Data access level*

Digital twin is to replicate specific physical objects in a digital way to simulate the actual behavior of objects. Especially in the field of transportation, in order to realize the close interaction between physical transportation and digital transportation, it is necessary to create more supporting technologies as support, and it is necessary to go through more stages of evolution, so as to achieve the shaping of physical transportation in the digital space. Therefore, the data access level must be fully valued.

In the data access level, the digital twin transportation clones collect, transmit, synchronize, and enhance the physical entity information in the real transportation through the Internet of Things and other technologies to form general data that can be used in the business. In this process, two methods, static modeling and dynamic perception, are used to obtain the transportation data in the real world, so as to prepare for the creation of the corresponding mirror model.

In the process of static modeling, the system mainly uses the way of static perception, and uses CAD, CAE, BIM, CIM, and other technologies to model the static part of the transportation objective environment.

The dynamic perception process is the key part of the data access layer. The perception and access of transportation data status can be realized in a more comprehensive and timely manner with the advantages of Internet and Internet of Things technologies, such as 5G and even future 6G technologies.

The data access layer belongs to the "input" part in the composition of the transportation digital clonal. The more comprehensive, real-time, and accurate the data information flow obtained at this level, the higher the value of transportation digital clonal will be.

11.4.1.2 *Computing simulation layer*

Digital twin technology needs to build a digital model of physical transportation and conduct simulation analysis on it to obtain a virtual transportation model of digital space version, for the purpose of making full use of data resources. This process requires the physical object to be represented as a digital model that can be identified by a computer network. Through modeling, digital twin can eliminate more uncertainties in physical transportation with less energy, so the computing simulation layer is the core level of digital twin.

Simulation refers to the dynamic simulation and prediction of the physical world in digital space. This requires not only the digital space to express the external image of the physical world but also the integration of physical laws into the digital space. The calculation and simulation layer should not only establish the digital model of the physical objects but also analyze and predict the future state of the physical transportation objects according to the real-time state and through the physical laws and mechanisms.

Therefore, the calculation and simulation layer includes two aspects: mining analysis and simulation prediction. The former is mainly to conduct deep-level mining on the primary model of the acquired data and create a higher-level information model. The latter mainly uses these information models to realize simulation calculation and predict the future according to the actual transportation conditions.

11.4.1.3 *Output application layer*

The core application of digital twin is mainly to assist decision-making and control. The digital twin technology can determine the current state of the transportation system based on the existing large amount of information in the output application layer and support users to use a proprietary analysis platform to find the best strategy to achieve their expected goals based on this state.

It can be extrapolated in different transportation development situations to verify whether the current possible transportation decisions are reasonable through the application of digital twin transportation system. Then, different intelligent optimization technologies are further used to improve the application iteratively, and the implementation strategy and decision-making of the control system affect the real transportation.

11.4.2 *3 Intelligent applications of digital twin in transportation*

The valuable value of digital twin in the field of transportation lies in the difference between it and simulation technology. Simulation technology focuses on "imitation", to reflect the characteristics and participation of physical transportation by mainly relying on the correct model, complete information, and hardware and software. On this basis, digital twin has the ability to analyze and optimize, and interact with physical transportation. Due to this, digital twin has a broader space for intelligent application in the field of transportation.

Digital twin has penetrated into different segments of the transportation industry and plays an increasingly important role driven by the technological growth of the industry and the continuous promotion of corporate entities. There are three main intelligent applications of digital twin in the field of transportation.

11.4.2.1 *Urban transportation construction and upgrading*

In the virtual space of digital twin manufacturing, urban transportation decision-makers and managers use the collected data to carry out model construction, and carry out simulation according to the actual needs of urban transportation, and automatically calculate the road network and trunk network density of urban transportation roads. By virtue of the intelligent application of digital twin, not only can the existing situation of transportation construction be analyzed and evaluated but also the solution can be automatically optimized to obtain the optimal effect, improve the transportation efficiency, and finally achieve the overall control of urban transportation.

11.4.2.2 *Application of smart expressway*

The intelligent construction and management of expressways is also an important application of digital twin. Currently, the demand for intelligent construction of expressways in China is concentrated in many aspects, such as the demand for toll collection at the entrance and exit of highways and the demand for overload control detection, such as the demand for road transportation and improvement of road conditions, as well as the demand for highway tunnels and bridge scenarios.

Some enterprises began to actively use digital twin technology to develop the all-weather transportation system of smart highway in response to these needs. The system has sensors at both ends of the vehicle and road to collect real-time data of the highway and process them through digital twin technology, and then display them on the visual display screen of the vehicle end and the management end, so as to help drivers and management personnel learn about the vehicle and road conditions and make early warning judgments in a timely manner.

11.4.2.3 *Collaborative application of vehicle and road*

The relevant agencies can build an automatic driving model based on the data collected from real-time perception and combined with the three-dimensional map through the application of digital twin technology. In such a model, the real transportation world can be restored to the virtual scene according to the original proportion, and intelligent algorithm training can be carried out in accordance with the operation rules of the real transportation, so as to improve the safety and operation efficiency of vehicle road collaboration and automatic driving.

The application of vehicle road collaboration is the focus of the application and construction of digital twin technology currently. Large Internet technology companies such as Baidu and Tencent are increasing investment in this field, with a view to achieving results as soon as possible.

Digital twin technology originated from the sense control system and rose due to the innovation of its application to transportation practice. It is believed that digital twin technology will continue to promote the integration of real and virtual transportation and provide a new model for the development of transportation, cities, and even the whole society.

11.4.3 *Digital cloning to improve the safety and efficiency of unmanned driving training*

The clonal transportation world created by digital twin technology is constantly integrating with the real transportation. Digital twin technology will penetrate into everyone's life, create a bridge to the new world, and make everyone's life travel more efficient and convenient in the near future. The application of digital twin technology to unmanned driving training will be the most promising way to achieve this goal as soon as possible.

China's transportation industry continued to develop and developed urban and rural transportation network was established after the founding of the People's Republic of China. Especially after the reform and opening up, driven by the economic take-off, the speed of population urbanization accelerated, the explosive growth of automobile ownership, and the pressure on the transportation system continued to increase. We must be committed to the development of unmanned driving with the aim of meeting people's growing travel needs and bringing a better travel experience. Exactly in the field of unmanned driving, enterprises can build digital twin space in a three-dimensional virtual environment based on real scene data, and complete the innovation breakthrough of this technology with low cost and high efficiency.

At present, digital twin-related research institutions are actively carrying out the application research of intelligent unmanned driving training system. In related projects, the real transportation system can be reproduced in digital space, creating various scenarios for unmanned driving vehicles. The virtual transportation space created by digital twin technology is a testing ground for unmanned driving vehicles in the "cloud". Digital twin technology can use a large number of real road data to train a transportation AI system, which can refer to the logic of human transportation behavior in the real world and generate transportation flow simulation model with high authenticity. At the bottom of the model, a high-precision three-dimensional map depicts the real urban environment.

The world constructed in this way will be able to form a variety of test scenarios to provide comprehensive support for the test of unmanned driving technology.

The transportation world created by digital twin technology is "virtual" rather than real, which can start and run a large number of test scenarios simultaneously with the support of cloud computing. These test scenarios can also be combined to form city level transportation scenarios. Enterprises can deploy thousands of autonomous vehicles to conduct continuous testing in such a large scenario, from which they can quickly find the transportation scenarios that are difficult to handle by the autonomous vehicle algorithm, and optimize the algorithm. Similarly, the R&D personnel can freely adjust the parameters, simulate and meet various types of unmanned driving needs, and evaluate the actual effect with the scenario at this level. What's more, the company can also conduct experiments with the application of technologies such as unmanned driving technology and test vehicle road collaboration, or even switch between different weather conditions, evaluate the impact of its changes on unmanned driving, and formulate corresponding plans.

11.4.4 *Hyper-spatio-temporal reduction, accurate analysis of the cause, and process of transportation accidents*

The digital twin technology can also be applied to the hyper-spatio-temporal reduction of the transportation accident situation environment and participants' tracks, and then observe the abnormal transportation accident process from different perspectives, and precisely analyze the specific causes of the accident, so as to take corresponding measures to reduce the occurrence of similar disasters.

As we all know, transportation accidents do not always occur under the cameras. For those transportation accidents in the blind area of monitoring, the traditional means of transportation supervision often have no response, and it is difficult to find the cause of regularity. Even for those transportation accidents that are monitored and recorded, they also cannot be recorded, analyzed, studied, and summarized from multiple perspectives under the traditional regulatory system. However, the corresponding technology can easily collect on-site data, simulate the accident scene, map it into a visual effect, model, and analyze from different perspectives such as drivers, pedestrians, and road conditions in the digital twin intelligent transportation system, so as to trace the source of the accident.

Digital twin technology can not only restore transportation accidents but also further develop its own potential to make up for the lack of human ability in the driving process. People mainly rely on visual observation of the transportation environment in the field of vision when driving vehicles, and rely on the vehicle itself tools such as turn signal, tail light, whistle, and external transportation signs to transmit information to each other. In addition, human attention is difficult to concentrate for a long time, and the response speed is insufficient, which is also exposed in transportation accidents. The Internet of Vehicles constructed by the vehicle sensor and the vehicle road collaborative technology will achieve sufficient information sharing among vehicles through the continuous improvement of the digital twin system, so as to solve the problem of personal misjudgment of drivers, and avoid dangerous operations such as drivers' fatigue, impulsion, and illegal driving.

When a transportation accident occurs, the digital twin system can enable all parties to recover the whole process of the accident by means of decomposition steps, explore each loophole in it, and find the implementation possibility of each prevention scheme. From the macro level, digital twin technology can also realize real-time detection of transportation flow based on the data shared by various types of roadside sensor equipment and users. For example, the digital twin can quickly identify the abnormal congestion caused by transportation accidents and other events, remind the transportation regulatory authorities to make decisions, and also remind drivers to avoid them.

11.4.5 *Parallel world simulation to achieve transportation deduction, prediction, and full management*

Research and development institutions simulate the urban transportation system by using digital twin technology, and optimize the management and control of the transportation system through deduction and prediction, which is an important contribution of digital twin technology to intelligent transportation.

11.4.5.1 *Control*

The digital twin system can transfer the real transportation information to the twin world, so as to obtain and control more information around the real transportation system to complete the closed-loop. Through this powerful capability of digital twin, the whole process control of the transportation system can be realized. The important value of this control ability is especially reflected in the specific scenarios, from which the digital twin system can easily find key problems and deal with them in a timely manner, while the traditional data analysis system is powerless. For example, the transportation condition near the city stadium does not cause congestion in the daily situation. Once a large-scale event is held, a large-scale transportation congestion will occur. The traditional means of transportation control can only be predicted in a broad range and cannot be accurately responded to. Through the twin digital platform, we can master the data information such as the holding time, number of participants and weather conditions of the concert in advance, and then control important information related to transportation.

11.4.5.2 *Deduction and forecast*

The biggest bottleneck in finding solutions to transportation problems is deduction and testing. Modifying the road environment in the real world, or conducting field tests, will cost a considerable amount of money, while the transportation scenes created by using digital twin technology can be easily achieved.

Using the digital twin system platform, research and development institutions can make different vehicles test on different roads, and can also use different combinations of conditions, such as lane line design, steering design, transportation light design, as well as transportation sign design, to test in the simulator and obtain the optimal solution of the deduction.

11.4.5.3 *Evaluation and optimization*

Transportation projects are increasingly complex and dynamic, and no means of transportation control can be perfect, which requires rapid and flexible assessment and optimization of the scheme. Transportation regulators and facilities operation departments need to understand the problems of existing strategies in a timely manner and explore the possibility of more “parallel world”. Especially at the meso and micro levels, the digital twin system can help them plan the allocation of transportation resources, integrate the updated data into more efficient models, and understand the efficiency level of the existing control measures.

The biggest advantage of the digital twin platform is that it can evaluate various transportation control methods in different "parallel world" by utilizing the large-scale parallel computing capacity of machines. Through the comparison and reference of the evaluation results, different plans are urged to learn from each other to obtain the most perfect solution.

Case: Guiyang pilot digital twin transportation system

Digital twin transportation system is a typical project of deep excavation and integration of big data in urban construction, which means that the accumulation of digital changes from quantity to quality. The formation of digital twin transportation system not only points out the advanced mode of intelligent transportation, but also constitutes a key part of digital city construction.

Starting from the beginning of 2021, Guiyang actively promotes the pilot project of digital twin transportation system. The digital twin transportation system of the city mainly uses the in-depth integration of the existing video monitoring resources in the transportation system and the MMW radar to make full use of the hologram perception of various transportation elements such as motor vehicles, non-motor vehicles and pedestrians in the transportation system, so as to fully create data integration and import the real transportation information into the digital version of the transportation simulation system.

For instance, the construction project of transportation intelligent facilities in Guiyang National Economic and Technological Development Zone has formed a transportation situation awareness system covering regional level, lane level, and node level for 23 urban roads in the economic development zone based on the deep integration of transportation data with intelligent algorithms. The system excavates and analyzes the real-time transportation situation data and can independently detect and distinguish sudden congestion on the road, and precisely correspond to the information management means according to the analysis of the causes of congestion. On this basis, transportation participants and managers can combine high-precision maps to locate, depict, and analyze individuals, motor vehicles, non-motor vehicles and roads, judge the significance of their activity tracks, comprehensively analyze the relationship between them, and finally solve various transportation problems.

The digital twin transportation system of Guiyang version still needs to be mature, but the system has shown the characteristics of accurate mapping, simulation, virtual reality interaction, intelligent intervention, etc. and is continuously promoting the innovative development of urban management methods. Guiyang is also expected to integrate the application of transportation law enforcement control, urban space analysis, urban construction simulation, urban feature portrait, urban emergency plan, etc., with the development of the system, to deeply reflect the law of urban development and support the implementation of precise policies.

Chapter Summary

The importance of digital twin technology plays an irreplaceable role from intelligent transportation to intelligent transportation, which has become a full consensus of the relevant industries around the world. The "what you see is what you get" feature of digital twin technology can ensure that intelligent transportation can obtain a more simulated operation system foundation. At the same time, digital twin will inevitably face faster, more stable, and more comprehensive industry requirements in the process of promoting the upgrade of intelligent transportation to intelligent transportation, which will in turn promote the rapid growth of digital twin technology.

Chapter 12

New Transportation Equipment Designed in Virtual Space and Applied in the Real World

Transportation development is an important indicator to measure the modernization of a country. The new transportation equipment can empower the traditional transportation infrastructure, promote the development of intelligent transportation, and improve the transportation development. Among them, simulation technology is an important means to develop new transportation equipment. Therefore, the traditional transportation equipment design process requires high-cost research and development matters, which can be quickly implemented in the virtual space and quickly applied to the real world.

12.1 Digital Twin: the Booster of Automobile Industry Development

As an important midstream manufacturing industry in the transportation industry, the automobile industry is a typical capital-intensive and technology-intensive industry that undertakes raw material enterprises such as steel and rubber upward and connects the demand side downward. Various sub-sectors in the automobile manufacturing industry present a pattern of monopoly competition due to the high concentration of the industry. With the overall development of the industry, digital twin technology has been actively involved in it, copying the physical objects of automobile factories in a digital way, simulating the behavior of automobile factories in the real environment, and starting to carry out virtual simulation of automobile design, process, manufacturing, and even the operation of the entire factory.

Digital twin, combined with automobile industry, can effectively improve the research and development capacity and production efficiency of automobile industry, reduce production loss, and predict the possibility of error in advance. The mutual mapping between virtual space and real factory will boost the automobile industry to develop in a rapid way.

12.1.1 *From digital twin to the trinity of vehicle, road, and cloud*

Digital twin, by means of digital means, can construct virtual things identical to real things in virtual space. The operator can control the virtual model, observe and study the changes of the virtual model in the digital space, so as to simulate and predict the specific performance of real things in the real environment. At present, the simulation level can not only replicate a city in a few hours but also clearly see the stone characters on the road. This is the specific performance of the trinity of vehicles, roads, and clouds.

12.1.1.1 *Virtual reality combination, two-way control of moped, road, and cloud coordination*

This city-level scale simulation capability can integrate multiple data, map, and learn from each other through machines, continuously understand the overall situation, and provide solutions for R&D and production of various industries. Digital twin technology can automatically collect multi-dimensional time and space and perception data of vehicles, roads, and clouds through the construction of an information management system for the interconnection of virtual models and intelligent devices especially in the field of transportation, so as to realize the virtual real interaction and two-way control between the virtual model and the real scene, and to achieve real-time scheduling and intelligent decision-making, and achieve the integration of vehicles, roads, and clouds.

The relevant perception data mainly comes from vehicles, roads, UAVs, and other equipment in the trinity collaboration. After data collection, the data is synchronized in real time through the cloud database, and then the intelligent Internet-connected vehicles and intelligent transportation are enabled. Digital twin not only drives the rapid development of information software but also continuously improves the technical requirements for hardware equipment via the coordination of vehicle, road, and cloud, especially the basic configuration of intelligent equipment and server.

12.1.1.2 *Specific application of digital twin technology in vehicle, road, and cloud*

The current development prospect of data twin is concentrated in the field of transportation, and its development direction is clear. The data twin platform can provide automatic driving simulation platform, cloud computing simulation platform, sensor, and intelligent computing platform in the environment test system, and multi-vehicle hybrid driving simulation system at the vehicle end. At the end of the road, the data twin platform can provide various scene editors and dynamic simulation components to achieve high-precision virtual scene construction. In the cloud, the data twin platform has various communication simulation components and testing systems, which can realize real-time state simulation and information synchronization.

In general, digital twin has the capability of providing technical support for the integrated management and control of vehicles, roads, and clouds and has great potential in the field of transportation. For example, the research and development, testing, and verification of autopilot are carried out by virtue of its high precision and high simulation characteristics. The data and experience obtained from this are conducive to the development of autopilot technology. The test platform built by digital twin is more diverse and efficient, and can meet the needs of customers in different scenarios in comparison with traditional research and development testing.

12.1.1.3 *Vehicle, road, and cloud collaboration to promote the development of intelligent Internet-connected vehicles*

Digitalization is the key to the transformation of many industries. As a key element to promote the development of smart vehicles, vehicle, road, and cloud collaboration has been highlighted in a number of policy documents issued around 2021. As the driving force of the development of intelligent connected vehicles, the vehicle, road, and cloud collaborative system can promote the research and development of intelligent connected vehicles, so as to form an independent, controllable, and complete industrial chain, and make the intelligent connected vehicles, intelligent transportation and smart city deeply integrated.

12.1.2 *Visualization + simulation + prediction, supporting the full element simulation of automobile research and development*

There are a few sub-industries in the automobile manufacturing industry, and the difference mainly lies in the product end, i.e. improving the market competitiveness through the difference of sub-divided products such as sedans, cross-country vehicles, and business vehicles. New vehicles are being updated and upgraded more and more rapidly in order to occupy the market share. This year, the configuration of million luxury cars has become the standard configuration of an entry-level car next year. Therefore, in addition to a small number of automobile manufacturers with high technical barriers, most automobile manufacturers are investing more and more in automobile research and development. The visualization, simulation, prediction, and other technologies provided by the digital twin technology can reduce the R&D costs for automobile manufacturers, save R&D time, and help automobile manufacturers to bring new products to the market more quickly.

The main steps of automobile research and development are market research, conceptual design, and engineering design. The R&D personnel will come up with a large number of overall automobile design schemes after understanding the needs of users, and enter the engineering design stage after the scheme is reviewed and approved. Engineering design is the most complex stage, which includes general layout design, body shape design, engine engineering design, chassis engineering design, and other processes. This stage involves a large number of parts and components testing in the traditional research and development process, which takes a long time. The technical support of digital twin will greatly reduce the experimental time at this stage, thus improving the overall research and development efficiency.

Based on the real production line, digital twin constructs factories in the virtual space and synchronizes with the real production line. Each equipment on the production line will be embodied in the virtual scene in the form of 3D modeling, and then through data modeling and simulation, the correspondence between the real production line and the virtual production line will be realized. Similarly, digital twin can also build different application scenarios to facilitate sample vehicle simulation tests.

Digital twin can not only describe the specific characteristics of real things but also accurately reflect the internal changes of real things. For example, the strength and durability of vehicles in different environments are different, and the changes of various material parts in different environments can be fully reflected through digital twin technology.

The R&D personnel can collect real things data and also drive the virtual space equipment to perform the same action, so as to realize the real-time linkage between the real environment and the virtual space. In the virtual space, all things are visual. The R&D personnel only need to observe the production line or test scene in the virtual space in the control room to understand the real-time R&D progress status of the sample vehicle.

In addition, with the 3D visualization effect, the driving status of the sample vehicle and the operating status of the parts can also be accurately predicted. The digital system will intuitively display the defective parts in the form of 3D visualization, which will facilitate the R&D personnel to quickly identify hidden dangers, solve the possible problems in advance, and improve the R&D efficiency.

The research and development process is presented in virtual space in digital form, and simulation experiments are carried out by virtue of the characteristics of digital twin visualization and simulation, which can predict the possible problems in the research and development process and formulate solutions in advance. The whole process of automobile research and development is more intuitive and provides strong support for the simulation of different elements through the combination of visualization, simulation, and prediction.

12.1.3 *Virtual industry chain of automobile industry: full simulation of R&D, production, and testing*

Automobile production needs to go through the process of research and development, production, and testing, and simulation test is an indispensable step for the smooth progress of the above process. With the maturity of digital twin technology, high-precision digital twin simulation of industrial chain is widely used in the automobile industry.

The core of simulation technology is the establishment of digital model. In the automobile industry, the traditional modeling is mainly based on physical methods. Through data driven, various modeling behaviors and modeling methods are comprehensively applied. Digital twin technology modeling can not only depict the external shape of products but also reflect the internal changes of products through data forms and even predict the future direction of products in comparison with traditional product modeling.

Data and algorithms are closely related to modeling. Data is the basis of the entire virtual space operation. On the one hand, it comes from the real-time feedback of various high-precision sensors, and on the other hand, it comes from the historical experience of keeping the system running well. The algorithm is the core factor to ensure accurate matching between the virtual space and the real scene. In order to achieve higher precision and higher density simulation, the algorithm needs in-depth learning process, continuous analysis and processing of historical data, and continuous establishment of models reflecting the development logic of the real world.

The simulation technology is the most efficient in the three links of research and development, production, and testing in the virtual industry chain of the automobile industry. Its essence is that the data generated from the three links has established the internal processing logic, which is commonly known as the model. Therefore, it will make a judgment based on the internal logic of the system when anything in the virtual industry chain receives the data transmitted by the sensors, so as to provide a decision-making basis for the actual research and development, production, and testing. Of course, such internal logic is not unchanged. The digital world of virtual industrial chain will be more consistent with the physical world of the actual industrial chain under the continuous in-depth learning algorithm. Only in this way, digital twin can guarantee the high-precision simulation of digital space and physical world. The following is the specific application of digital twin in the three stages of R&D, production, and testing.

12.1.3.1 *Research and development stage*

With the aid of digital twin technology, R&D personnel can model, verify, and optimize their product design in the virtual space. As the virtual space highly matches the data of the real world, the 3D visual results obtained through the virtual space can achieve rapid and low-cost product iteration, which is caused by relatively low iteration cost time and human and material resources, and the R&D personnel can carry out higher precision tests to develop products with higher performance level.

At present, many famous automobile brands have achieved substantial results by using digital twin technology. In particular, the three companies, BMW, Tesla, and TOYOTA, have developed a lot of vehicle components through 3D simulation software. These vehicle components can effectively improve the linearity of vehicle motion flow and reduce air resistance during driving.

12.1.3.2 *Production stage*

The staff only need to observe the workshop operation status of the virtual space in the control room to understand the actual production line working status with the aid of the digital twin platform. Workers can grasp the actual production situation on site through the screen without visiting the site. The digital twin platform can display all its states in the form of 3D visualization, whether it is personnel management, equipment management, operation and maintenance management, project management, or energy management, and enterprise operators can optimize production plans based on these data, so as to improve the overall production efficiency and realize flexible production.

Digital twin can not only integrate the production and consumption process of the production line in the virtual space but also present it in the form of comprehensive analysis report, enabling the decision-maker to obtain all information intuitively. With the change of production plan and continuous self-learning of algorithm, digital twin technology can predict the production capacity of one week, one month, or even longer and help operators to achieve production scheduling management.

12.1.3.3 *Test phase*

The scene editor and various dynamic simulation components in the digital twin platform can realize product testing in different scenarios and under different states. This virtual space simulation technology has brought great optimization to the test of automobile products. The tester can edit various scenarios anytime and anywhere and observe the actual performance of the experimental objects, especially in extreme scenarios. In contrast, traditional testing methods are constrained to a variety of scenarios or states, and the results are not universal. In order to get the most objective results, staff often need to find or even build a variety of scenarios, which requires a long time cycle. On the contrary, the most intuitive advantage of the testing process is that the development node moves forward, the development time is shortened, and the development cost is reduced under the support of digital twin technology.

Digital twin technology can make the research and development, production, and testing of the automobile industry more convenient. Among them, the visualization simulation technology plays a core role. With the continuous upgrading of simulation technology under the in-depth learning of algorithm, the resulting virtual industry chain will also provide more efficient solutions for the automobile industry.

12.2 Digital Twin Aircraft Popularizing Digital Twin Airport

The digital twin aircraft can receive the operation status information of each component and each structure of the real aircraft in the virtual space. When there is any problem with the aircraft, the digital twin system can be perceived in advance, so as to avoid the occurrence of various accidents. Digital twin aircraft can not only collect and analyze the real data of aircraft, but also serve as the basis of digital twin airport, providing strong support for the construction of intelligent airport.

12.2.1 *Boeing aircraft ignited digital twin*

From the end of 2018 to the beginning of 2019, within less than half a year, there were two major air accidents of Boeing aircraft. A number of airlines around the world announced the suspension of flights of this aircraft, including China Southern Airlines and China Eastern Airlines. This focus event has aroused the concern and discussion of the whole society on the safety of aircraft flight. As a widely used means of transportation, aircraft freedom and high efficiency are incomparable with other means of transportation. However, it is also limited by the environment, and people hope to improve the safety of flight through scientific and technological means. Subsequently, the digital twin technology related to aircraft received great attention. Similarly, the purpose of digital twin technology was initially proposed to reduce the probability of flight accidents.

Digital twin was first proposed as a concept beyond reality, which aims to judge the safety of flight by conducting virtual space simulation flight based on the data provided by real aircraft. In this scenario, twin refers to two identical sets of data, one of which is derived from real flight parameters, such as flight height, visibility, remaining fuel, number of passengers, as well as channel settings. The other group of data is generated by transferring the above data to the corresponding software, and through specific algorithm analysis, it can provide parameters such as the safety of each flight. In the actual development and application, digital twin technology involves far more than two sets of parameters, including high-precision sensors, artificial intelligence equipment, and computing chips, in the aircraft industry. These infrastructures need complex frontier knowledge and technology as technical support.

The flight of aircraft generates a large amount of data, and the safety of aircraft is related to the safety of each passenger's life. This is why digital twin technology is so important. The National Aeronautics and Space Administration has published a paper on digital twin, pointing out that digital twin technology will become one of the key factors driving the development of future aircraft. With high precision simulation as the core, digital twin technology can map each other in the real world and virtual space, and predict the possibility of error in advance. Since then, digital twin has taken on the major responsibility of ensuring people's livelihood and transportation safety.

Although digital twin requires forward-looking scientific and technological support, users are not deterred by the difficulty of its use. After recent years of practice, the current digital twin technology is not limited to the aircraft industry but is widely used in all walks of life. At present, the visibility, predictability, and interoperability of digital twin are not comparable to other technologies. Decision-making is often more effective and accurate and has been recognized by the majority of users through digital twin technology.

12.2.2 *Digital twin gives birth to intelligent airport*

Airport is an important part of traditional transportation infrastructure. Its development not only is related to people's livelihood but also promotes rapid economic growth. Therefore, major countries in the world attach great importance to the construction of intelligent airports and take scientific and technological innovation and intelligent transformation as the main direction of intelligent airport construction. The following are some key elements of digital twin technology to support the intelligent transformation of airports.

12.2.2.1 *Transformation road of intelligent airport*

From 2020, under the impact of the new crown pneumonia epidemic, the airport as an aviation hub will be greatly affected. Under the severe situation, the airport model constructed by the digital twin through the virtual space and the real airport are mapped to each other, and the operation track of the real airport business cycle can be predicted. With the help of digital twin technology, airports can carry out various innovative attempts and reforms to test or predict the feasibility of decision-making, thereby greatly reducing the cost of trial and error in real airports. Similarly, with the help of deep learning of the algorithm, airport management decision-making will be more efficient and feasible.

The information construction of airports in China has undergone tremendous changes in the past years and is gradually moving toward digitalization, IOT, and intellectualization. The wide application of digital twin has provided core technical support for the construction of intelligent airports. Today, people often point to “two airports” when they talk about airport construction, one is an increasingly new modern civil airport, and the other is a digital twin intelligent airport. The technology department collects the operational status, geographical space, facilities and equipment, business processes, production factors, and other data of the real airport through sensors, and directly imports these data into the airport in virtual space, so as to form an airport highly matching with reality. Staff can see real-time flight, equipment, ground handling, and other data in this virtual space of the airport. In addition, these data will be presented in the form of 3D visualization, which is conducive to the airport to reduce management costs and improve operational efficiency.

12.2.2.2 *Construction content of intelligent airport*

The construction of the airport under the digital twin technology mainly includes three aspects: data input, data perception, and data decision-making.

Among them, data input refers to the implementation of high-precision physical simulation by arranging sensors inside and outside the airport, on the ground, and combining the geographic information system and simulation technology to import all the data of the real airport into the airport in virtual space. Data perception refers to the use of Internet of Things to sense the operational status of key facilities and equipment due to the large amount of airport data. Data decision-making refers to the system can make scientific and accurate decisions according to the actual needs by receiving all the above data. The algorithm will continue to conduct in-depth learning in the airport of virtual space, and the decision-making direction provided by the virtual airport will become more and more scientific and feasible.

The construction of intelligent airport is essentially to empower the decision-making level, management level, operation level, and executive level of the airport, and its ultimate purpose is to function in three aspects, namely security, operation efficiency, and service experience. The intelligent airport can operate efficiently. In the final analysis, the virtual airport can quickly and effectively receive data information in the real scene, such as passenger data, flight data, as well as equipment status data.

12.2.2.3 *Construction object of intelligent airport*

In contrast, there are limitations in the traditional airport management system, such as information non-interoperability, data difficult to understand, as well as scene segmentation, and these problems are solved in the intelligent airport under the digital twin. For example, the airport in virtual space can feed back the real geographical coordinates of the aircraft in real time when the aircraft takes off and lands, open up the data of various departments of the airport, and improve the safety of the aircraft taking off and landing. In addition, the ground ferry, tractor, and other vehicles are also the targets of digital twin technology in addition to the relevant data of aircraft flight. Any movement of them will be transmitted to the virtual airport through sensors, and the staff can understand the working status of these vehicles at any time, so as to make more flexible linkage management. What's more, the ground and terminal are also the key construction objects of the intelligent airport. Flexible control of different airport transportation can be achieved through the introduction of heat and scene simulation technology.

12.2.3 *Development direction of aerospace equipment: solve the present and face the future*

Digital twin technology is based on modeling, sensor, Internet of Things, data analysis, decision-making, and other systems. Through real-time monitoring and online optimization, digital twin can accurately describe and forecast the real scene. Compared with other simulation models, digital twin has the characteristics of complete visualization and dynamic optimization.

Due to the above characteristics, digital twin has shown great potential in the development of aerospace equipment. The development efficiency of aviation equipment data analysis and business management has been significantly improved by using digital twin technology. At present, digital twin technology can provide a visual solution path based on historical experience and existing data when aerospace equipment researchers encounter problems. Relevant personnel only need to follow this path, which has a positive role in promoting the development of aerospace equipment.

The research and development of aerospace equipment involves high-precision technology; therefore, it is a huge and long project, and its program iteration, product manufacturing, sample, and finished product trial often require a huge cost. Facing such a difficult problem, digital twin gives researchers confidence. Its main direction is to adopt digital strategy and help the design, research and development, manufacturing, assembly, and operation and maintenance of aerospace equipment through high-precision simulation. The following is the specific application of high-precision simulation in different stages of aerospace equipment.

12.2.3.1 *Design and development stage*

The twin of aerospace equipment is established in the virtual space through 3D modeling, and then the testing and verification of various scenes and states are directly carried out, and the defects found are timely modified. This ensures that a complete set of tests and verifications has been completed before the physical objects are processed, avoiding the high cost and long cycle brought by repeated iterations.

12.2.3.2 *Manufacturing and assembly stage*

The staff only need to look at the virtual production line in the control room to understand the real production situation. The system can integrate the resources of all production lines, especially the personnel management and equipment operation status, which can be presented in a 3D visual manner, so as to facilitate the decision-maker to allocate resources and make reasonable production scheduling.

12.2.3.3 *Operation and maintenance stage*

Digital twin can visualize the quantity, status, life, and other parameters of all aerospace equipment and combine with intelligent algorithm to estimate the loss and life of equipment. The staff can maintain or replace the equipment according to the forecast report, so as to maintain the best performance of the aerospace equipment in the use phase.

The application of digital twin in aerospace equipment has solved a large number of problems, which proves that the prospect of digital twin in aerospace industry is particularly promising. Of course, aerospace equipment involves multi-disciplinary, multi-departmental, and multi-stage collaboration, and the digital twin itself is based on a variety of technologies, such as digital modeling, data analysis, simulation, and prediction. The comprehensive combination and application of the two require the joint efforts of staff in various industries. It is reasonable to believe that the development of aerospace equipment will have more possibilities with the continuous expansion of relevant theoretical technology and the continuous upgrading of twin model and simulation technology.

12.3 Digital Twin Model Activating High-Quality Development of Rail Transit

Rail transit is an indispensable way for people to travel. Rail transit has a high density of passenger flow, and its transportation place also has its particularity. It is not only necessary to ensure the personal safety of passengers, but also to ensure the comfort of passengers. The real-time monitoring and monitoring of rail transit is particularly important. Through the digital model, the linkage between physical reality and virtual space can be realized, enabling high-quality development of rail transit.

12.3.1 *How does the digital twin track system solve the security problem of 17.59 billion passengers*

Digital rail transit features with large scale and dense flow of people. Faced with such a huge dynamic database, the traditional rail transit management system has great limitations and cannot establish a perfect real-time security mechanism. For example, it is unable to present all the data in a visual manner, and it is unable to make intelligent analysis and decision on all scenarios. In view of the above problems, the combination of digital twin technology and rail transit system is on the horizon, and the digital twin rail system came into being.

Digital twin track system, mainly for the security of rail transit, flow control, environmental control, and other needs, analyzes the characteristics of physical objects. With the help of various sensors, the R&D institutions collect various data in the design and construction stage, operation and maintenance stage, and historical operation. Through high-precision 3D modeling, the interactive mapping between the physical entities of rail transit and the virtual space is realized, and the corresponding transportation management is ultimately served. The reason why the digital twin track system can solve the security problem of 10 billion people depends on the advantages of its architecture and the universality of information integration, and introduces its architecture and system advantages.

12.3.1.1 *Overall structure of digital twin track system*

The overall architecture of the digital twin track system is divided into perception layer, convergence layer, data layer, service layer, application layer, and display layer. The perception layer collects information through security inspection equipment, monitoring equipment, face recognition, mobile phones, and other devices, and then all information enters the data aggregation layer of each station and finally summarizes to the rail transit security big data center. In these stages, the data is only transmitted to each other, and there is no real data analysis. Until the service layer, the data has been used. The most intuitive application scenario is the twin service of 3D geographic information, followed by the integrated communication service, data analysis service, data sharing service, etc. These services are mainly applied in the four aspects of service command, industry supervision, intelligence early warning, and police coordination, which together provide a strong guarantee for the security of rail transit. Finally, all the data will be presented in a visual form, and the staff can understand all the data of rail transit through the large screen display system, desktop browser, or even mobile terminals and make real-time work scheduling to ensure the smooth implementation of rail transit security decisions.

12.3.1.2 *Advantages of digital twin rail transit system*

The advantage of digital twin rail transit system is three-dimensional visualization, which integrates the information of rail transit from construction to operation and maintenance stage and provides practical support for predictive decision-making. Therefore, even if it is more crowded and large-scale, digital twin rail transit can realize the digitalization of people, things, and other elements, and through the process of data collection, data transmission, data storage, and data processing, realize the precise mapping between the physical rail transit and twin rail transit, and assist the decision-makers to make better decisions.

12.3.2 *Construction of real twin bases for rail transit*

The real twin base of rail transit is built on a three-dimensional geographic information platform, on which the staff can observe the operation of the rail network and a single station from any perspective, and even realize hierarchical browsing within the station.

The technical support of the real twin base of rail transit comes from the integration of digital twin and remote sensing images, electronic maps, and spatial data technology. The base is matched by 3D video map scene, and the technology department constructs scene data index in the virtual space. The basis of the twin bases of the real scene is data. The more comprehensive the data acquisition is, the closer the virtual space rail transit is to reality. Specifically, the source of base data is from four aspects in the digital twin track system, namely GIS platform, video platform, monitoring platform, and historical operation data of virtual rail transit. Among them, the information obtained by GIS platform is city map, transportation map, and key location identification. The real-time monitoring video and playback video obtained by the video platform are important information for algorithm learning in virtual rail transit. The information obtained by the monitoring platform is real-time event monitoring and early warning data. In addition, the historical operation data of virtual rail transit also plays an important role in reference.

The business scenarios that can be supported by the real twin base of digital twin rail transit are very rich. The widely used three-dimensional geographical information scene browsing of rail stations, three-dimensional scene linkage movement of video monitoring probes, three-dimensional scene visual scheduling and command of police business, three-dimensional scene mastering and application of monitoring and early warning, and three-dimensional geographical scene virtual and real integration of key station areas are three dimensions. The above business scenarios can basically achieve the digital management of physical rail transit.

Based on the real twin base of digital twin rail transit, rail transit can achieve its maximum value. Plan, design, and build rail network through the use of digital twin rail transit and three-dimensional geographic information platform, and use digital twin rail transit system during the operation period; thus, the management level of operators can be greatly improved, and passengers can directly experience better comfort and safer ride protection.

12.3.3 *Promote the integration of the virtual and the real in rail transit*

Digital twin technology ensures that staff can switch between physical space and digital space at will according to their needs. Users can not only watch the physical space in the digital space but also enter the digital space through the physical space. With the development of sensors, 3D modeling, Internet of Things, database, artificial intelligence, and machine learning, data can be exchanged in real time. In addition, the algorithm of virtual space can also use these data for deep learning, thus promoting the efficient use of data resources.

In the digital twin rail transit system, physical entities and virtual spaces are more fully integrated. For example, the traditional video monitoring can only see partial pictures, while in the digital twin rail transit system, the monitoring pictures can be combined with the virtual model, with the help of deep learning algorithm, to analyze and collect the behavior images of passengers. In addition, based on historical passenger data and three-dimensional virtual model, different algorithms can be used to configure air conditioning system in the virtual space to analyze the site comfort in the study of air conditioning temperature and passenger comfort, so as to save energy and reduce energy consumption.

The application of digital twin technology has promoted the integration of virtual and real in rail transit. After data collection, data transmission, and data analysis, the digital twin track system can make a visual prediction, which is convenient for managers to accurately measure and fine manage.

At present, the digital twin rail transit system has been able to achieve real-time twin in the whole region. Staff can sense all the data in the rail transit through various display devices in a three-dimensional way. In case of emergencies, the digital twin rail transit system can achieve coordinated linkage in the whole region, and dispatch resources near to ensure the safe operation of the rail transit.

12.3.4 *Construction of three-dimensional twin model for rail transit*

The efficient operation of digital twin track system relies on three-dimensional virtual model and real-time data transmission. Both data analysis and decision prediction of the latter must also be based on the three-dimensional virtual model.

The purpose of the three-dimensional modeling of rail transit system is to build a three-dimensional twin model. Therefore, the one-to-one mapping of physical entities in rail transit is required. The most basic physical entity of the rail transit three-dimensional twin model is the building part, including walls, ceilings, and drainage pipes. Then, there are some physical entities serving the application target, such as monitoring equipment, security inspection machine, and import and export gate. The extensive application of the three-dimensional twin model in rail transit depends on the following characteristics.

12.3.4.1 *Intuitive image*

The primary feature of the three-dimensional twin model is its intuitive image. The traditional regional plan display information is limited and not intuitive, and the three-dimensional twin model focuses on the pursuit of authenticity at the beginning of modeling, so as to give full play to the role of digital twin in the rail transit system, and effectively enhance the practicability and versatility of the three-dimensional twin model.

12.3.4.2 *Real-time feedback on physical changes*

The visual three-dimensional twin model can bring great convenience to the staff. The ultimate purpose of building the three-dimensional twin model is to match with the existing rail transit operation and maintenance experience. Through continuous iteration of the data collected by the physical entity, the physical entity changes in various properties, states, logic rules, and other data under external conditions are gradually fed back to the three-dimensional twin model to achieve information integration in specific scenarios according to the needs of research and application.

12.3.4.3 *Reduce storage costs and maintenance costs*

The three-dimensional twin model receives and uses data throughout the whole life cycle of rail transit. Whether the construction starts or the subsequent operation and maintenance, the data can maintain consistency and integrity, which is different from the traditional mode of separate storage. The three-dimensional twin model greatly reduces the storage cost and maintenance cost in terms of data.

While the three-dimensional twin model continues to receive information, it is also constantly optimizing itself. When the continuously optimized three-dimensional twin model is further applied to the theoretical and practical research of the industry, the track three-dimensional twin model can promote the development of various research and guide the industry to make further changes due to its own storage of many experience data in planning, optimization, decision-making, and other aspects.

12.4 Digital Twin: Boosting High-Quality Development of Water Conservancy Modernization

According to the widespread application of digital twin, it has been proven that the digital twin is no longer just a technology but the beginning of a new development model of the transportation industry, which can also be said to provide a new path for the transformation of the traditional transportation industry. The combination of digital twin and water conservancy industry will also form an important driving force for the reform of water conservancy industry as a carrier of digitalization and intelligence.

12.4.1 *Beijing 51 WORLD digital twin technology is fully implemented in the water industry line*

Water treatment includes river, lake, reservoir pipe network, and other scenarios, which involve massive data processing and complex quality system, and its simulation is very difficult.

In 2020, Beijing 51 WORLD, together with top partners in the water industry, carried out water system modeling in the virtual space and visualized the data and information of natural water systems and urban pipe networks based on the full element scenario digital twin technology, overcoming many difficulties, bringing new management methods to the water industry.

At present, Beijing 51 WORLD water digital twin industry line has been continuously established in many provinces and cities in China and overseas cities, mainly applied in three aspects of water supply, disaster resistance, and ecological governance. The water management department can intuitively obtain water monitoring and ecological and environmental protection data through the two-way mapping between the web end and the full element three-dimensional scene, which is convenient for decision-making and simulation demonstration. The following is the specific application of digital twin technology in three aspects of water supply, disaster resistance, and ecological governance.

12.4.1.1 *Water supply*

In the process of solving the problem of urban water supply and drainage, the digital twin industry line of water affairs mainly starts from three aspects, i.e. the division of water supply areas, the layout of water plants, and the pipe network system. The division of water supply area is to restore all areas that can supply water, such as water source, water pump station, and water supply point, by virtue of the existing virtual city map. In this way, when water supply and drainage are needed in the city, the optimal route or location can be found immediately to facilitate the work scheduling of the water management part. The layout of water plants is similar to the principle of water supply area division, but the layout of water plants is to restore the location of water plants in the form of POI points. In addition, the water digital twin system can also display all parameters related to water sources, such as turbidity, total bacteria, and other data. In some key water source protection areas, UAVs can be equipped to update water source data in real time to ensure the safety of urban water use. The pipeline network system combines real urban data to restore the pipeline network location trend. Once there is a problem with the local pipe network, the management department can choose whether to repair or replace the line according to the prompt of the pipe network system, so as to restore the smooth pipe network more quickly.

12.4.1.2 *Flood control and disaster reduction*

Flood control and disaster reduction is another important application direction of water digital twin industry line. In order to minimize the disaster, the water digital twin system can predict the disaster area in real time according to the flood data and the urban virtual water pipeline network structure, and calculate the flood evacuation route intelligently. "Whether a city is advanced depends on its drainage system", and the application of water digital twin will provide greater technical support for the construction of smart city.

12.4.1.3 *Ecological governance*

At present, the public values more and more about the treatment of water pollution and water quality safety as the country accelerates the construction of ecological civilization. The role of water twin technology in this area is also very prominent. Water source data and detection are the basis of digital twin of water affairs. The water affairs system in virtual space collects a large amount of water network information and carries out real-time monitoring. Therefore, in the future, relying on intelligent models can even achieve automated governance. It is a representation of the construction of water conservancy information infrastructure system and water conservancy network security system in China ushering in faster development with the wide application of digital twin technology in the water conservancy industry.

12.4.2 *The water and the sky are the same, and the digital twin technology opens up a new path for port and shipping management*

Water transportation plays an important role in both domestic and foreign trade transportation, and the management level of ports and shipping directly affects the efficiency of water transportation.

Port and waterway management not only involves ship management but also includes port infrastructure, inland waterway planning, coordination of transportation of key materials, and other aspects. The traditional port and shipping management mostly refers to historical experience to maintain the order of operation due to its wide coverage. However, the traditional management methods are inadequate with the improvement of the national and local requirements for port management. Therefore, it is necessary to learn from the experience of water conservancy management and introduce digital twin technology.

Digital twin technology can digitalize the elements of port and shipping management. Through the modeling technology, the same port and shipping model can be built in the virtual space, and its most prominent feature is visualization. Whether it is the port infrastructure or the administrative region of port and shipping management, it is clear at a glance in the virtual space. In the virtual space, all ships are important sources of data. The system can analyze according to the collected data, optimize the planning of inland waterway, and make the ships more efficient when entering and leaving the port. In addition, the digital twin technology superimposes the mapping of digital space into the displayed port management based on the complete collected data, forming a new route of port management with the combination of the virtual and the real and the twin interaction.

The application scenarios of digital twin technology in port and shipping management will become more and more abundant. Digital twin will be committed to digitalizing the physical projects related to the port and shipping, and calculate, analyze, and optimize the digital port and shipping management information in the virtual space. Its data display is more intuitive, operation is more convenient, and management efficiency is higher compared with the traditional management methods.

Case study: Digital twin technology helps Italy recover road system quickly

Italferr is a large international infrastructure project construction company in Italy, mainly engaged in the construction of ordinary railway and high-speed railway, urban and road transportation and port facilities.

In August 2018, the collapse of the Morandi Bridge destroyed many buildings and caused a large number of casualties, raising concerns about the condition of bridges in Europe, since there are studies showing that due to erosion and deterioration of structural performance, many bridges have similar safety hazards. Italferr, a leading company in infrastructure construction, was commissioned by Pergenova Consortium to build a 200-meter-long viaduct section to replace the original function of the Morandi Bridge. However, in order to restore the road to normal transportation status as soon as possible, the construction period has been seriously shortened, and the new bridge deck standards have to be met.

Italferr realized that if traditional processes were used to carry out construction, there would be no way to complete the tasks on time at all. Since the construction period is already tight, the new standard still needs to be designed and tested, which takes a lot of time.

After comprehensive consideration, Italferr has created a digital model of the viaduct section, in which the design team can carry out simulation collaboration and testing at any time. This greatly saves time. In addition, before the commencement of construction, simulation construction can be carried out to understand the dynamics of each construction stage and the progress of the project. In the formal construction, some potential difficulties have been eliminated. In the end, Italferr completed the project more than expected and the road system in Italy was soon restored.

Italferr uses digital twin to speed up design decisions, improve design quality, reduce design costs, minimize uncertainty risks, and improve data accuracy and project team collaboration efficiency. Using digital twin technology, Italferr has been able to design durable structures in virtual spaces by collecting existing bridge data, which has helped to rebuild the technological reputation of bridges in Europe and greatly restored public confidence.

Chapter Summary

The application of digital twin technology in the field of transportation has created more possibilities for urban transportation management. In the digital model, through the construction of urban three-dimensional space model and the visualization of dynamic operation data, the transportation management department can not only achieve real-time monitoring, but also carry out dynamic intervention.

In addition to the use of transportation management, digital twin is also playing a huge role in urban ecological governance, which indicates that smart city construction is constantly moving from laboratory to reality, from mapping to manipulation, and from small scenarios to broad industrial and urban governance space.

Chapter 13

Empower the Future of the City and Give Higher Wisdom to Transportation

The upgrading and transformation of urban transportation industry has been put on the agenda with the rapid development of digital and information technology. The application of intelligent technology based on digital twin has become more and more extensive, and the urban transportation system has ushered in digital transformation.

13.1 Digital Twin Enabling Urban Transportation from Intelligentized to Intelligent

Virtual figures are constantly integrated with reality in the development of urban transportation system. As the cutting-edge technology of intelligent transportation, the key task of digital twin is how to find the appropriate application scenarios in combination with the current practical needs, so as to enable urban transportation more efficiently.

In recent years, China's digital construction has been continuously strengthened under the guidance of the strategy of "digital power" put forward by the 19th National Congress of the Communist Party of China, and digital twin technology has been effectively promoted in this trend and promoted China's urban transportation from intelligentized to intelligent.

For example, in February 2022, the National Development and Reform Commission, the Cyberspace Administration of the CPC Central Committee, the Ministry of Industry and Information Technology, and the National Energy Administration jointly issued a note, agreeing to start the construction of national computing hub nodes in 8 places, including Beijing–Tianjin–Hebei, the Yangtze River Delta, Guangdong–Hong Kong–Macao Bay area, Chengdu–Chongqing, Inner Mongolia, Guizhou, Gansu, and Ningxia, and planning 10 national data center clusters.

This project is known as “east digital and west computing”, with the main goal of enabling Chinese cities to accelerate the pace of intelligent development. This project is another national super project after “South-to-North Water Diversion Project” and “West-to-East Power Transmission Project” in China.

In fact, before the official commencement of the “east digital and west computing” project, many cities in China had already started to carry out the construction of related facilities. In 2021, more than 20 cities in China rushed to build urban intelligent computing centers. By the beginning of 2022, the number of smart computing centers in China had reached 27, of which 12 are located in the eight hubs.

The intelligent computing center not only brings the improvement of computing power to the city but also accelerates the construction of the city’s smart brain. According to the statistics of China’s authoritative departments, the calculation power of each intelligent computing center in China starts from 100P (1P represents 1,000,000 billion calculations per second); this unimaginable computing capacity is enough to support all kinds of smart construction in China’s major cities to complete various intelligent construction. The empowerment of digital twin by intelligent computing center will play a greater role in urban transportation management and regional transportation development, and even the upgrading of China’s comprehensive three-dimensional transportation network has injected strong momentum.

13.1.1 *Digital twin: Intelligent brain formed in transportation system*

The implementation and application of smart city governance have greatly improved urban transportation, all of which are inseparable from the digital twin. Digital twin replicates the complex transportation network system in reality by using Internet of Things, 3D modeling, cloud computing, and other technologies and constructs the digital twin transportation system through real-time data transmission, so as to optimize urban transportation and ensure transportation safety.

With the aid of visualization, the twin transportation system of the city is easy to perceive and use. Managers can systematically see the travel path of millions of vehicles in the whole city, from where to start, through which intersections, and finally where to park. In addition, all transportation facilities, road networks, and parking lots in the whole city are objectively sensors of the digital twin transportation system. Managers can restore the real transportation operation status of the whole city on the screen through the real-time data transmission of the Internet of Things and the Internet.

From the macro level, the research on the transportation system operation of a large number of vehicles provides valuable data. Facing the ever-changing urban transportation situation, the algorithm can dynamically optimize the solution of urban transportation problems through in-depth learning based on historical data and real-time data. For example, it optimizes the signal light timing scheme according to the transportation flow and reduces the red light waiting time. In the face of emergencies or holidays, urban managers can also implement corresponding time limit, current limit, transportation limit, and other measures through historical data and vehicle travel rules to improve road transportation efficiency.

The application of digital twin technology in the transportation system has reshaped the operation mode of urban transportation and promoted the intelligent development of urban transportation management. This intelligent direction has two levels of significance in the technical system. On the one hand, real physical entities are allowed to form data in the virtual space through sensors, and then restore the state of the entire urban transportation operation, and then present it through the visual effect, which is called "cognitive intelligence". On the other hand, it is "intelligent in decision-making". The twin transportation system of virtual space can conduct comprehensive perception, intelligent analysis, accurate research and judgment, collaborative command, and emergency response to the real transportation, producing important strategic significance to alleviate transportation congestion, reduce transportation accidents, and reduce transportation pollution. Therefore, the empowerment of digital twin in transportation system provides strong support for the upgrade of intelligent transportation system and services.

Currently, the digital twin transportation system is piloted in major cities in a hierarchical and sub-regional manner to meet the penetration of different urban needs and help build urban intelligent brain. In addition to the use of digital twin in the transportation system, it can also be used in the field of social security prevention and control. For example, when carrying out epidemic prevention and control, digital twin can carry out key personnel management and control, trace the operation track of personnel or vehicles, intelligently analyze risk areas, and create a new pattern of urban grassroots management.

13.1.2 *Multi-dimensional interpretation of the difference between intelligent transportation and intelligent transportation*

Intelligent transportation and intelligent transportation are different stages of the twin digital transportation system. Although the concepts of the two are similar, a clear distinction between the two is conducive to a better understanding of the hierarchical structure and technical connotation of the digital twin technology in the transportation system, to prevent difficulties in promoting due to the excessive demand for preliminary development, and to avoid the lack of forward-looking and limited iteration of the system. In a narrow sense, intelligent transportation is biased toward the iteration of technology and pursues more high-tech applications. Intelligent transportation is an upgraded version of intelligent transportation, which is closely related to "people" and solves the problem of human–machine system coordination. In addition, there are the following differences between the two.

13.1.2.1 *Technical application*

Intelligent transportation integrates information technology, sensor technology, communication technology, and other modern electronic technologies. On this basis, intelligent transportation integrates high-tech IT technologies such as Internet of Things, cloud computing, and big data.

13.1.2.2 *Data processing*

Intelligent transportation focuses on the integration of various transportation application information and emphasizes the real-time application of data. Intelligent transportation is inclined to build digital models and data decision-making, emphasizing the overall application scheme.

13.1.2.3 *Analysis capability*

Intelligent transportation can sense the environmental state through various sensors and automatically perform corresponding operations relying on the collected variable data in terms of analysis capability. Intelligent transportation is more advanced. On the original basis, it adds the "human brain" simulation ability, which can learn from the existing information and make better decisions by using the unique learning ability.

The external factors of urban transportation are not unchanged, and their management countermeasures should also be continuously upgraded. The transportation system used to only need to solve the problem of people's travel, but now it undertakes the task of guiding the change of lifestyle. Therefore, the relatively extensive management measures are difficult to adapt to the current complex transportation management system, resulting in the transformation of refined and precise countermeasures, and their underlying logic is the upgrade from intelligent transportation to intelligent transportation.

Intelligent transportation can face the complicated factors in the transportation system with independent adaptability. It can adjust the behavior mode and action strategy of the twin system in the virtual space through in-depth learning of the algorithm according to the changes of the environment and the behavior subject, so as to provide guidance for the real world. Although intelligent transportation has highly automated procedures and can make many different responses according to a variety of different situations, it is still difficult to fully adapt to the complex system because its physical logic can only achieve intelligent control and cannot adapt to the new control rules system, let alone the evaluation, research, and correction of the new system.

Intelligent transportation belongs to a more single technology management system, while intelligent transportation belongs to a management system combining society and technology. The previous simple technical system is often overstretched in the current application due to the complex and dynamic nature of the transportation system, and the establishment of a system integrating society and technology is more in line with the needs of the development of the times. It is impossible for the future transportation system to stop at intelligent transportation, and it is inevitable to move towards intelligent transportation.

13.1.3 *Transportation resource management system originated from digital twin*

Urban congestion is becoming more and more serious, the incidence of transportation accidents is increasing year by year, and vehicle exhaust emissions cause serious air pollution with the improvement of economic level and social progress. Countries gradually found that it is difficult to fundamentally solve the problem by simply adding transportation infrastructure in the corresponding long-term governance process, and improving the efficiency of transportation resource management is the core to solving the problem. As a result, countries have turned to the perspective of reasonable allocation of transportation resources to seek solutions.

To establish and improve the transportation resources management system is the premise of realizing the reasonable allocation of transportation resources. However, the transportation resource data has a huge amount of information, a wide range of sources, complex types, difficult integration, and efficient use, and has always been the bottleneck for countries to find solutions. The emergence and maturity of digital twin technology have solved this problem perfectly.

Based on digital twin, the transportation resource management system belongs to a new type of resource control system. It collects the information of urban transportation resources and digitalizes all the information in virtual space by using the network transmission system and intelligent application system, so as to realize the mutual mapping of real transportation resources and digital transportation resources, and promote the digitalization of transportation resources, visualization of transportation information, intelligent transportation management, and personalized transportation service.

Digital twin transportation resource management system integrates a large number of advanced technologies, including artificial intelligence, information collection, communication technology, and so on. A large number of data can be presented in a visual way when the staff needs to understand the urban road transportation information, specifically to the transportation volume, transportation time, road network efficiency, etc. When the regulatory authorities need to solve the congestion problem of a certain road, the transportation resource system of virtual space can use intelligent algorithm to change the existing transportation control to solve the congestion problem, such as through the control methods of transportation light duration or tide lane.

Most of the traditional transportation resource management systems only belong to the data collection center; facing massive data, they cannot carry out transportation coordination and linkage control. As a result, the problem of congestion on this road is often solved, and the next road is blocked to the ground. This is because it is difficult for traditional systems to carry out global analysis and make decisions to solve only the current problems. In contrast, the transportation resource management system, which originated from the digital twin, can carry out dynamic, diversified, and global transportation control. Its approach is to collect resources to the maximum extent, then analyze the optimal scheme in digital space, and finally implement it in the real world, so as to make the most efficient use of infrastructure such as road networks and transportation lights.

The biggest advantage of digital twin transportation resource management system is real-time and dynamic data analysis. Therefore, the acquisition of data is particularly important. The data mainly comes from the road transportation system, the network toll collection system, the urban intermodal transport management service system, and the urban information management system. With these systematic data support, the transportation resource management system can not only easily cope with the daily transportation scheduling work but also play an important role in emergency command.

The digital twin transportation resource management system uses the data mapping of virtual and real space to provide information management support for the dynamic transportation system and improve the efficiency of transportation resource management, producing important theoretical value and practical significance for the intelligent and intelligent process of the transportation system.

13.1.4 *Urban transportation from data management to digital management*

In the third ten years of the 21st century, people's travel demand and the number of trade transactions have increased significantly. The single line of traditional urban transportation can no longer meet the demand. Therefore, urban transportation construction began to develop into a network and diversified complex line network project, forming a comprehensive transportation system such as subway, light rail, bus, and tram.

Extensive experience-oriented management has been used by urban transportation management for a long time, which is relatively weak in project management, such as safety assurance, quality inspection, progress control, work plan, and personnel arrangement, and cannot meet the social expectations and the needs of rapid development of the industry. Therefore, the transformation of urban transportation management is imminent.

There are two stages in the transformation of urban transportation management. The first stage is datamation, and the second stage is digitalization. The following is its specific performance in the two stages.

13.1.4.1 *Datamation stage*

Digitalization refers to the system construction carried out by the transportation management department in order to manage various business processes, business data, infrastructure, etc., and the service objects of these systems are mainly the inherent work links. For example, the road network transportation data mainly reflects whether the road is congested, and the transportation light timing scheme mainly measures the overall waiting time. As for the problems that occurred, how to give full play to the value of data based on the actual situation requires more manual intervention and relies on historical experience to solve.

13.1.4.2 *Digital phase*

Digitalization is different. In this stage, the management system will express a lot of complex and difficult to estimate information in the virtual space through simulation modeling and other ways, and build an intelligent decision-making system that can be perceived, multi-scene, and coordinated, and then rely on the management mode of the virtual space to provide strong data support for the business in the real transportation world.

With the support of digital twin technology, information technologies such as artificial intelligence, Internet of Things, cloud computing, and 3D modeling are integrated into urban transportation and deeply penetrated into each network. The digital clonal urban transportation and real urban transportation data are mapped to each other, which makes the digital information become the basis for decision-making, actively explores the digital road of urban transportation, strengthens urban transportation management, and lays a foundation for the transformation of digital city and smart city.

Of course, digital transformation is not a direct simplification of data. In fact, the relationship between data management and digital management is upstream and downstream. Without the basis of data, there will be no digital construction, and the digital construction must finally return to the data itself and present it in a visual manner. Therefore, the two are interrelated and mutually driven, realizing the whole life cycle information management of urban transportation management.

13.1.5 *Warehousing and logistics system from digitalization to automation*

Under the background of intensified market competition, online and offline integration and omni-channel marketing have become the first choice for merchants to occupy more market share. The new Internet business model has brought more sales opportunities and challenges to merchants. Especially in the field of consumption, the traditional warehousing and logistics system has been difficult to meet the personalized, grid, and customized needs of consumers. The warehousing and logistics system urgently needs transformation under the dual pressure of market environment and consumers.

The traditional warehousing and logistics management is relatively extensive, which adopts manual collection, handling, and stacking management. Most of the management rules follow the historical experience, resulting in a series of problems such as high error rate of warehousing and logistics, disorderly goods placement, inventory difficulties, and lack of process tracking. The overall low efficiency of warehousing and logistics management has seriously restricted the development of warehousing and logistics industry, and directly affected the trade between regions and even countries.

Digitalization has become the only way for the warehousing and logistics system with the transformation and upgrading of traditional industries and the rapid development of information technology. Digital twin provides ideas for the transformation of warehousing and logistics system.

From the perspective of the management of the warehousing and logistics system, first, the operation process of warehousing and logistics should be digitalized to ensure more transparent information of logistics operation, improve the efficiency of management, and provide data support for the decision-making of warehousing and logistics operation. Second, we should extract operation data for analysis, and directly reflect the current warehousing and logistics capacity by analyzing data such as picking efficiency, warehousing efficiency, loading efficiency, inventory volume, outbound volume, and amount to be expired, so as to lay a foundation for continuous upgrading. Finally, through the core technologies such as Internet of things, big data, cloud computing, as well as sensors, the management party shall conduct automatic data collection for all operation links from the arrival inspection to the delivery of goods and then present the data in the warehousing logistics in a visual manner, so as to facilitate the management to accurately grasp the warehousing information and timely track and manage customer orders, and maximize the efficiency of warehousing and logistics.

Once upon a time, the digital warehousing and logistics management method once solved the pain point of traditional warehousing and logistics, and upgraded the warehousing and logistics from semi mechanized state to mechanized warehousing. However, in recent years, the rapid development of market competition has put forward a higher test to the warehousing and logistics management. In front of the large volume of commodity inventory, it is difficult to load the mechanized warehousing, and it must further move to automation through the update and iteration of digital tools.

With the rapid development of artificial intelligence technology, managers began to focus on the exploration of equipment. The emergence of intelligent transportation equipment such as electric forklifts, automatic guided vehicles, shuttle cars, and automatic three-dimensional warehouses marked that the warehousing and logistics system began to enter the era of automation. Based on the digital warehousing and logistics system, after connecting with customers, the operator only needs to publish the project plan on the system, and the automated warehousing and logistics system can realize the whole process automation from production planning to output and input of the warehouse. The automatic mode not only improves the space utilization rate of the warehouse by 2~5 times and greatly reduces the inventory cycle and cost but also reduces the capital turnover speed, and promotes the transformation of the warehousing and logistics industry from labor-intensive to technology-intensive.

The emergence of automated warehousing and logistics systems not only makes the warehousing and logistics industry develop rapidly but also provides new opportunities for the digital twin empowerment of warehousing and logistics systems, which plays a positive role in promoting the development of upstream and downstream enterprises. Beyond all question, the digital twin system will speed up the popularization of automated warehousing and logistics systems, and more and more enterprises will enjoy similar technical dividends.

13.2 Digital Twin Accelerates the Construction of Urban Intelligent Transportation

The construction of urban intelligent transportation is the basis for future cities to enhance their comprehensive competitiveness and achieve sustainable development. At present, problems such as high data collection threshold, insufficient information infrastructure, low standardization of platform models, and insufficient application breadth of relevant industries are still seriously restricting the development of urban intelligent transportation construction. Digital twin emerged as the times required. As an important carrier of urban comprehensive governance and development of digital economy, it is an important historical mission to empower urban intelligent transportation, realize the integrated application of new generation information technology, and drive the rapid development of urban intelligent transportation construction.

13.2.1 *Build a digital twin city and upgrade urban management*

Many problems will be exposed in the process of urban development, such as difficult parking, transportation congestion, as well as insufficient public facilities, which have a serious impact on the quality of life of urban residents, reflecting the deficiencies in urban management. These problems are gradually being solved by the application of digital twin technology.

Before 2018, digital twin was only a simple technical concept. On the basis of the development of big data, cloud computing, Internet of Things, simulation and modeling, and other technologies, digital twin has rapidly evolved into a new industry upgrading path. All walks of life are exploring how to use digital twin technology to achieve change. As the carrier of economic development, it is also necessary to use digital twin technology to achieve the upgrading. By copying twin cities in digital space one by one through information technology, the management and order of the city which is difficult to be directly measured will become apparent. All urban management elements will be visualized to help improve the level of urban management. The main improvement lies in the following aspects.

13.2.1.1 *Easy for analysis and control*

Digital twin city is to completely describe the real world of the physical city in the digital space with the use of digital twin technology and visualization technology, and through visual means, it conducts simulation, monitoring, analysis, and control of the physical city, so as to help people more conveniently cognize and manage the city with digital twin visualization technology.

The digital twin city is not traditional urban planning and construction, but promotes the intellectualization and digitalization of urban management. Digital twin cities directly show the urban problems in a virtual form, help the future direction and purpose of urban governance more clearly, and help further improve the quality of public services.

13.2.1.2 *Provide decision support*

In the past, in order to solve urban problems, decision-makers and management teams often needed to visit the site in person. After a series of steps, the plan could be implemented. Now, with the digital twin system, the improvement plan can not only be seen but also be operated directly. The system will give feedback quickly. With the in-depth learning of the algorithm, the feedback of the system will act on the actual management decisions and provide strong support for the upgrading of urban management.

13.2.1.3 *Extensive management direction*

Digital twin cities involve a wide range of management directions, including urban management, power, high-speed, weather, water affairs, etc., and ultimately serve all residents of the city. Almost all of the digital twin cities under construction have listed the expansion of public services as key projects, taking into account social development and the needs of public life, including the improvement of urban ecology, education, and other hot issues.

The infrastructure construction capacity and management capacity of physical cities have been significantly improved by relying on the empowerment of digital twin cities. Urban economic construction, political construction, cultural construction, social construction, and ecological civilization construction are coordinated, and the ultimate goal of urban management, i.e. smart city, is moving forward steadily.

13.2.2 *Take in everything at a glance: The three-dimensional panoramic presentation of comprehensive urban transportation situation*

Urban development is inseparable from urban transportation, and the planning and construction of each city is inseparable from transportation planning. Urban transportation management problems are gradually exposed with the rapid growth of transportation demand, including a large number of equipment difficult to manage, a large amount of intangible energy waste, passenger flow monitoring lag, single fault information difficult to trace, and other issues.

After the promulgation of the "14th Five-Year Plan", the national management of urban transportation has been put on the agenda, and intelligent transportation has become the ultimate development goal of urban transportation. However, the system applications cannot be displayed on the same platform due to the large number of system applications and scattered data of urban transportation, making it difficult to achieve correct decisions based on data, and the construction of intelligent transportation has been stagnant.

The emergence of digital twin technology has enabled urban transportation in many aspects and satisfied complicated and diverse urban transportation scenarios. Digital twin technology combines various technologies such as BIM and three-dimensional GIS to make urban transportation visible at the same time. It replicates the real urban road network transportation through modeling technology, presents the overall situation of the whole city in digital space, collects data in real time, and synchronizes transportation operation conditions, so as to provide experimental space for transportation model deduction and realize data-driven decision-making.

In the digital twin urban transportation system, all transportation elements are clearly visible, and the transportation network, vehicle operation track, and transportation infrastructure are all presented in the form of visualization. In the virtual space, the real transportation operation state of the whole city reappears, from the micro vehicle travel law to the macro urban transportation system operation state, realizing the full scene coverage.

The basis of intelligent transportation construction planning is the digitalization of urban transportation elements. Only when it is digitalized, physical entities and digital space can effectively communicate with each other in the form of data, and build a three-dimensional panoramic transportation system that can be displayed, measured, and managed. All transportation elements are complete and rich in information in such a system. It can not only replicate the real world one-on-one but also overlay dynamic transportation travel status information and implement transportation management simulation.

The overall situation of urban transportation can be clearly seen through the three-dimensional visual panorama of urban transportation. Urban managers can not only organize the remote inspection and maintenance of transportation infrastructure but also dredge the congested roads thousands of miles away. These services greatly reduce the consumption of personnel and significantly improve the efficiency of management.

Digital twin makes use of a variety of modern science and technology, integrates transportation information resources, coordinates the operation status of urban transportation, creates a perfect visual system of urban transportation operation monitoring, realizes the data exchange and sharing between physical space and digital space, and improves the level of urban transportation management, playing a great role in promoting the smart city.

13.2.3 *Responding immediately once there are calls: Efficient command of urban transportation emergency*

Transportation emergency system is the basic guarantee of urban transportation. Transportation congestion, high accident rate, transportation pollution, and other sudden problems seriously affect the quality of urban life in the moment of transportation scene complexity. Therefore, the voice of emergency management model reform is growing.

Urban transportation emergency management needs real-time, accurate, efficient, and visual support platform. Through the platform, the management department can obtain the transportation operation conditions at any time, such as which transportation lights are congested, which road section has transportation accidents, and how to travel smoothly. All kinds of transportation information can be obtained through the platform at any time, which is not only convenient for people to travel but also greatly reduces the pressure of the transportation management department in emergency command.

The emergency management platform based on digital twin technology should not only be able to solve the travel problem but also provide advanced warning and establish relevant plans for major events such as natural disasters, accident disasters, public health events, social security events, and major activities that may affect the normal operation of transportation. As the emergency management system has integrated the relevant resources required for transportation safety supervision and emergency support in advance, it has realized the information management of the resources related to emergency command. Once an emergency occurs, the command and scheduling of personnel, materials, and technology under emergency state can be highly coordinated, so as to minimize casualties and property losses and ensure urban transportation safety.

The reason why the urban transportation emergency command is so efficient is because of the unique advantages of digital twin technology. Real-time and detailed data command can provide sufficient decision-making support for the emergency management department when the emergency command is started, not only the transportation management department.

Another important role of the urban transportation emergency system is the early warning function. The system can process the historical data and analyze the conditions or factors that may cause a dangerous situation. Once the city has a situation of contact with the police, all operations such as operation deployment, resource allocation, and order issuing can be carried out efficiently, so as to prevent the situation from happening.

With the help of three-dimensional modeling, GIS, Internet of Things, cloud computing, and other technologies, the timeliness of transportation facilities management and control and operation supervision is becoming more and more strong. In addition to presenting the transportation operation status in a three-dimensional scene in a visual and dynamic manner, it can also analyze the urban transportation operation data through in-depth learning of the algorithm. The coverage of urban transportation emergency command has gradually expanded with the upgrading of digital twin technology, and both the execution efficiency and service quality have been improved. As a result, digital twin technology should be closely combined with urban transportation emergency to be the core content of the development of intelligent transportation.

13.2.4 *Be clear at a glance, scientific analysis, and decision-making of urban transportation*

Digital twin technology is highly integrated with urban transportation data resources, which can monitor and visually analyze the core indicators in various fields of urban transportation, assist transportation management departments in mastering the urban operation situation at a glance, and improve the supervision and administrative efficiency.

Transportation management departments can carry out comprehensive transportation situation monitoring through the digital twin urban transportation system. Transportation comprehensive situation monitoring includes road condition monitoring, alarm receiving monitoring, and video inspection monitoring. Road condition monitoring mainly monitors real-time transportation index, congested road sections, and other information. The algorithm will conduct scientific analysis with historical data to provide scientific decision-making support for transportation management and command. Alarm receiving monitoring mainly focuses on the rapid positioning of alarm events, intelligent access to surrounding monitoring and police resources, and provides decision-making support for alarm processing to achieve efficient emergency command of urban transportation. The video inspection monitoring is mainly used for early warning. The data is collected through sensors in the real city to intelligently analyze the vehicle operation track, road congestion points, high accident sections, predict abnormal points, and schedule in advance.

The digital twin city transportation is based on the visualization technology of time, space, data, and other dimensions, which is convenient for the transportation management department to integrate the data resources of the existing information system, covering multiple business scenarios such as daily monitoring and emergency command. With the core technologies such as cloud computing, big data, Internet of Things, the virtual and real spatial data are highly integrated.

With the help of the three-dimensional scene comprehensive situation of urban transportation, the information between various businesses of urban transportation is fully integrated to ensure the efficient docking and display of massive data and the normal operation of urban transportation, providing a scientific basis for decision-making of urban transportation scheduling and optimization.

13.3 Digital Twin Guarding Urban Transportation Safety Development

Digital twin technology brings new opportunities and challenges to the development of urban transportation safety, and new transportation control and communication technologies create new possibilities for transportation management. The mapping between the real world and digital space can not only realize real-time monitoring but also carry out dynamic intervention when necessary, which greatly promotes the development of urban transportation safety.

13.3.1 *Why is digital twin regarded as a new concept of transportation safety management?*

With the increasing complexity of domestic urban transportation scenarios, people pay more and more attention to the safety of road transportation, the stability of vehicle driving, the convenience of daily travel, and other issues. The design of effective urban transportation safety management solutions has become an integral part of urban planning and construction.

The traditional urban transportation safety management is still in the two-dimensional plane stage, which is far from systematic management and difficult to cope with complex and dynamic transportation projects. Therefore, urban transportation management departments urgently need a set of fast and flexible solutions.

The emergence of the concept of digital twin gives the city more space for development. The application of digital twin technology in the field of urban planning has also brought new opportunities for urban transportation safety management. In the technical system, the real-time feedback exists between the real assets, infrastructure, and vehicles related to urban transportation, in which the transportation management department can realize the transportation flow modeling and complete the simulation of the overall transportation operation and infrastructure construction. The simulation test in digital space can also form different solutions by optimizing parameters and implementing options. Compared with the traditional experimental method, it is safer and more cost-effective.

Nowadays, with the rapid development of 5G industry and intelligent driving, urban transportation management departments have to bear more safety responsibilities. With the power of digital twin technology, as the most basic guarantee system, urban transportation safety management has begun to realize synchronous visualization and model deduction, which provides experimental space and decision-making support for urban safety management. It is more convenient for the transportation management department to carry out safety management due to the high-precision real scene presentation. Based on historical experience and existing data, this system can intelligently analyze and early warn, and improve urban transportation safety problems in advance.

Digital twin has the ability of macro vision and global planning, which can effectively solve the waste of transportation resources and quickly respond to safety management. In its actual application process, it can greatly reduce the human cost and improve the management efficiency. Therefore, it is regarded as a new concept, new tool, and new path of transportation safety management.

13.3.2 *What is "digital twin + security infrastructure"?*

With the wide use of Internet of Things, cloud computing, big data, and other technologies, people ushered in the digital era. In this historical environment, the core of national construction strategy is shifting to digital infrastructure.

Digital infrastructure has two major development directions, namely industrial Internet and smart city. In the future, countries, cities, and industries will not only exist in the real world but also build a twin on the basis of the Internet and software. Just like the transportation digital clones, the entire industrial production chain, detailed to every part and every process, is highly digital, promoting the overall infrastructure into the digital twin era.

In the digital twin space, all elements are created by physical entities through software editing or modeling. Here, digital drives everything and software defines everything. However, as long as it is software, there must be a vulnerability. When more and more physical entities are mapped into digital space, once the vulnerability is attacked, the loss to the physical entity will be immeasurable, which may be a building, a community, or even the whole city or country.

In the virtual space, the surface appears to be calm, but in fact, the risk of secret war has been continuing. The forms of war are not guns and rockets, but similar forms of hacker attacks. By threatening the digital twin clones, they pose a threat to the data and information of cities, transportation, and industry and even ultimately destroy this virtual space. Therefore, for the construction of the digital twin system at the city and national levels, it is not necessary to simply improve the security capability of software products, but to build sufficient security facilities while digital infrastructure is being built, so as to fundamentally improve the security performance. For example, it can create a vulnerability analysis platform even though no single product can solve all the vulnerabilities in various kinds of clonal bodies, which allows users to have the ability and technology to discover vulnerabilities, and combine the in-depth learning of algorithms to continuously discover and remedy vulnerabilities, so as to solve security problems from the root.

At present, the development of the network security industry in China is not sound. The industry is often not really aware of the harm caused by security issues when it vigorously develops digital transformation. When vigorously developing digital twin technology, we must attach importance to network security protection means and form a real "digital twin + security infrastructure", so as to make the digital infrastructure more stable and reliable.

13.3.3 *Deepen the connotation and extension of urban transportation and forge intelligent transportation safety roads*

Since the reform and opening up, the process of urbanization has accelerated, a large number of people have transferred to the city, the rapid development of the city has also ushered in a dilemma, and the shortage of resources, transportation congestion, environmental pollution, and other problems have become increasingly serious, seriously affecting the quality of life of urban residents. As an important part of the construction of smart city, the rapid development of intelligent transportation and the gradual improvement of urban transportation infrastructure provide a solid guarantee for urban transportation safety, with a focus on the promotion of the connotation and extension of urban transportation.

The connotation of urban transportation cannot be separated from safety. With the rapid increase in the number of motor vehicles in cities, the incidence of transportation accidents has increased significantly. In the progress from intelligent transportation to intelligent transportation, safety is the most basic guarantee. How to eliminate transportation safety hazards is a concern for all people. The empowerment of digital twin makes the transportation safe, orderly, and smooth and realizes the refined and intelligent management of the connotation of urban transportation safety.

The expression of intelligent transportation on the connotation of safety includes the distinction between macro and micro. Among them, micro objects refer to factors closely related to transportation, such as passenger and freight cars, transportation infrastructure, road operation and management; macro objects refer to government decision-making, urban comprehensive planning, public information services, and other factors.

13.3.3.1 *Application extension at micro level*

The application extension of digital twin technology at the micro level is to reduce the overflow of queuing at the intersection, increase the average speed, and reduce the rear-end accident caused by the sudden stop of vehicles by optimizing the timing of transportation lights.

In addition, the pedestrian crossing warning system can remind the pedestrian to pass in an orderly manner through the LED lights of the pedestrian crossing and can automatically detect the pedestrian running the red light and record the illegal acts.

13.3.3.2 *Application extension at macro level*

On the macro level, the urban transportation management system can realize the data integration of the three-dimensional scene of digital space and the actual transportation operation status. For example, the system will monitor the panorama of urban transportation in real time. When there is a warning event, the system can intelligently analyze and dispatch the nearby personnel and passenger flow or vehicle diversion routes, so as to achieve efficient emergency command and management and reduce safety risks and impacts. In addition, the construction and simulation of safety management model can be carried out in the digital space to provide decision-making support for the actual urban transportation safety management.

The connotation of urban transportation is to improve the quality of urban life, and the extension is to continuously expand the application of digital twin technology in the field of transportation. By playing a role in two aspects of connotation and extension, digital twin technology will promote the improvement of intelligent transportation construction and the formation of smart city.

13.4 Deep Exploration of Internet of Things Potential by Digital Twin

In the future, the Internet of Things will provide technical support for the digital transformation of all walks of life. Although the Internet of Things is still under development, the empowerment of digital twin technology can give full play to the potential of the Internet of Things to a greater extent, complement each other's strengths, and realize the mapping of reality in virtual space and the upgrading of Internet of Things technology.

13.4.1 *Why digital twin is known as the core technology of the future Internet of Things*

Digital twin technology constructs a virtual copy in the digital space, but it must rely on the real-time data transmission of the Internet of Things in order to realize interoperability with the real data.

Internet of Things is a ubiquitous network formed based on the Internet. It can be integrated through wired network and wireless network to transmit information and data to the end of the network in real time. The information used by the digital twin is first collected by sensors and then transmitted through the Internet of Things. The correctness and timeliness of data must be guaranteed through various different networks and protocols due to the delay and packet loss in data transmission. Therefore, digital twin can achieve indicator measurement and even accurately predict the future through Internet of Things technology.

Indicator measurement refers to the prediction of indicators that cannot be directly measured through machine learning after the sensor collects data with the help of historical experience and big data, which greatly reduces the time and cost of the test, and is often used in simulation experiments.

Accurate prediction is a more intelligent prediction. The existing digital technology can rarely achieve accurate prediction, and some problems hidden under the surface are easy to ignore, often leading to deviation in decision-making. Digital twin can carry out panoramic modeling and analysis in combination with a variety of modern information technologies, including Internet of Things technology in particular, so as to realize systematic assessment of current data, diagnosis of problems in the past, and simulation of various possibilities at any time, to achieve full prediction of future trends and shape decisions with a global outlook.

Digital twin integrates multi-disciplinary, multi-physical quantity, multi-scale, and multi-concept simulation process, which also provides a guarantee for the information security of Internet of Things. However, even if the Internet of Things takes all security measures, the system almost does not provide any control, so it is easy to trigger a variety of network attacks. Therefore, digital twin is also known as the core technology of the future Internet of Things.

In the future technology matrix, digital twin is bound to be in a core position, fully coordinating and using various technologies to optimize the deployment of the Internet of Things. The combination of Internet of Things and digital twin technology can help it establish a corresponding digital model, restore the nature and state of the real scene, and achieve intelligent decision support even in a more complex system.

Digital twin not only makes up for the shortcomings of the Internet of Things but also complements and coexists with each other, providing solid technical support for the digital transformation of all walks of life.

13.4.2 *Visual design and digital testing to comprehensively improve the development of Internet of Things*

The implementation of digital twin technology, to a certain extent, depends on the significant progress of Internet of Things sensor system. With the continuous improvement of Internet of Things equipment, digital twin is not only applied to urban construction, transportation development, industrial transformation, and other macro level, but for micro objects, it can also predict its development results through variable data. In this process, digital twin visual design plays an important role.

The practical significance of visual design lies in the construction of visual business decision-making model by copying the real world and massive data information. Digital testing, by contrast, is a core use of digital twin technology, which was born to test what might happen in the future.

The combination of visual design and digital testing refers to the establishment of respective parameters on the basis of visualization, and the user can simulate the possible scenarios according to the given parameters, so as to evaluate the real operation status and property changes of the tested physical entities. This kind of data testing based on physical model is widely used and has achieved good results.

The higher-level digital test is based on the analysis model. It is to create a prediction model through the acquisition of big data and machine learning. The advantage of this method is that it does not need to introduce designers. The R&D personnel only need to create a prediction model based on the product shape so that it has been processed before the problem occurs.

The integration of visual design and digital testing is particularly beneficial to the development of the Internet of Things. In recent years, the Internet of Things has developed rapidly and also exposed a lot of problems, including the non-uniform technical standards, complex platform management, high cost, security loopholes, and other problems, which affect the further development of the Internet of Things. Through the virtualisation of physical entities, the visual design of digital twin can reflect the whole lifecycle process of corresponding entities, facilitate the research and development of the Internet of Things and users to make decisions, and improve management efficiency. Digital testing uses various technologies to build the analysis model, which will simplify the problems to be solved and save manpower and material resources. Therefore, the Internet of Things can ensure the balance between informatization and security since visual design and digital testing enable the Internet of Things on the technical level. On this basis, the development of the Internet of Things will be more systematic and help more industries to solve core technical problems.

13.4.3 *Digital twin: Precise prediction technology for future development of Internet of Things*

Digital twin has now been widely regarded by the Internet of Things industry as a new tool for human to deconstruct, depict, and understand the physical world, providing a new direction for the change of the existing management model.

The most commonly used digital means of digital twin are Internet of Things, virtual reality, and simulation tools. Through these means, people can understand physical entities in virtual space. Especially those technical means that cannot be operated in real physical entities can be easily realized through virtual space.

Before designing and manufacturing any products in the real world, relevant institutions need to calculate and measure their attributes, operating status, and other parameters to ensure accurate analysis and optimization. However, the traditional measurement methods are time-consuming, costly, and biased, resulting in unsatisfactory results after the products are manufactured. Compared with traditional measurement methods, digital twin technology can achieve measurement in virtual space. Data acquisition is fast and direct, which is not only efficient but also accurate.

In addition, digital twin also has comprehensive analysis and prediction capabilities. It realizes the analysis and diagnosis of the current physical entity state through the data collection and big data processing of the Internet of Things, and predicts the future development trend. Of course, this speculative technology is based on machine learning, so the more historical data, or the more speculative items, the more reference value the decision-making generated by digital twin.

With the rapid development of Internet of Things in recent years, countries have invested huge human and material resources in research and development, which is sufficient to prove that the Internet of Things has a positive role in promoting all walks of life, and even the whole country. However, the main role of the Internet of Things is actually to achieve information exchange and communication between items. Simple understanding is to obtain data to achieve intelligent identification, positioning, tracking, monitoring, and management of items. In contrast, its original system is especially lack of accurate prediction ability. Digital twin can make up for this shortcoming. Therefore, digital twin technology is known as the accurate prediction technology for the future development of the Internet of Things.

Whether it is digital twin or Internet of Things technology, its prediction ability is based on the data it has. When forecasting by Internet of Things, it needs expensive sensors, collection systems, and detection systems. They actually face the same problem in comparison with the traditional measurement methods, i.e. both of them will have limited measurement coverage due to the high threshold, resulting in the failure to achieve accurate prediction.

With the use of digital twin technology to enable the Internet of Things, people no longer need to explore different ways to improve the process nor need to terminate the current Internet of Things formation process due to the need for testing, only by running simulation scenarios in digital space can we understand the advantages and disadvantages of the new model and can find the best results through repeated tests, so it can be applied to the real world. Therefore, digital twin provides solid technical support for the development of Internet of Things, enabling Internet of Things technology to play a role in more areas.

Chapter Summary

Driven by digital twin and Internet of Things technology, urban transportation information data is horizontally connected and exchanged and shared. Integrated with social comprehensive management and social public service resources, the digital twin transportation system will be able to coordinate the operation of various functions in the city in a digital and automatic manner, and create a highly integrated intelligent transportation system with more thorough, in-depth, and comprehensive technology.

Chapter 14

Digital Twin "Intelligently" Contributes to Sky Science and Technology

With the continuous integration of digital and information technology, intelligent manufacturing has become a key development area of aerospace technology industry. Machine learning and new energy materials are closely following the pace of the times to help the development of aerospace technology. The application of digital twin technology promotes the manufacturing of aerospace technology into the digital and intelligent era.

14.1 Digital Twin "Intelligently" Contributes to Aerospace

As a key technology to solve the physical integration of intelligent manufacturing information and to practice the concept and objective of intelligent manufacturing, digital twin plays a great role in promoting the multi-dimensional design and practice of aerospace tools through the mutual mapping of physical entities and digital space, and makes the aerospace industry enter the "intelligent" manufacturing era.

14.1.1 *Open up the channel of the universe and accelerate the pace of the era through virtual space*

The concept of digital twin was first applied to the Apollo project at NASA (National Aeronautics and Space Administration). NASA manufactures two identical space vehicles. One of the "twin" space vehicles is mainly used for simulation tests and simulation experiments to provide data support for the operation of space vehicles. At this time, the twin aircraft is still a physical entity, which has the problems of high cost, long time-consuming, and high maintenance cost and has great limitations in simulation experiments, which limits the development of the aerospace industry. Aerospace researchers put forward the idea of putting twin aircraft in digital space with the development of digital and information technology. In the subsequent Apollo project, the physical twin aircraft began to become virtual twin aircraft, and the digital twin aircraft began to be officially applied.

Digital twin aircraft is a digital model of aircraft created in digital space by applying modeling technology and then realizes the mutual mapping of virtual and real space through Internet of Things, big data, cloud computing, and other technologies. The virtual model not only needs to be consistent with the physical entity in appearance but also needs to ensure that its nature changes, state changes, and logic laws are consistent with the reality under the superposition of the external environment, and as a result, the application of digital twin has moved the test with complex original process and low intelligent level to digital space. Through the twin model simulation test, the failure prediction and health management of the whole lifecycle of the aerospace vehicle can be realized, and the remote monitoring and status assessment can be realized, which greatly improves the safety of the aircraft and plays a positive role in promoting the development of the aerospace industry.

Digital twin not only provides a more accurate description and forecast for the manufacturing and R&D of aerospace products but also continuously assists space exploration with its technical strength.

In the launch mission of carrier rocket, the digital twin technology visualized the five steps of rocket launch, booster ignition, second-stage rocket ignition, fairing stripping, and entry into orbit for comprehensive and full-cycle monitoring, which is convenient for ground staff to obtain flight data at different stages. The intelligent control of rocket launch monitoring management, route management, operation data, and equipment status is realized.

The digital twin simulation and visualization technology provides guarantee for the operation of space flight and ensures the safety and reliability of each flight. The simulation test and prediction deduction in digital space indicate that human beings have begun to open up the channels of the universe and entered the era of space "intelligence".

14.1.2 *New energy replaces human aerospace wings, digital twin upgrades aerospace intelligent brain*

The requirements for aircraft are becoming higher and higher with the continuous exploration of the aerospace industry. The extensive use of new energy has improved the performance of space flight devices. With the addition of digital twin technology, new intelligent brain has been embedded into the development of aerospace industry.

The intensification of energy crisis and environmental damage has made the world aware of the importance of sustainable development. Therefore, the voice of low-carbon development of aerospace industry is growing. There is a consensus from all walks of life that new energy materials have the characteristics of clean and environmental protection, which can save energy and reduce the weight of aircraft, and is conducive to the optimization design of intelligent structures and integrated devices, and can improve the overall performance of aircraft and help human aerospace exploration. Flying is becoming quieter and more environmentally friendly with the use of new energy materials, and mankind is beginning to enter the era of green aviation.

In combination with new energy materials, there is also a technological system related to digital twin. Digital twin technology has the characteristics of online operation and maintenance of physical entities, reducing structural design redundancy, low maintenance cost, and can dynamically update the digital model according to the data collected in real time. The system's diagnosis, evaluation, and prediction capabilities are continuously improved and finally applied to the real world through in-depth learning of the algorithm. Due to these characteristics, digital twin has shown great application potential in the field of aerospace "smart brain".

First, in the design and development stage, the trial and error of virtual space has been completed before the products come out, which greatly reduces the cost and cycle of research and development. In the aerospace vehicle manufacturing and assembly phase, the efficiency of resource utilization is maximized by collecting the data and status of each component. In the operation and maintenance stage, digital twin can monitor the whole cycle of aerospace products, and finally present in a visual manner, which is conducive to the maintenance of the staff, and give full play to the maximum performance of each component to match with new energy materials, so as to make aerospace industry more environmental protection.

Digital twin provides an experimental platform for product design and R&D, and various application concepts emerge in an endless stream, bringing many products from theory to practical application. The mutual mapping between physical entities and virtual space provides more possibilities for the development of the aerospace industry. With the help of the simulation of twin model, it assists aerospace equipment in multi-dimensional design and practice. Thus, the digital capability and intelligent level of aerospace equipment design and manufacturing can be improved. In addition, the visual monitoring, evaluation, and prediction of digital twin facilitate the R&D personnel to formulate more reasonable maintenance strategies, which greatly improves the safety and stability of space equipment. The digital twin technology upgrades the aerospace intelligent brain and plays a huge value in the aerospace field, laying a solid foundation for the digital and intelligent development of the aerospace field.

14.2 Airbus Digital Twin Technology Helps Aviation Production

Aerospace manufacturing industry has the characteristics of many parts and components, complex assembly process, etc. Aerospace manufacturers always face problems such as long cycle and insufficient precision in production. Through the use of digital twin technology, Airbus aircraft research and development institutions enable manufacturers to improve quality and quality, effectively reduce costs, and achieve more economic aircraft production. With the extensive use of digital twin technology, aerospace production ushered in a better development opportunity.

14.2.1 *Why digital twin is ranked as the top technology by Sikorsky Aircraft*

The rise of digital twin is derived from the development of Internet of Things, big data, and virtual modeling technology, and is feeding back into the development of high-end technology industry. The combination of digital twin and aviation manufacturing industry has enabled famous aircraft manufacturers such as Sikorsky Aircraft and Boeing to enjoy technical dividends. The most intuitive result is the improvement of production efficiency and the completion of delivery beyond expectations, so it is ranked as the top technology. Among them, the most significant are the following characteristics.

14.2.1.1 *Reduce iteration costs*

The essence of digital twin is to use artificial intelligence, machine learning, sensor data, and other technologies to complete the digital mirror process of physical entities in virtual space. Mirror image is a digital model, which can interact with physical entities in real time and provide decision support for the whole lifecycle of products. After the digital twin is modelled based on the three-dimensional scene, it is not only similar to the physical entity in appearance but also has the same property changes under different scenes and conditions as the actual situation. Taking advantage of this unique advantage, aviation manufacturers can test the possible problems of products in advance during the production cycle, complete the iteration of products in digital space, and reduce the cost of component replacement in reality.

14.2.1.2 *Develop simulation plan*

In addition, digital twin can also simulate the development of activity plans. For example, after collecting real-time data with digital twin, the simulation plan can be completed in digital space, and the simulation results and solutions can be "seen". The more complex the activity, the more data and dynamic factors it can collect, and the higher the cost, the more obvious the advantage of digital twin technology.

The application of digital twin in the production optimization of F-35 model gives new inspiration to the researchers. For a long time, even if the model has been abandoned in the real production line, the data obtained by adding sensors in the locations of interest is still valuable for the production and research and development of products. In this way, the F-35 model uses three-dimensional modeling for real-time monitoring, the experience of production and manufacturing is updated combined with data collection of inferior products, and the multiple processes are improved accordingly, so as to achieve faster and more accurate defect analysis and decision-making, and greatly shorten the overall production time.

14.2.1.3 *Improve detection level*

As is known to all, problems in various components of the aircraft will threaten the safety of flight. Therefore, it is essential for the detection and supervision of aircraft structure, but the number of parts and components of aircraft design is huge, and it is difficult to detect their life span, which is not a small hidden danger for flight safety. Although the traditional detection method is safe enough, it is difficult to be persuasive based on experience and surface testing. After the introduction of digital twin technology, digital twin can collect all components or structural data, assess risks through modeling and simulation tools and probability analysis methods, and establish product lifecycle monitoring and planning. The advantage of this method is that it can predict the safety performance of all the important structural components in the aircraft, timely repair in case of abnormal conditions, reduce maintenance costs and time, and improve flight safety.

From the perspective of the application level of the aircraft manufacturing industry, the digital twin has entered the system level from the equipment level. For both parts suppliers and aircraft manufacturers, the application of the digital twin not only reduces costs but also improves production quality. For a high-tech integrated industry, the market scale prospects are promising. Therefore, digital technology is worthy of Sikorsky Aircraft's reputation as the top technology.

14.2.2 *Future prospects of aviation science and technology, mixed virtual and real, human–machine coordination*

The overall goal of aviation technology manufacturing is to achieve intelligent manufacturing. The most prominent feature of intelligent manufacturing is the combination of virtual and real, and human–machine collaboration. All of these cannot be realized without digital twin technology. Digital twin technology is the mutual mapping between the real world and the virtual space through Internet of Things, cloud computing, big data, and other technologies. At this time, the boundary between human and machine is further blurred; in other words, both human and machine are in the process of data collection, data transmission, data analysis, data decision-making, and other processes, namely virtual and real mixture and human and machine collaboration.

The virtual and real mixture is to build a twin visual environment in the digital space through modeling technology and combining the scenes of the physical world. In this digital space, everything is restored through 3D simulation, so you can control it more intuitively. For example, the R&D personnel only need to understand the production situation of the aviation factories and production lines through the large screen, and carefully consider the idle status, inventory quantity, and process stacking status of the equipment. Based on these data, the production scheduling pre-judgment can be carried out, and the visual management of the production process and operation status can be carried out without leaving home, which greatly improves the management efficiency.

Human and machine collaboration is not only an important application function of digital twin technology but also the reason why it is widely used in various manufacturing industries. It emphasizes that machines should cooperate with people's work independently, adapt to environmental changes independently, and even make decisions superior to people's through in-depth learning, which is also the goal of intelligent manufacturing. Especially in the research and development stage, through machine learning, experiments that are difficult to achieve in the real world can be realized in the digital space. In the manufacturing stage, digital twin can spontaneously complete some complex operations according to the instructions or historical data, and spontaneously allocate materials for production according to the production plan.

Compared with other manufacturing industries, aviation manufacturing industry has the characteristics of long R&D and manufacturing cycle, complex process, and high cost, so many aircraft manufacturers introduce the concept of digital twin. This means that the production and R&D process of aviation vehicles will gradually move to digital space, and complete product testing and iteration in digital space. Since digital twin can monitor the whole lifecycle of products through modeling, the R&D personnel can predict the possibility of error in advance, so as to provide data support for the iterative upgrading according to the change of product status or nature.

Once upon a time, the aviation manufacturing industry required a large amount of cost and time to test a certain component, which seriously hindered the development of aviation technology. However, all components are a digital model through the digital twin technology. The R&D personnel only need to test the digital model, and the system will describe the state and changes of the product in the whole lifecycle, which accelerates the development of the product, reduces the time of replacement or maintenance due to product iteration, and greatly reduces the cost. For the aviation manufacturing industry, which costs hundreds of millions of US dollars frequently, the extensive application of digital twin technology will promote the industrial upgrading of the entire aviation manufacturing industry and make it further on the road to intelligent manufacturing.

Case study: How digital twin technology will change Russia's aviation industry in 2024

In March 2020, Engine Alliance of Russia worked with dozens of enterprises, industry research institutions, and universities to introduce digital twin technology into the research work of aviation engines, and it was expected to complete the plan by 2024.

According to Ivan Timofeev, a senior expert of Russia Engine Alliance, digital twin can describe the whole lifecycle operation of products in digital space, including design, testing, production, and operation. This is difficult to monitor and perceive in the real world. Therefore, this technology will greatly reduce the cost of engine research and development, promote the development of new products, and increase the competitive advantage of domestic Russian engines.

Engine Alliance said that the research of digital twin on aeroengine is so important, on the one hand, because it can describe a series of technical and mathematical models of product lifecycle operation, and on the other hand, it can integrate all component standards to enable it, including all factors that can affect the product, such as product characteristics, economic and environmental indicators, technical documents, the standard of the product and its components. Due to this unique digital integration capability, Timofeev believes that the digital twin technology will drive the development of Russian engine technology and gain a larger market share in the global market.

The basis of digital twin is digital model, and different digital models can be designed according to the product lifecycle. For example, the engine data center uses 1D and 3D format in the design phase, while the complex 3D model is used when the full-scale parts need to be tested.

The use of digital twin often needs to be combined with artificial intelligence. For example, a large number of real-time data about engines can be obtained through digital models and sensors. These data can detect the operating status and service life of engines. Combined with artificial intelligence, some possible problems can be predicted to achieve early prevention.

The use of digital twin in the aerospace industry has been fully recognized by many Russian experts.

Sarnikov of the central Aeroengine Research Institute of Russia believes that the research framework of Engine Alliance does not solve the modeling problem of large products or complex products, because for the aircraft, it involves thousands of parts, and each part requires special mathematical models and methods, which is not a small difficulty at present, and after all, digital twin will be applied to all aeroengines in the future.

Kirill Paateni, head of Engineering Analysis Department of Engine Alliance, proposed that the use of digital twin can significantly reduce the cost of engine use, shorten the research and development cycle to 5~6 years, and reduce the overall design cost by one-third, which is a shocking upgrade.

The requirements of modern aircraft on engines are increasing, resulting in more complex research and development. In addition, after each research and development, a large amount of time will be spent on fine-tuning and certification, which seriously limits the development of the aerospace industry. The digital twin exactly solves this problem. The design, development, and certification time of the new aero-engine is shortened to 5~6 years.

From the experience of Russian aerospace industry, it is necessary to ensure that the test value basically conforms to the real situation in order to shorten the overall development time, and the verification model and calculation level need to be improved, so as to improve the quality of testing and reduce the number of test projects. In the framework of digital twin, the digital model is not immutablel; it needs to be constantly checked and improved in the calculation and testing, and combined with historical data and real-time data of the engine, a more comprehensive maintenance of the engine and its components is conducted.

Due to the large and complex data involved in the aerospace industry, the collaboration between multiple departments is limited. Digital twin technology optimizes the collaborative process, digitizes all the information, data, and documents designed in the whole process, and presents them in a visual manner so that the staff will no longer have a sense of whether to start with complex information. The practice of Russian aviation industry shows that all project participants can clearly obtain the information they want by using digital twin technology, which greatly improves the team cooperation ability and plays a positive role in the research, development, and production of aviation products.

Chapter Summary

The digital twin technology is deeply integrated with the aviation manufacturing industry and vigorously promotes the digital transformation of the aviation industry. Its core is the digital expression of aviation products, digital design and virtual experiment, digital manufacturing, and whole lifecycle data management. In the future, digital twin will make greater contribution to reshaping the aviation industry capacity system and building intelligent aviation road facing the higher requirements of aviation product development.

Chapter 15

Digital Twin "Intelligently" Empowers the Stronger Earth

In the manufacturing industry, a virtual environment which is not different from the real environment is built in the computer network through the digital twin technology. The virtual objects that need to be tested are injected into the virtual environment for testing, and the test results are analyzed and evaluated, so as to simulate the optimal solution to adapt to the complex real environment. As a result, the manufacturing industry has opened the door to "smart" manufacturing.

15.1 Digital Twin Modeling of Unmanned Driving Vehicles

Unmanned driving is a technology based on advanced technologies such as computers, navigation, and positioning, as well as sensors; it is an AI application in the field of transportation. Researchers have been conducting research and tests on this basis and have achieved certain results since the National University of Defense Technology developed China's first unmanned driving vehicle in 1992. The concept of digital twin has been mentioned and clarified, and the construction of digital model has enabled unmanned vehicles to carry out digital testing and training before practical application after entering the 21st century.

15.1.1 *Digital twin drives unmanned driving vehicles into explosive phase*

With the emergence and development of digital twin technology, the connection among the three aspects of vehicles, roads, and users has become more close, which has promoted the explosive development of unmanned driving vehicles and further strengthened the collaborative work between virtual and reality.

15.1.1.1 *Principle of unmanned driving vehicle under digital twin technology*

We can use digital twin technology to build virtual unmanned driving vehicles and driving scenarios for virtual driving test and training in the process of research and development and production of unmanned driving vehicles. The test results are analyzed, evaluated, and optimized to obtain the optimal results in accordance with the driving data of the virtual unmanned vehicle. The tester inputs the data into the real vehicle system and uses the real vehicle in the real scene to reproduce the virtual scene after determining the optimal result, thus affecting the real production and manufacturing.

15.1.1.2 *Development of unmanned driving technology under digital twin technology*

In the original concept of unmanned driving vehicles, it was to build a kind of intelligent vehicle that could free hands. However, digital twin technology gives it a new development direction with the development of science and technology.

Transportation management departments can use digital twin technology and unmanned driving technology to reproduce the scene of transportation accidents in dealing with transportation accidents. The whole process of the accident was comprehensively analyzed in combination with the accident scene environment and the driving track of the driver, so as to find out the specific cause of the transportation accident and the main responsible person for the transportation accident.

In terms of transportation monitoring, transportation management personnel can monitor the ground transportation conditions in real time through digital twin technology, so as to detect and deal with problems in a timely manner. For example, timely control of vehicle density. In addition, digital twin technology can also monitor weather conditions in real time. When a road section is covered by heavy fog, the visibility is low. The technology can timely transmit the signal to the driving system to remind it to pay attention to safety.

In terms of transportation prevention, as the digital twin technology is the data virtual body mapping of physical entities, transportation management personnel can simulate the development of ground transportation through digital twin technology according to the feedback information of the real environment, and predict the possible results. In short, that is, to predict the future through realistic data. On the basis of the forecast results, the possible problems are prevented.

15.1.2 *Digital twin, making it possible for vehicles to stay online forever*

The vehicle is always online, that is, the unmanned vehicle is always kept online, the information is fed back to the digital twin system in real time, and the state of itself is changed at any time according to the instructions of the digital twin system through the control of digital twin system. This is the development direction of unmanned driving vehicles, and also the research focus of Chinese researchers.

Although the level of science and technology in China is in a leading position in the world, the research process of unmanned driving vehicle eternal online is still slow.

In 2020, digital twin technology was developed rapidly with hundreds of leading enterprises in the world bringing digital twin into the ranks of core technology research, and with the development of Internet of Things and Internet technology. The state perception of digital twin system to the real environment is becoming more powerful. The virtual model simulated by the digital twin system is more and more close to the perfect physical entity. Under the scenario of a digital model, extreme virtual scenarios can be constructed to conduct in-depth testing and training of unmanned vehicles. For instance, extreme weather, extreme ground environment, sudden accidents, etc. ensure that the unmanned driving vehicle can find the optimal solution in the complex virtual situation, and improve the safety guarantee ability and continuous work efficiency of the real unmanned driving vehicle with the goal of eternal online.

The spatial virtual and reality of digital twin have begun to merge. In the future, virtual data is likely to become a physical entity in the present reality, which is also the development direction of automobile life in the future. At the same time, the rapid development of artificial intelligence and Internet of Vehicles has improved the real-time performance of digital twin, and it is possible to break through the limitation of high-precision map and model the real scene in real time. This situation shows that the digital twin system has more powerful control over the unmanned driving vehicle, and can ensure the status of the unmanned driving vehicle more timely and accurately.

At the same time, the core application of digital twin technology lies in decision-making and control. In terms of unmanned driving vehicles, the virtual driving conditions are affected by the real environment, and the real driving conditions are controlled by the virtual data. Both of them can be connected and interacted dynamically in real time continuously, so as to build a closed-loop flow between the real and the virtual. In this case, the digital twin system can control the unmanned driving vehicle at all times so that the vehicle can obtain the status of online at all times. It is through the digital twin system that the eternal online ideal of automobiles will come true sooner or later.

15.2 Entry of Digital Twin Technology into Autonomous Navigation Logistics Robot Field

Since the concept of digital twin was clearly put forward, digital twin technology has been in a state of sustainable development and rapidly integrated with advanced technologies such as AI and VR. At this stage, digital twin not only plays a role in the field of autonomous vehicles but also has a huge driving effect in the field of logistics robots.

15.2.1 *3 Major features of future logistics: digitalization, automation, and intellectualization*

Since the 21st century, the huge population pressure has resulted in the shortage of resources with the continuous improvement of science and technology and the continuous development of economy. Under the background of the development of the times, the user demand has changed, which makes the global logistics industry face more severe challenges. While facing challenges, the future logistics industry is facing new development opportunities. In order to enhance the core competitiveness of the future development of the logistics industry, logistics enterprises must constantly learn new technologies and improve innovation capabilities. Therefore, the future logistics industry is gradually showing the characteristics of digitalization, automation, and intelligence.

15.2.1.1 *Digitalization*

Digitalization refers to the transformation of a large number of complex and changing information into measurable figures and data, and then the establishment of appropriate digital models based on these figures and data. In the future logistics industry, digitalization refers to the digitalization of logistics information such as goods and vehicles with the help of computers, and the construction of a clear and intuitive digital model based on the converted data for deployment and integration.

Digitalization is an important direction for the development of logistics industry. The application of digital technology in the logistics industry has promoted the transformation of goods distribution from fast logistics to precise logistics. Particularly, digital twin technology is used to establish digital models for airports and ports which occupy an important position and monitor the cargo transportation in real time. Or make digital tracking of valuable goods, real-time monitoring of the status of goods, so as to ensure the safety of goods.

15.2.1.2 *Automation*

Automation is the most prominent feature of the future development of logistics industry. Logistics automation means the automation of logistics equipment and facilities in the logistics operation process without direct participation, including warehousing, loading and unloading, transportation, sorting, packaging, and a series of work. The logistics equipment and facilities are summarized and integrated through the computer network, and the overall situation is controlled by the same system, so as to achieve efficient and flexible logistics work.

15.2.1.3 *Intellectualization*

In December 2011, "The 12th Five-Year Development Plan for the Internet of Things" issued by the Science and Technology Department of the Ministry of Industry and Information Technology of the People's Republic of China clearly pointed out that the development of Internet of Things in the future will mainly focus on several major areas, including intelligent logistics.

Intelligent logistics is to make the logistics system imitate human intelligence through the use of integrated skilled technology, with the ability of thinking, perception, learning, reasoning and judgment, and solving logistics problems by itself. The logistics industry makes full use of intelligence to improve efficiency and reduce costs.

The application of intelligence in logistics is mainly reflected in the storage, transportation, monitoring, and other aspects. With the development of the times, it will continuously be given new content.

15.2.2 *Great change of logistics robot, from intelligent equipment to digital twin*

Logistics robot, as the name implies, refers to the robot applied in the logistics process. Common logistics robots include logistics transportation robot, logistics sorting robot, logistics stacking robot, and logistics operation robot.

In January 2016, Citibank and Martin College of the University of Oxford jointly proposed the "Technology in the Age of Work 2.0", indicating that the development of advanced technologies such as artificial intelligence and robotics will have different degrees of impact on employment in developed countries and even developing countries. In particular, single tasks such as loading and unloading, carrying and packaging are more easily replaced by robots. Therefore, the logistics industry is the first to bear the brunt and has become the forefront of "machine replacement".

The level of logistics intelligence is also deepening with the introduction of digital twin technology into the logistics robot field. The artificial intelligence integrated with digital twin technology has a stronger ability of perception, understanding, and execution of the surrounding environment when compared with the traditional logistics robot. In the next few years, after the integration of digital twin technology and artificial intelligence technology, the application scope in the logistics industry will also be expanded.

In terms of logistics and warehousing, different types of logistics robots can be connected through the digital twin system, and the best scheme of efficient and low-cost work of logistics robots can be tested by using the simulation calculation of the digital model, which mainly includes the optimal action of the robot, the optimization of the robot's operation track, the robot's load balancing, etc. In such a logistics warehousing system, the online digital twin system sends instructions to coordinate and control the offline execution of different types of logistics robots, so as to achieve online and offline integration, efficient collaboration, and the purpose of improving efficiency and reducing costs.

In terms of logistics and transportation, artificial intelligence, such as UAVs, unmanned express vehicles and unmanned distribution robots, has entered the test stage and is gradually put into use. For example, in September 2013, Shunfeng Express carried out a test of UAV express delivery in Dongyuan. In May 2016, Jingdong launched an open platform for intelligent logistics, which put UAVs, autonomous vehicles, and other artificial intelligence into the platform. In the future, through the use of digital twin technology, the logistics robots such as UAVs, unmanned express vehicles and unmanned distribution robots, as well as the actual environment of a specific region will be incorporated into the digital twin system, and these enterprises can carry out remote control through virtual simulation of physical entities, and use the system to track the whole process to prevent Identify and solve the problems that may occur in the logistics process.

15.3 Digital Twin Simulation of Optimal Urban Public Transportation System

With the continuous development of economy and the continuous growth of population, the process of urban construction in China is accelerating. The huge population and complex urban environment have also brought certain pressure to China's transportation. Therefore, the concept of "green travel" is constantly mentioned, and the innovation of public transport system is imperative.

Digital twin technology has shown great advantages in the construction of urban public transport system. Professional institutions use digital twin technology to build urban virtual transportation, and then simulate the optimal urban public transportation system to solve most of the public transportation problems. It is of great significance to promote the development of intelligent transportation and even smart city.

15.3.1 *Build a "mirror system" of smart public transportation in a virtual city*

As an important part of the urban transportation system, urban public transport plays an important role in the daily travel of citizens. The development of modern network technology has promoted the transformation of traditional public transport system to intelligent public transport system after entering the 21st century.

The emergence and development of intelligent public transportation is the result of the integration of traditional public transportation system with GPS (Global Positioning System), GIS (Geographic Information Technology), AI (Artificial Intelligence), and other technologies. The emergence of smart public transportation has realized real-time monitoring, dynamic perception, safety protection, and other functions, and improved the operation efficiency of the transportation system. For instance, the combination of traditional bus cards and convenient mobile payment makes daily travel more safe and fast.

The intelligent public transportation system has been further improved with the application of digital twin technology in the field of transportation, and the "mirror system" of the virtual city intelligent transportation has begun to be built. At this stage, the construction of "mirror system" has begun to take shape.

The term "mirror image" can be understood as "copy", which is another expression of digital model. The construction of the "mirror system" of the intelligent public transportation in virtual city mainly uses digital twin technology to digitalize the information of the public transportation tools and roads in the whole city, and feed back the collected data to the computer network. The virtual model of the physical entities of the public transportation system is constructed by using digital twin technology. In short, the construction of the "mirror system" of the public transport system is to copy all the factors of the physical public transport system into the virtual space.

If the "mirror system" of smart public transport is built in a virtual city, the smart public transport in a real city can also be connected to build a complete city public transport perception network in the network. Through the "mirror system", urban public transport management departments can monitor the operation status, driver status, as well as regional vehicle density, etc. of intelligent public transport in real time, so as to avoid the danger of drivers due to visual blind angle, fatigue, interference and other emergencies during the driving process. Even if there is danger, the "mirror system" of smart bus can quickly simulate the best solution to the problem and intelligently take over the bus in a timely manner.

In addition, the management organization can also set up an intelligent platform that interacts with the intelligent bus in real time, and broadcast the information, such as the estimated arrival time of the bus and the number of passengers in the bus in real time, so as to provide passengers with real-time information of taking a bus and make them ready in advance.

15.3.2 *Running out the optimal solution of urban public transportation management in virtual space*

At the "HUAWEI Smart City Summit 2019" in 2019, HUAWEI proposed "5G + AI, open a new era of smart city twin". As pointed out in this slogan, in the future, the best way to use virtual space to simulate urban public transportation management needs to combine 5G, AI, and digital twin technology to promote the online and offline integration of bus and virtual space model and seek the best solution conducive to urban transportation management.

Among the problems of urban public transport management, transportation safety and low passenger flow are more obvious.

15.3.2.1 *Urban public transportation safety*

In recent years, reports on bus safety accidents have emerged in an endless stream.

On 28 October, 2018, the No. 22 bus in Wanzhou, Chongqing, crashed into the river at the second bridge of the Yangtze River in Wanzhou due to the interruption of normal driving by passengers, and all 15 people in the bus were killed. On 11 February 2019, a No. 12 bus in Ningqiang

County, Hanzhoung, Shaanxi Province fell into the Jialing River due to driver's illegal operation, killing 2 people. On the first day of the college entrance examination in 2020, a No. 2 bus in Anshun, Guizhou Province, fell into a lake due to drunk driving, killing 21 people, including 5 students. Similar bus safety accidents have occurred for 3 consecutive years, which has caused a very bad impact on society.

The reason for bus safety accidents is largely affected by people. Whether it is the interference of passengers, the driver's own problems, or other external vehicles, it is because in a certain period of time, emotions drive out rationality, and irrational factors occupy the dominant position. And unmanned driving technology can solve such public transport risks.

The relevant agencies can establish the urban public transport model in the virtual space, master the vehicle conditions remotely and in real time, detect abnormalities in time, and take over the bus quickly before the danger comes. At the same time, the vehicles and personnel are checked before departure to solve the problem at the source through the immediate response of the model.

15.3.2.2 *Low passenger flow of urban public transport*

Compared with the relatively sparse public transport vehicles such as subway stations and light rail stations, the dense bus stations obviously have more advantages. However, the passenger flow of most city buses is lower than the above two. The reason is, on the one hand, due to the complexity of ground transportation, and on the other hand, buses are often delayed due to road congestion, resulting in weaker accessibility. Therefore, taking the bus will cost more time.

In view of this, the urban public transportation management department can master the road conditions remotely through the digital twin system, optimize the bus operation route, and achieve timely departure and arrival, so as to improve the operation efficiency of the bus.

15.4 Logistics Supply Chain Based on Digital Twin Technology

With the development of Internet of Things, cloud computing, artificial intelligence, virtual reality, and other technologies, digital twin technology has begun to accelerate its development and has been gradually valued by major industries. Among them, the logistics industry is one of the most widely used areas of digital twin technology, and has a close relationship with the logistics supply chain.

15.4.1 *Modern intelligent logistics warehouse of "digital twin + AI + IOT"*

Zhang Yong, CEO of Alibaba, once proposed that logistics in the future must be from digitalization to "digital intelligence". "Digital" refers to digital twin technology, and "intelligence" refers to artificial intelligence, i.e. AI. Gu Xuemei, CTO of Cai Niao network, said at the 2019 global intelligent logistics summit that IOT (Internet of Things) also plays an important role in the whole logistics chain, including warehousing, transportation and distribution. When the digital twin encounters AI and IOT, the model of modern intelligent logistics warehouse has jumped into people's eyes.

In addition to affirming the importance of IOT technology, Cai Niao Network has established an open logistics IOT platform with the industry based on digital twin technology, AI and IOT technology. The construction of the platform will provide technical support for enterprises to build a modern intelligent logistics warehouse. It can be directly dispatched by intelligent scheduling to work in the warehouse after the digital warehouse (virtual warehouse) on the computer network is connected to the logistics IOT open platform.

In order to enable warehouse staff to receive digital warehouse information in a timely manner, Cai Niao has developed a minimal PDA (Intelligent handheld operation terminal) - CAINIAO LEMO for the logistics IOT open platform, which is used to connect digital warehouse and physical warehouse. Due to its small size, the PDA can be worn on the wrist by warehouse staff. When the digital warehouse releases the task, PDA will guide the relevant staff to complete the task through lighting, vibration, and other ways, so as to make the work more efficient and concise.

The practice of Cai Niao network shows that if an enterprise wants to build a modern intelligent logistics warehouse, it cannot do without the integration of the three technologies, namely digital twin technology, AI, and IOT technology.

Among the relationships of digital twin technology, IOT and AI, IOT plays a very important role.

On the one hand, IOT technology promotes digital twin technology and AI. With the development of IOT technology, digital twin technology and AI will develop accordingly. On the other hand, IOT and AI are complementary and interdependent. IOT can obtain a large amount of data through sensors, and AI can analyze the data, both of which are indispensable.

In the end, the remote real-time monitoring function of data twin technology is the real-time data transmission based on IOT technology. The future prediction of data twin system through the construction of data twin model is also realized with the help of IOT technology.

It can be expected that it will be soon to complete the modern intelligent logistics warehouse under the common development of the complex of the three.

15.4.2 *Why DHL digital twin intelligent warehouse technology has gradually become mainstream*

Li Jianghua, CEO of Cai Niao, once said, "The pace of urban life is getting faster and faster, and there is a 24-hour express demand. Overnight delivery is based on meeting the needs of more people, bringing more service options and better service experience". Due to the progress of the times and economic development, the living standards of residents have gradually improved, and there is also a higher demand for personalized logistics services.

"Ordinary people will retreat if they do not advance". The same is true for enterprises. Enterprises in any industry should constantly learn new science and technology, improve innovation ability, provide high-quality services to users for the purpose of occupying a favorable position in the fierce market competition, and thus enhance their core competitiveness. Therefore, DHL digital twin intelligent warehouse technology has gradually become the mainstream in the logistics industry.

The so-called DHL digital twin intelligent warehouse technology is developed and proposed by Deutsche Post DHL, a world-famous postal and logistics company, and is a plan and scheme for building a new type of intelligent warehouse. The reason why DHL actively promotes the logistics reform and wants to build a new global logistics ecology is largely due to the fact that a competitor "stole" parcels from Chinese corporate customers. The occurrence of the incident, including DHL and other major international logistics companies, is in a crisis of trust. In order to regain customers' trust and improve service quality, DHL is striving to build a new global logistics ecosystem. DHL digital twin intelligent warehouse technology is an important step.

Similar to the modern intelligent logistics warehouse of "digital twin + AI + IOT", DHL digital twin intelligent warehouse technology also uses digital twin technology to build a model connecting virtual and reality, and integrates it with Internet of Things technology and data analysis to create a new intelligent warehouse solution.

The internal situation of the warehouse can be monitored anytime and anywhere through DHL digital twin intelligent warehouse technology, and the operation coordination can be achieved at all times. In addition, the six key factors of DHL's smart warehouse plan are also the reasons why smart warehouse is highly sought after in the logistics industry. The key factors of DHL intelligent warehouse plan are as follows: ① Reduce congestion and improve efficiency, ② control area alert, ③ temperature monitoring system, ④ complete transportation visualization, ⑤ real-time operation data, and ⑥ enhance the safety of mechanized handling equipment.

With the support of digital twin technology, DHL digital twin intelligent warehouse technology can optimize the present and forecast the future, simulate various possibilities, and provide users with more comprehensive solutions. Therefore, DHL digital twin intelligent warehouse technology has gradually become the mainstream of intelligent warehouse construction in the current era.

15.4.3 *Digital twin makes logistics supply chain approach to unmanned mode*

In recent years, countries around the world have gradually entered a new era of intelligence with the development of Internet of Things technology, artificial intelligence, GIS, big data, cloud computing, and other technologies. The economic society has entered an explosive development period under the influence of the smart era, and the industrial structure has been reshaped. Affected by the global epidemic, the logistics industry in various countries has accelerated the digital upgrading of the supply chain.

Matthias Schutger, senior vice president of DHL, said, "The digital twin market is expected to grow by more than 38% per year, breaking the USD 26 billion mark by 2025. Digital twin provides unparalleled tracking, monitoring and true short assets capabilities to transform traditional supply chains by facilitating data-driven decision-making and collaboration, simplifying business processes and new business models". The Log Research Institute also proposed that the logistics supply chain supported by digital twin technology is the development trend in the future. China's Fifth Plenary Session of the 19th CPC Central Committee proposed to include "improving the modernization level of the industrial chain supply chain" into the "14th Five-Year Plan".

The rapid development and application of digital technology provide a broader space for the transformation and upgrading of traditional logistics supply chain. Digital twin can analyze the past according to historical data on the basis of the function of digital twin technology, make use of historical information that was originally lack of value for predictive maintenance and scheduling, and can monitor the current logistics supply chain in real time through the connection between reality and virtual model, so that the logistics supply chain has a higher efficiency. In addition, digital twin can provide a model to analyze the possibility of the future and better ensure the security of the entire supply chain link. Therefore, the digital twin's whole grasp to the past, the present, and the future is an important factor to enable the logistics supply chain to achieve full automation.

The future prediction function of digital twin data can provide business guidance to users whenever necessary and help users make decisions quickly. For example, when a logistics company needs to change the logistics supply chain link due to changes in its operation and management, it can test and analyze all possibilities through the model to quickly find the optimal logistics supply chain link.

Digital twin technology has been involved in all aspects of logistics supply chain development. From the smart warehouse, through the unmanned express car, UAV, to the unmanned distribution. Digital twin technology, with its rich application functions, combines with artificial intelligence, Internet of Things technology, and other technologies in the new era of intelligent logistics, thus creating an unattended and fully automated logistics supply chain.

Chapter Summary

Digital twin technology is so disruptive that it is worthy of the name of "game-changing technology". In recent years, the construction of smart city, intelligent transportation, smart logistics, and other surface systems has made great achievements with the continuous follow-up and innovative application of digital twin by Chinese researchers. However, these applications are still only digital twin "small trials". By relying on the huge advantages of digital twin, digital twin is destined to show its skills in the transportation-related manufacturing industry, and "wisdom" will create the strongest transportation manufacturing power on the surface in the future!

Chapter 16

Digital Twin Creates a Master's Work

Water conservancy project is an integral part of water resources management. Digital twin technology has been gradually applied in various fields of water, land, and air since digital twin technology has been gradually valued by various countries. The construction of the strongest transportation system on the surface has begun to take effect. With the enhancement of China's scientific and technological strength, digital twin technology has gradually entered into water conservancy, and is determined to create a large, comprehensive, and sophisticated intelligent water supply of river channels.

16.1 River Intelligent Management System Based on Digital Twin

In the traditional period, it was often unable to find or deliver information in time due to the backwardness of river management methods, resulting in inability to do anything about accidents. The real-time monitoring and simulation prediction function based on digital twin technology can realize the visual management of river dynamics with the support of modern technology, and ensure the safety of people's lives and properties to a greater extent.

16.1.1 *Flexible simulation, flexible management method of smart river channel*

Water is the basic condition for human survival and development, and is a basic natural resource and strategic economic resource. Without any water, the human body can persist for 3~10 days at most. The Ministry of Water Resources of China also adheres to the most stringent water resources management system in order to protect water resources.

However, it has become a global water shortage phenomenon due to the growth of the population, as well as the damage and pollution to the environment caused by human development, which has also been demonstrated in China. Taking water pollution in China as an example, there are more than 1,700 water pollution accidents every year in China in recent years according to the statistics of the environmental supervision department. Countries around the world are looking for effective solutions.

After the 21st century, China has gradually started the construction of smart cities with the development of science and technology, especially the practical application of digital twin technology, virtual reality, artificial intelligence, GIS, and other technologies. Urban river courses, as part of water resources, are included in the construction of smart river courses. The water system, river, and river course are simulated and directly displayed on the computer. The relevant departments can dynamically grasp the changes of water level, flow rate, flow rate and water quality of the river through real-time monitoring. At the same time, the regional precipitation will be monitored, and early warning information will be issued. The emergence of smart river system has realized the effective management of water resources.

At this stage, the construction of smart river has begun to take shape. The relevant management departments tend to use flexible management methods, so as to combine the watershed with the region, coordinate the urban and rural areas, and coordinate the development, utilization, and conservation and protection. "Soft" is in line with the characteristics of "water conservancy and all things are indisputable", especially it can more scientifically and comprehensively correspond to the characteristics of different regional environment. The application of flexible management to intelligent river management is more conducive to promoting scientific and intelligent management of water conservancy projects.

Flexible management refers to a humanized management method based on the theory of flexible management and promoting people-oriented management. Flexible management is generally applied to people, which is contrary to the traditional rigid management methods. For instance, the flexible management of enterprises will not impose more restrictions on the work of workers, and will not establish strict commute time and complete tasks in accordance with regulations but more advocate human autonomy.

In terms of water conservancy, digital twin technology can be used to integrate large and small river courses nationwide into the computer network. They are allowed to develop according to the natural law without being interfered with by other factors. If necessary, the problems reflected by the digital model will be comprehensively adjusted and controlled in a small area. Smart river channel realizes overall control, while flexible management improves operation efficiency and reduces operation cost while implementing overall control.

16.1.2 *Management and control of the whole line and segmented management method of water transportation*

Water transportation is an important part of transportation. Although with the progress of the times, the proportion of road, railway, aviation, and other modes of transportation has gradually increased, but low-cost waterway transportation still has no small advantage in the transportation of large-scale goods.

However, waterway infrastructure construction is less and more vulnerable to natural factors compared with other modes of transportation. For example, due to natural factors such as water level, climate, as well as season, the uncontrollability of waterways is greater, which has great hidden danger to the safety of people's lives and properties. Therefore, the pace of building smart water transportation is urgent.

Objectively, China has a vast territory, with a distance of about 5,500 kilometres from north to South and about 5,200 kilometres from east to west. In addition, the domestic river network is densely distributed, and the rivers are criss-cross. From west to east, due to the great difference between the east and west topography, the upper section of the river has a large drop, rapid water flow, and narrow river channel. The middle section is located at the junction of mountains and plains, with many waterfalls or rapids. In the lower section, the flow rate is relatively slow, and the river channel is wide, which is also the main location for water transportation. Through the simulation of digital twin technology, the real and complex water transportation lines can be translated into the twin space, and the water transportation can be monitored in real time.

Based on the differences in location and climate of water transportation, and the limitations of digital twin technology, when choosing the management mode of water transportation, we can choose the segmented management mode. While ensuring efficient supervision, it also reduces management costs.

Segmentation refers to the division of a whole into several different parts. Segmented management, as the name suggests, is a management method of segmenting the rivers within the scope of water transportation according to the city, topography, climate, etc. Through the segmented management, the key river basin is strengthened to ensure the normal operation of water transportation. At the same time, it can also carry out global control from point to area.

In addition, the relevant departments may also consider contracting a certain watershed to individuals or groups, who will manage it according to the rules by using the digital twin system, so as to expand the water management team. Realize the dual supervision of virtual and reality, so that the management is more strict and the operation mechanism is more smooth.

16.1.3 *Gain a decisive victory a thousand miles away, digital space controlling river operation research*

With the development of modern Internet, the future river channel is the integration of virtual space and the real world. Through the Internet of Things technology, the data of the actual river is digitalized and informationized; the digital twin technology is used to model the river channel model and ensure that the river channel model can keep pace with reality. Finally, according to the digital model of the real world of the river channel, the visualization of the river channel through digital space is realized. In the virtual digital space, the river channel is under full control.

As for the practical application of digital twin technology in this aspect, it was first born in Hangzhou's "urban brain" in April 2016. In recent years, Hangzhou Shangcheng District Urban Management Bureau has closely followed the pace of digital development of Hangzhou, started from the control of river channels, and built "Hangzhou City Brain Intelligent River Platform" on the basis of "city brain".

The platform consists of four parts, i.e. smart river network, command center, data center, and business management. On the platform, the distribution of rivers, water quality, flow rate, and water level in the Shangcheng

District can be seen. In addition, river treatment projects and river-related approval projects can also be displayed on the platform. Through the platform, the management personnel can carry out comprehensive management and control of the river channel in the region without leaving home.

Digital space controls the whole process of the river. On the one hand, the safety of the river can be managed through the simulation function of digital space. Through the access to online monitoring data, we can timely detect the emergency situation in the operation of the facilities, quickly trace, track, and early warn of the accident, assist the management department to prevent the accident in advance, and improve the early warning and processing capacity of the river emergency situation. On the other hand, digital space is used for real-time monitoring. Through water environment simulation modeling, the water level, water quality, flow, infrastructure, etc. in the covered watershed are effectively analyzed. On the basis of the analysis results, it predicts accidents, disasters, and other emergencies for the managers to form a decision-making plan so that the managers can make decisions in a timely and accurate manner.

In terms of the improvement of the water environment in the river channel, the past and present water environments are analyzed through digital space, so as to conduct a comprehensive assessment on the treatment plan of water resources according to the results.

In short, the real world and virtual space have begun to break through the boundaries with the arrival of the digital economy era. Both the river channel and even the entire human survival environment will be affected by digital space with the help of digital space. And the planning of a human survival environment will also be more consistent with the laws of nature.

16.2 Digital Twin and Smart Water

"Technology changes life and technology makes the future" has penetrated into all aspects of social life. In the era of rapid changes in information and digital development, the intelligent development of water management to a higher stage has become an inevitable trend of the development of the times.

Urban intelligent construction is rapidly carried out with the rapid development and in-depth application of Internet of Things, mobile Internet, big data, cloud computing, and other technologies, and urban drainage system is constantly updated with the development of the city.

Especially with the development of cities, the problem of water pollution is becoming increasingly severe, and the supervision and control of water has become a major problem that plagues the development of cities.

The development and in-depth application of digital twin provides a new idea for the control and governance of water resources in a timely manner, i.e. the use of digital twin and other technologies to build automated, digital, and intelligent water affairs.

Smart water has three core concepts, namely perception, collaboration, and intelligence. Perception is based on the application of sensors, which can be embedded in each stage of natural water and social water cycle to complete data collection. Collaboration refers to the use of virtual platforms to complete the control and treatment of water resources and enable each link to operate efficiently and collaboratively. Intelligence is the prediction function of the model to provide scientific decision-making. These three core concepts are also a strong driving force to support the construction of smart water.

The management department can fully apply the Internet of Things, the Internet, big data, cloud computing, and other new technologies to the comprehensive management of water affairs, rely on sensors to obtain water information, and build a virtual reality management platform of water affairs through the analysis and sorting of big data and cloud computing and through digital twin simulation, so as to complete the real-time monitoring of the water quality, water level, water pressure, flow, precipitation, weather, and other data, so as to realize the refined, intelligent, and standardized management of water resources.

In terms of smart water construction, the combination of digital twin and water construction has already begun to take effect. The Yangtze River Protection Project in Jiujiang City, Jiangxi Province, adopted the new thinking of "Internet + water affairs" and started the construction of intelligent water affairs. A smart water service platform with "one cloud", "one network", "one map", "one book", and "five centers" as the overall design framework was established. According to the platform, Jiujiang has realized real-time monitoring and dynamic management of the operation of drainage facilities.

Case: digital modeling, saving RMB4 million cost of Jiujiang water affairs

According to Jiujiang News Channel, the Yangtze River Protection Project takes Jiujiang, Jiangxi Province as a pilot city to build a smart water platform. The project involves sewage treatment plant, supporting pipe network, water ecological restoration, river regulation, pump gate, hub project, ecological transformation of sponge avenue, etc.

The project was jointly carried out by Shanghai Survey and Design Institute and Yangtze River Ecological Environmental Protection Group, with a construction period of 3 years and a total investment of approximately RMB7.7 billion. The water information of Jiujiang city is managed in a multi-dimensional, multi-scale, and multi-faceted mannerL:

(1) BIM design for the whole urban area. In terms of model construction, we integrated all relevant professional BIM models, adopted tilt photography technology, and used UAVs to complete the image acquisition within the scope of the project. After the completion of image acquisition, the 3D real scene modeling software of collision check is used to construct the digital model. With the support of Internet of Things and GIS, the data of underground pipe network within the scope is sent to the real model to realize the storage and management of underground pipe network.
(2) On the basis of BIM design, build a smart water platform. The platform is based on the overall design framework of "one cloud", "one network", "one map", "one account", and "five centers".
"One cloud" refers to the Yangtze River Ecological cloud.
"One network" refers to the monitoring perception network.
"One map" refers to BIM + geographic system map.
"One account" refers to the performance indicator system.
"Five centers" refer to the engineering management center, intelligent perception center, water application center, decision-making support center, and exhibition and publicity center.

The platform framework design includes infrastructure layer, service sharing layer, smart application layer, and security layer. In addition, the platform will open external interface to facilitate data exchange with the outside world.

Smart engineering can be realized after the construction of smart water platform is completed. With the aid of data twin model, the construction progress and quality of the Great Yangtze River Protection Project are monitored in real time.

(3) The digital model is used to collect the operation information of drainage households, pipe networks, storage tanks, pump stations,

sewage treatment plants, and other facilities in real time. According to the collected data, the emergencies in the project operation are found in time, and correct actions are taken in time to avoid the occurrence of problems.

It is understood that by virtue of the digital model, the team reduced design errors by 80%, saved 800 design and proofreading working days, and saved RMB4 million in cost during the model design. During the construction of the Great Yangtze River Protection Project, the construction period was shortened by 4 months and the construction cost was reduced by 20%.

The construction of the platform has realized the transformation and upgrading from traditional extensive water management to modern standardized, intelligent, and refined management.

16.3 New Path of Integrated Development of Digital Twin and Land Water Transportation

Digital twin is a new direction for future transportation. The dynamic simulation of real entities by this technology can solve the problems of real space and simulation space, and integrate the two dimensions of digital model and real land and water transportation to bring a new path to learn from the advantages and make up the disadvantages, which is of great significance to promote the development of intelligent land and water transportation.

16.3.1 *Collaborative management of the earth's surface transportation and integration of water and land transportation*

In recent years, we have built a modern, high-quality, national comprehensive three-dimensional transportation network to support the construction of a modern economic system and a strong socialist modern country in order to speed up the construction of a strong transportation country. In February 2021, the Central Committee of the Communist Party of China and the State Council issued the "Outline for the National Comprehensive Three-dimensional Transportation Network", which clearly put forward the requirements for promoting the integrated development of comprehensive transportation. This also puts forward a new topic for the digital twin technology to promote the coordinated integration of land and water transportation.

In waterway transportation, port is one of the important transportation infrastructures and the hub of waterway transportation. Under normal circumstances, any domestic or foreign trade involving two areas of water transportation and land transportation will be transshipped through the port. However, the land water transportation system has always been a relatively independent individual due to the regional characteristics of ports and the influence of seasons and climate.

With the global transportation entering the digital and information era, water transportation and land transportation should not be separate individuals but should become a whole of mutual connection and coordination. Therefore, we should have a sense of overall situation and systematic thinking and make the relationship between water transportation and land transportation coordinated in the management of water transportation, so as to achieve complementary advantages and disadvantages, resource sharing, and common development, and build a sustainable and dynamic multi-win situation.

The integration of water and land transportation is a systematic process. Strengthening the coordinated management of ground transportation requires the joint efforts of all regions and departments in China. Nowadays, the development of modern technology has played an important role in the coordinated management of land and water transportation. With the help of modern technology, especially the application of 5G, GIS, AI, Internet of Things, digital, and other technologies, it provides a strong technical guarantee for the coordinated management of land and water transportation.

We should make full use of modern new technologies such as digital twin to strengthen the ability of coordinated management of land and water transportation with the aim of promoting the coordination of land and water transportation, so as to improve the quality and efficiency of the development of land and water transportation.

We believe that we can use digital twin technology to translate the real transportation into twin space in the near future, and predict the changes of transportation flow according to the changes of land and water transportation rules, and then predict the specific situation of land and water transportation conversion and connection, so as to provide scientific basis for the decision-making of management personnel. The integrated transportation system will be further improved, and the coordinated development of land and water transportation will be promoted. In addition, a comprehensive three-dimensional transportation network can also be used to facilitate the connection of vehicles and vessels, and to avoid the threat to the safety of drivers due to weather and road conditions.

16.3.2 *Explore new development path of digital twin and water integration*

The development of modern science and technology has had a profound impact on the world's economic development and the transformation of production and lifestyle. With the progress of Internet of Things technology, 5G, AI, and other technologies, the development field of digital twin and water integration is blooming with new vitality.

16.3.2.1 *Digital twin leads the "China Intelligent Manufacturing" of water treatment industry*

In 2015, China first proposed "Made in China 2025", aiming to change the "big but not strong" situation of China's manufacturing industry and improve the competitiveness of China's manufacturing industry. In "Made in China 2025", China proposed the transformation from "made in China" to "created in China", "speed in China" to "quality in China" and "products in China" to "brand in China", so as to complete the task of "made in China" from "big to strong". Reflected in the water treatment-related manufacturing industry, digital twin and AI and other technical support intelligent manufacturing is one of the main directions.

From 2016 to 2020, China plans to increase the scale of sewage treatment facilities by 50.22 million cubic meters per day, but the water quality after treatment is still not optimistic. In fact, in terms of sewage and wastewater treatment, there are higher operational requirements on the operation and management process of key elements such as water treatment membrane system due to the limitation of technical requirements. Therefore, the water treatment industry is one of the industries that need to be changed. With the help of modern technology, relevant departments and enterprises can collect data through the Internet of Things, conduct big data analysis with the help of AI, and finally predict and make decisions by digital models. The indicators that cannot be directly measured by the machine are intuitively and clearly simulated dynamically to achieve high efficiency, high quality and low cost of water treatment.

16.3.2.2 *Digital twin helps water conservancy modernization*

On 11 June 2021, the 333rd China Engineering Technology Forum - Digital Twin and Water Technology Innovation Forum hosted by the Chinese Academy of Engineering was held in Wuhan. It is jointly undertaken by the Department of Civil, Hydraulic and Architectural Engineering, Chinese Academy of Engineering, and the Changjiang Institute of Survey, Planning, Design and Research (CISPDR).

With the theme of "digital twin and water, help water conservancy modernization", the forum carried out paper collection and exchange around digital twin technology, intelligent water area and water engineering joint scheduling, intelligent construction throughout the life of water conservancy and hydropower projects, urban intelligent construction and water affairs comprehensive governance, so as to promote water conservancy modernization and digital advice and suggestions.

During the exchange, Niu Xinqiang, President of the Changjiang Institute of Survey, Planning, Design, and Research (CISPDR), believed that an efficient and three-dimensional digital platform should be built through new technologies such as digital twin. By using the dynamic perception of the digital platform, the comprehensive management and control of water conservancy is carried out, and the ability of "prevention, pre-control, early warning and forecast" of flood and drought disaster prevention is realized.

"Digital hail", a leading visual digital twin enterprise, has released the smart water digital twin IOC system in terms of digital twin and water integration. The system covers water environment monitoring, urban water supply monitoring, drainage operation monitoring, water ecological monitoring, water engineering monitoring, water safety monitoring, and other business areas. With this system, the comprehensive operation situation of water conservancy management can be monitored in real time and mastered with one key.

Nowadays, the trend of digital twin and water integration is inevitable. Digital twin is an important means to support the high-quality development of the "great water conservancy" industry in the future.

Chapter Summary

It is the focus of China's scientific and technological innovation to enhance the strategic scientific and technological strength of water conservancy. Digital twin technology is an important tool to realize the modernization of China's water conservancy governance system and governance capacity. Based on the model constructed by digital twin technology, China's water resources information can be more comprehensive, clearer, and more macro display and provide a strong boost to the modernization of water conservancy in the future digital era.

Chapter 17

Digital Twin Connects the Real World and Digital Space

As early as 2011, the U.S. Air Force Research Laboratory proposed the prototype of digital twin technology and established the model of real aircraft in digital space. Since then, the real world and digital space have begun to integrate. In recent years, with the development of VR, AR, and other technologies, the distance between the virtual world and reality has been further shortened. In the future, digital twin technology is bound to usher in greater changes, and the complete integration of digital space and the real world will no longer be a fantasy.

17.1 Digital Twin: Connecting Reality and Digital Space

From WeChat to Alipay, from sharing bikes to unmanned driving cars, the impact of digitalization on people and society has been pervasive. The maturity of the concept of digital twin and the development of technology break the boundaries between the real world and virtual space. This has made it possible for human beings, who were originally limited by their physical structure, to go to the sky and enter the sea freely. The emergence of digital twin provides an increasingly clear exploration route for human imagination.

17.1.1 *Digital twinL: the infrastructure of future technology*

The intersection of virtual reality and reality makes digital twin a hot spot in the global economic development, which has attracted extensive attention from countries around the world, and has listed digital twin as a key technology application.

In September 2020, the State-owned Assets Supervision and Administration Commission (SASAC) of China issued the "Notice on Accelerating the Digital Transformation of State-owned Enterprises", which clearly proposed to accelerate the promotion of digital twin and accelerate the application of digital twin in various fields.

Nowadays, digital twin technology has become one of the largest technology development trends in the future. Not only that, but the digital twin is also building the infrastructure of future technology.

Based on the real-time monitoring, simulation, and prediction of digital model, most of the practical applications of technology need to be based on digital twin technology. For example, the logistics transportation robot, logistics sorting robot, logistics stacking robot, and logistics operation robot, which are widely used in intelligent logistics, are put into use on the basis of the twin integration of logistics network, AI, and digital. In addition, the construction of smart city and intelligent transportation also requires the integration of GIS, GPS, Internet of Things, and digital twin technology.

In some manufacturing industries, robots are currently unable to do more sophisticated work. In particular, when milling, industrial robots need to use high-speed rotary milling tools to cut the required shape on the roughcast. The powerful impact produced in the process of cutting will lead to inaccurate movement. However, inaccurate movements of the robot in the milling process will be calculated by digital space in real time to predict the feasibility and reliability of the robot's future motion path with the application of digital twin in the field of artificial intelligence, so as to ensure that the robot can complete fine motion on the accurate path.

Digital twin can bring virtual presentation of physical entities, and not only the surface of physical entities, but also their internal structure and operation will be presented in real time through a high simulation model. It also benefits from the support of big data, cloud computing, and other technologies. More importantly, digital twin technology is still at the stage of development and can continue to evolve. Digital twin evolution will also promote the development of other technologies. Therefore, digital twin technology is not only the infrastructure of science and technology but also the innovation pillar of future science and technology.

17.1.2 *New development direction of virtual reality in the future*

As early as 1929, the idea of virtual reality was contained in the dynamic simulation of sound. Around the 1980s, Jaron Lanier of VPL Research proposed the concept of "virtual reality" and developed the first set of VR glasses and gloves. By relying on the solid foundation of electronic information manufacturing and broad market, virtual reality technology has gradually become mainstream after entering the 21st century.

The emergence and application of virtual reality technology have changed the original way of life of human beings and have become an important force leading a new round of global industrial reform. In the future, VR may become an important force to promote the development of biological, medical, social, educational, commercial, military, and other fields. Although most of the applications of virtual reality are still in the stage of idle theorizing. The relationship between virtual space and the real world is also a simple interaction of "plane, passive, and one-way". However, driven by digital twin technology, virtual space will have more and more development space in the future.

In 2021, the press conference of the Yunfeng Meeting of the world VR industry conference was held in Nanchang City, Jiangxi Province. At the press conference, Li Wenqiang, vice president of China Academy of Electronic Information Industry Development, believed that the new development direction of virtual reality itself in the future is three-dimensional, active, and interactive. In the future, the model data acquisition and understanding will tend to be hologram.

For example, some e-commerce enterprises have envisaged virtual reality, the main idea of which is a "Buy +" system to change the existing online shopping methods. Users do not need to understand the products through pictures and evaluation; they only need to wear helmets and can shop like shopping in physical stores.

In addition, the future development of virtual reality will gradually tend to be convenient, natural, and lightweight. Virtual reality equipment can bring more immersive experience with the progress of virtual reality technology. With the high simulation and prediction of digital twin technology, accounting calculates the best proportion of manufacturing and production of virtual reality products, so that the production cost of virtual reality products decreases, the production efficiency increases, and the price becomes cheaper.

The integration of virtual reality and industry is also expected. The development of virtual reality will be guided by integration and innovation, promote the integration of virtual reality and key industries, promote the construction of industrial application demonstration base, and promote the coordinated and joint development of various regions in the country. In the meantime, the government departments are actively cultivating new models and new types of virtual reality industry to expand the development space of virtual reality industry.

17.1.3 *VR and ARL: the first key to the virtual and real world*

Unlike VR, which is dedicated to making people enter the virtual space and can provide immersive experience, AR is to put virtual objects into the real world and is the interaction between reality and virtual. Therefore, the Chinese translation of VR is "virtual reality", and the Chinese translation of AR is "augmented reality technology". VR is the "key to the virtual space" and AR is the "real world auxiliary equipment".

Virtual reality is widely known for its practical application in the entertainment field, but augmented reality is rarely mentioned. Even some people who do not often contact with technology will confuse VR, AR, AI, MR, etc. In fact, both AR and VR, as products of the development of the times, exist to promote the development of human society. AR and VR will also become the first key to open the door between the two in the integration of the real world and virtual space.

The so-called augmented reality is a technology that superimposes virtual objects into the real world through simulation and can be perceived by people using a certain medium with information that does not exist in a certain area of the real world. For example, Google Glass of Google is a medium for perception. In 2015, Magic Leap released the AR promotional film "Big Beluga Comes Out of Water", which was once a global sensation, which is integrating reality and virtual. Although ultimately Magic Leap company was exposed to releasing promotional videos which were all made in 3D, which greatly reduced the valuation of Magic Leap company. However, the imagination of the Magic Leap company for the future development of AR still brings people to think about innovation.

On 30 January 2021, following the Magic Leap, its former chief executive officer, Ronnie Abovitz, formed a new company called the Sun and Thunder. Ronnie Abovitz, in an interview with GameBeat, said that the reason for the establishment of the company is to continue to focus on research and development in the field of AR, and use AR to integrate intelligence, technology, and art, so as to show a new world of virtual and real interaction. Similarly, the Ministry of Industry and Information Technology of the People's Republic of China put forward in its "Notice on Promoting 5G and Accelerating Development" that the combination of 5G, AR, and VR should be promoted to promote the practical application of AR and VR.

For example, the design, research and development, and testing of new products in the past will incur huge time and money costs in the manufacturing industry, while the advantages of virtual reality and augmented reality possessed by VR and AR can reduce costs and improve production efficiency. Virtual models of products can be built based on the real world via the combination of VR, AR technology, and digital twin technology, and real-time adjustments can be made through virtual reality to determine the design scheme. This will further shorten the distance between the virtual space and the real world.

17.2 The Structural Awakening of Digital Twin

Since professor Michael Greaves proposed the concept of digital twin in 2003, it was not until 2017 that DTT, a well-known consulting company, clearly described the structure of digital twin in "Industry 4.0 and digital twin". Driven by the awakening of digital twin architecture, a large number of technology architects have sprung up in various industries. The use of digital twin architecture has played a huge role in enhancing the competitiveness of enterprises.

17.2.1 *Five levels of future technology of digital twin innovation*

Digital twin is of great significance in promoting future scientific and technological innovation. According to the "Technology Trend 2020 Report" issued by DDT, in the long run, all systems and data in the entire ecosystem need to be integrated in order to release the full potential of digital twin technology. This means that the continuous development and in-depth application of digital twin technology cannot do without the support of existing science and technology. On the contrary, the development and innovation of science and technology in the future also need the guidance of digital twin technology, which is an important driving force to promote the development and innovation of science and technology.

In order to promote the innovation of future science and technology, it is necessary to promote digital twin technology innovation. We can update the concept of digital twin from the five levels of planning, design, governance, operation, and development to provide innovative ideas for the development of science and technology.

17.2.1.1 *Planning*

"No unprepared battle, no uncertain battle, and every battle should strive to be prepared". To do anything, we must plan well in advance and have a strategic and overall view. In recent years, smart projects such as smart cities, smart logistics, and smart water affairs constructed in China are all integrated applications based on digital twin technology and artificial intelligence, Internet of Things, GIS, and other technologies. Before promoting the future digital construction, the relevant departments shall conduct a comprehensive inspection of the industry and region, and promote the innovation and development of science and technology according to local conditions and time.

17.2.1.2 *Design*

Innovation is a process of trial and error. People gradually enter the digital world with the development of science and technology. In the process of innovation, the unavoidable error cost has become a low cost that can be paid repeatedly. Therefore, the simulation model of virtual space can be used for variable parameters and repeated testing both in product design and in the practical application of new technologies. The existence of digital twin technology provides a low-cost trial-and-error path for future scientific and technological innovation.

17.2.1.3 *Governance*

After the Third Plenary Session of the 18th CPC Central Committee, China has also clearly proposed to promote the modernization of the whole country. Governance is not the simple application of a single technology, but the establishment of a larger scientific community. Therefore, in the process of governance, we should make full use of advanced technologies, including digital twin, build a new governance tool system, and promote the innovative development of modern, intelligent, and digital governance.

17.2.1.4 *Operation*

In the deep integration of technology and operation, technological innovation model is becoming the mainstream respected by major industries. Only by continuously improving the level of scientific and technological innovation and combining the operation capability of digital twin with the depth of scientific and technological innovation can we help various industries establish stronger competitive advantages.

17.2.1.5 *Development*

Innovation and development is an eternal theme. At present, digital twin technology has been fully developed with the support of other technologies, and its application has covered three major areas, namely water, land, and air. The deep application of digital twin technology can broaden the application space of other technologies. Under this virtuous cycle, the innovation level of science and technology application in various industrial fields will be improved. Therefore, it is necessary for us to make full use of digital twin technology to achieve innovative development of science and technology in the future.

17.2.2 *Four stages of future application of digital twin*

China is in the information and digital era nodes, Internet of Things, mobile Internet, big data, cloud computing, and other advanced technologies, providing a solid guarantee for the development of digital twin, making its application more extensive. From the current development trend of digital twin, the application of this innovative technology in the future will go through four major stages, namely virtual verification, two-way connection, intelligent decision-making, and virtual reality interaction.

17.2.2.1 *Virtual verification*

Before the formal application of digital twin technology, the design, manufacturing, testing, and other links of products, production lines, systems, and other real-world things not only need to pay a huge cost, but also their efficiency cannot be guaranteed. However, after the in-depth application of digital twin technology, through the construction of physical entity simulation model, repeated, variable parameters, and accelerated simulation experiments are made for design, manufacturing, testing, and other links, and the best results are obtained through repeated tests. This means that digital innovation in its virtual verification stage can improve the accuracy of products in design, research and development, manufacturing, and other aspects.

17.2.2.2 *Two-way connection*

Internet of Things is the link between the real world and the virtual space in the digital model. Through a single information source provided by the Internet of Things, the model can realize the two-way connection between the real world and the virtual space. The combination of Internet of Things and digital twin provides a large amount of real data support for the simulation of the model. Through the simulation model, the two-way connection is realized to maximize the utilization rate of data resources and promote the efficient coordination of design and manufacturing.

17.2.2.3 *Intelligent decision-making*

The intelligent decision-making stage of the development of digital twin technology mainly focuses on the combination with artificial intelligence. On the one hand, artificial intelligence can better receive and analyze the data information sent by the sensors and help the virtual model to improve its decision-making ability. On the other hand, the application of artificial intelligence is inseparable from the comprehensive control of digital model. In recent years, the integration of digital twin and artificial intelligence has been more and more widely spread; the integration of digital twin and artificial intelligence has been more widely applied from the fields of power, ships, agriculture, construction, and manufacturing, to the fields of medical, education, aerospace, etc.

17.2.2.4 *Virtual real interaction*

The core technologies of digital twin (including dynamic perception, data integration, model construction, and human–computer interaction) promote the transformation from "static mapped physical entities" to "dynamic collaborative physical entities" and realize the collaborative interaction between the real world and virtual space.

The development of digital twin application is from virtual verification to closed-loop optimization of virtual reality interaction. In the future, all major industrial areas will be deeply affected by the popularization of digital twin applications. Digital twin technology will become a new driving force for the transformation and upgrading of the industry.

17.2.3 *Digital space, real future*

In August 2021, the Ministry of Natural Resources of China issued the "Outline of Real 3D China Construction Technology (2021 Version)". Real 3D is a digital virtual space that reflects real, three-dimensional, and real-time real space and is a digital image of the real world. The release of the "Outline" marks a new climax in the construction of three-dimensional real projects in various regions. The development of digitalization is irresistible, and the integration of the real world and digital space has become a general trend.

As the data and information interface of the virtual and real world, digital twin is an important driving force to promote the digitalization of the real world and the development of digital economy. The curtain of the digital world is gradually opened with the support of big data, cloud computing, artificial intelligence, 5G, and other technologies.

Before 2001, a netizen named "Mr. Mao" made a video about the future digital space and expressed his wonderful imagination of the future digital space.

In Mr. Mao's video, the magic index of digital life is divided by star level. One star is a "carefree and healthy world" in medical treatment. Two stars is a "lively and interesting learning space" in education. Three stars is a "convenient and comfortable home life" in terms of household. Four stars is a place to enjoy the world in person and a place to eat and drink at will. Five-star is an "unmanned driving trip at will" in the field of transportation. In this video, digital space brings great convenience, security, and comfort to people.

Wang Tao, executive director of HUAWEI and President of ICT products and solutions, said, “most of today’s creation comes from the past wild imagination”. With the rapid development of AR, VR, AI, and other technologies, Mr. Mao’s imagination of future digital life is gradually becoming a reality.

In the future, whether it is commerce or industry or agriculture, it will be based on digital. People can directly enter the shop experience of the shopping platform with the help of virtual images when shopping. In the manufacturing industry, managers can directly carry out one-stop management of design, manufacturing, and testing through the virtual model in the computer network. In the planting industry, the virtual model can use sensors to analyze soil fertility, climate, crops’ habits, etc. and control the unmanned equipment to work around the clock according to the analysis results, and never delay the agricultural time.

In the future, more and more people will obtain digital identity, and people will be stored in the cloud in the form of digital. In the future, digital will be everywhere.

17.3 Technical Ethics and Trust of Digital Twin

In the future, collaborative interaction between the real world and virtual space is inevitable. The limitations of the real world make people have greater expectations for virtual space.

However, the unrestricted nature of virtual digital space may cause people to deviate from the basic moral standards of real society, thus promoting the generation of malice. In addition, in the future digital world, everyone’s information and data will be stored in cloud. At that time, it is also a bottleneck of the development of digital twin as to whether the security of people’s own information can be guaranteed. Therefore, there has been a heated discussion on whether the digital twin technology should be used to build a virtual space.

The issues that must be considered in the process of digitalization are as follows: Whether the digital twin technology will develop “good” or “evil”, how the virtual space constructed by it will impact the real society, how to balance the relationship between the real world and the virtual space and so on.

In 1970, Meseny, an American scholar, published “Technology and Society” on the social evaluation of technology. Through various investigations and speculations, Meseny divided the concept of technical ethics into three concepts:

(1) **Technology "good" theory:** In the book, Meseny wrote that technology is the driving force for all progress, can solve the fundamental problems of human survival and development, and can create an ideal society for people. For instance, the mobile Internet makes people's lives more convenient.
(2) **Technology "evil" theory:** The idea is that technology is the source of all evil. On the one hand, the development of technology reduces the employment rate, and the work with low technology content is often replaced by intelligence. On the other hand, the different levels of scientific and technological development among countries have resulted in different strengths of national science and technology, and some powerful countries often advertise themselves as the Savior, and carry out wars against the weak countries under the banner of rescue, which poses a serious threat to the safety of people's lives and property. Finally, the natural environment is more and more polluted with the development of technology. Therefore, some even believe that technology will lead to the destruction of mankind and the earth.
(3) **Technology "neutral" theory:** The idea is that technology is not good or bad in itself. However, there is a trial-and-error process in the development of digitalization, so the application of technology is full of great uncertainty. Meseni himself supports this view, as do the authors of the book on digital twin technology.

The same technology is used in the field of benefiting the country and the people, that is, good; if it is used in the field of calamity to the country and the people, it is evil. There is no good or evil in technology itself, but the use of technology affects people's evaluation of technology. Therefore, we need to actively balance the relationship between technology and human, and society and nature in the process of using digital twin technology, and promote the development of science and technology in a favorable direction.

Digital twin technology will not fail us, and we can trust it, as the law and time will prove.

17.4 Digital Twin AI Experience Platform

Before the deep integration of digital twin technology with Internet of Things, big data, cloud computing, AI, and other technologies, digital twin has demonstrated a strong synergy. The integration and innovation of digital twin and other technologies have also been continuously breaking through and fully applied to various fields through years of exploration. With the arrival of the Industry 4.0 era, the social production mode will gradually change from automation to autonomy based on the deep integration of digital twin and artificial intelligence.

17.4.1 *Future experience gained on AI platform*

In July 2021, the fourth World Artificial Intelligence Conference was held in Shanghai. The conference mainly focused on the application of AI in various fields. Among them, the use of AI technology to create a one-stop AI platform, optimize resource allocation, and achieve efficient use of resources has become a hot topic at the conference.

AI platform refers to a tool platform built on the basis of AI technology. The AI platform is built based on the characteristics of high precision, high efficiency, and high coverage of AI analysis and data processing. Users can make full use of existing resources for real-time data analysis, achieve efficient network management, and improve operational efficiency through the AI platform.

The development of digital world is inseparable from the support of the three aspects, namely data, computing power, as well as algorithm. However, the data is relatively disorderly due to the rapid development of mobile Internet, which makes it difficult to obtain and analyze the data. AI has the ability to search for laws and values from numerous data. It is an inevitable trend for the development of this technology to integrate the algorithm with the production service platform and make the AI platformization.

In May 2021, Hua Xiansheng, vice president of Alibaba group, clearly stated that AI platformization is the future trend of AI industry development. Hua Xiansheng delivered a keynote speech titled "Exploration and Practice of Large-scale Application of Artificial Intelligence Technology" at the "2021 Sohu Technology 5G & AI Summit". He believes that artificial intelligence technology has changed a lot in less than 10 years, and has made great breakthroughs in computing power, deep learning technology, vision, hearing, natural language decision-making, big data and other aspects. Therefore, the development of AI industry in the future will tend to the development of platform mode.

Shortly after that, at the Fourth World Conference on Artificial Intelligence, the participant, Transwarp Technology, launched a one-stop

AI platform “Transwarp Sophon 7.0”. According to the concept of the AI platform, Yang Yifan, the R&D director of the AI product R&D Department of Transwarp Technology, once said, “In the practical application of artificial intelligence era, the user’s scenarios are becoming more and more complex, and sometimes it has to maintain a variety of databases to meet different needs. At this time, the need to deal with multi-modality becomes urgent”. Therefore, based on the big data + cloud computing + artificial intelligence technology, the data can be analyzed and processed in real time, and the high-efficiency AI platform “Transwarp Sophon 7.0” was born.

AI platform is in the ascendant and will soon become a new stage of digital twin technology, creating a new experience for the whole society.

17.4.2 *AI + 5G drives human evolution and civilization transition*

On 6 June 2016, the Ministry of Industry and Information Technology of the People’s Republic of China issued the 5G commercial license, announcing that China officially entered the first year of 5G. Subsequently, the second World Artificial Intelligence Conference with the theme of “smart world, unlimited possibility” was held in Shanghai. The conference covered a number of areas, including AI, 5G, chips, and algorithms, and emphasized that human will enter the era of great changes in the technology industry after the deep integration of AI + 5G.

There is a complementary relationship between AI and 5G. The arrival of 5G provides AI with a broader space for development and further improves the data calculation and transmission speed of AI. AI’s strong ability to analyze and operate complex systems can solve most of the problems in the development of 5G and promote the development of 5G.

The deep integration of AI + 5G has added more power to the digital twin technology to promote the transformation of many industries.

17.4.2.1 *Communication industry*

From the white paper “5G Opens the Era of Wireless Connectivity and Intelligent Automation” issued by the Global Association for Mobile Communications in 2017, China Mobile had built nearly 500,000 5G bases by 2021, promoting the completion of large-scale application of AI + 5G, and the deep integration of AI + 5G has become the core technology leading the transformation of the communication industry.

17.4.2.2 *Medical industry*

With the application of AI + 5G in the medical field, medical imaging technology and other medical technologies have made breakthrough progress, and difficult and complicated diseases have been found. Not knowing what caused it is a thing of the past. According to media reports, in September 2021, under the guidance of the First Affiliated Hospital of Guangzhou Medical University, Foshan Second People's Hospital, and Qingyuan People's Hospital successfully carried out two remote operations. Remote surgery has high requirements for image recognition and real-time performance, while 5G ensures the quality of images and the speed of data transmission. With the support of 5G, the diagnosis of the patients was successfully completed by using the technology of bronchoscopes and automatic navigation.

17.4.2.3 *Transportation*

Since AI was put forward, people have been full of wonderful imagination for unmanned driving vehicles. With the comprehensive application of 5G and AI, unmanned driving has become a reality. 5G can provide high-precision and intelligent high-precision positioning services, and provide a solid guarantee for automatic driving of vehicles, thus satisfying the needs of building an intelligent transportation power in China and promoting the construction of strengthening country with strong transportation network.

In addition to the fields of communication, medical care, and transportation, AI + 5G has also been fully applied in industries, energy, education, culture and tourism, the party and government, finance, and other aspects. The comprehensive application of AI + 5G is of great significance to the development of human evolution and civilization.

17.4.3 ***The possible trilogy of XR achievements***

While many people do not know much about XR, XR has entered people's lives in a quiet way.

XR (Extended Reality) refers to a real and virtual combination and human-computer interaction environment generated by computer technology and wearable devices, which is collectively referred to as VR (virtual reality), AR (augmented reality), MR (mixed reality), and other forms of presentation. Under the support of XR technology, it can realize the perfect interaction and integration of virtual space scene, real-world scene, props, and users and provide users with highly immersive visual experience.

XR technology is generated to promote the transformation of social productivity. Hisilicon, a semiconductor company under HUAWEI, called XR technology "new entrance to digital space", and described XR technology as follows on its official website: "XR refers to VR/AR/MR and other related technologies, which are the next generation of experience revolution and computing platform, the advanced integration of digital space and physical world, and the revolutionary upgrade of computing power, connectivity and display". Products based on XR technology are regarded by Hisilicon as the entrance to the subsequent new world and "can better realize the connection, mapping and enhancement of digital space and the real world".

It can be seen that XR technology will have unlimited possibilities and can drive the development and innovative application of the three visual technologies of AR, VR, and MR. As a result, XR technology has been applied in construction, education, medical, entertainment, industrial manufacturing, and other industries. With the rapid development and in-depth application of digital twin, 5G technology, Internet of Things, and other technologies, it will not only promote the development of XR technology but also provide a new idea for the innovative application of XR.

For example, in the Tokyo Olympic Games, which were popular due to the fact that the judges still held their posts when they were ill, XR + 5G technology has been applied to sailing, swimming, and golf course halls. Viewers will be able to put on smart glasses and see relevant statistics about athletes. It can be predicted that over time, sports events will bring more rich experience, and the cost of organization and operation will also be greatly reduced.

Chapter Summary

At present, it has become a consensus of all countries to incorporate digital twin technology into the strategic development plan, and the trend of deep application of digital twin in various fields is irresistible. In the process of its application, the solution of "efficiency increase and cost reduction" in various industries has reached a new height of digital concept. The deep integration of digital twin technology and other technologies will also promote scientific and technological innovation and provide a strong impetus for China to seize the commanding heights of science and technology.

Chapter 18

"Mirror World" of Digital Twin Reproducing Transportation Equipment

Digital twin, as the name implies, is "twin in digital form". Unlike the real twin, one of the digital twin exists in the real world, and the other exists in the virtual "mirror world". Based on the high similarity in shape and structure, digital twin has a wide range of applications. It can also create fruitful results in the field of transportation equipment industry with the in-depth implementation of "smart power" and the deep integration of digital twin and transportation industry.

18.1 Application of Digital Twin in Transportation Equipment Layer

Through the innovative application of digital twin technology in the field of transportation equipment, China has made major breakthroughs in the design, manufacturing, operation, and maintenance of transportation equipment such as high-speed rail, intercity rail, and underwater tunnel. Moreover, the digital twin has achieved whole lifecycle coverage in the transportation equipment manufacturing process. As the first link of manufacturing, the equipment layer should take the lead to strengthen the in-depth application of digital twin in the equipment layer.

18.1.1 *How to give new life to transportation equipment in digital space*

In recent years, with the innovative application of digital twin technology, the virtual space constructed based on digital twin has given new life to transportation equipment, and also provided a strong driving force for the construction of China's "strengthening country with strong transportation network".

With the support of modern science and technology, China has completed the construction of a number of highly difficult transportation equipment projects. With the emergence of the Pearl River Mouth tunnel, the first undersea high-speed railway in China, means that China has made another major breakthrough in the field of transportation equipment.

The Pearl River Estuary tunnel is part of the high-speed railway built in connection with the Belt and Road Initiative. The Shenzhen–Jiangmen high-speed railway connects Shenzhen and Jiangmen and passes seven stations, including Xili, Shenzhen airport, Dongguan Binhai Bay, Nansha, Zhongshan north, Henglan, and Jiangmen. Among them, the Pearl River Estuary tunnel is located at the estuary of the Pearl River between Dongguan and Guangzhou, which is the only way for the Shenzhen River high-speed railway. It is a great problem for the high-speed railway to cross the Pearl River Estuary as the geographical environment near the Pearl River Estuary is very complex and the water corrosiveness is relatively serious.

With the advantage of digital technology, China Railway Engineering Equipment Group Co., Ltd. and China Railway Tunnel Group jointly developed the "No. Greater Bay Area" super large diameter slurry balance shield machine, which solved the problem. With the help of "No. Greater Bay Area", the project completed the construction of the Pearl River Estuary tunnel with a maximum underwater depth of 115 m and a maximum water pressure of 1.06 MPa.

Under the guidance of advanced technologies such as digital twin, China has also made major breakthroughs in the construction of high-speed trains, heavy loaded trains, urban rail trains, and other transportation equipment fields, enabling the transportation mode to truly achieve the omnipotent "go to the sky and enter the sea".

In the future, society will be fully digitalized, and digital transportation will be an important part of the digital economy. Therefore, the whole society should make full use of the advantages of digital twin technology, especially in the transportation equipment manufacturing industry, and actively integrate with the Internet, Internet of Things, artificial intelligence, big data, and other technologies, accelerate the digital and intelligent construction of China's transportation equipment, and promote the transformation and upgrading of the transportation equipment field.

18.1.2 *Whole lifecycle + real-time mapping + two-way interconnection*

The key to the deep application of digital twin lies in its three major features, namely whole lifecycle, real-time mapping, and two-way interconnection.

The whole lifecycle refers to the whole process of an organism from birth to death in biology. The whole lifecycle in the field of transportation equipment refers to the whole process of design, research and development, manufacturing, testing, operation, and maintenance of transportation equipment covered by digital twin.

Real-time mapping refers to the comprehensive real-time connection between physical entities and digital twin.

Two-way communication refers to the two-way flow of data and information between physical entities and their digital twin. During the whole lifecycle operation, the digital model will interact with each link of transportation equipment manufacturing in real time and continuously through sensors, artificial intelligence, operation history data, etc. and complete accurate mapping of physical entities in digital space so as to improve its own accuracy. At the same time, it forms a data flow closed-loop between the digital space and the physical world. In the end, real-time monitoring in various aspects, directions, and scales is carried out by the use of digital space model, providing scientific reference for each link of operation, and further optimizing and improving the work of all stages of the whole lifecycle of transportation equipment.

In view of the three characteristics of digital twin, digital twin has been applied in many aspects in the field of transportation equipment. For example, Pci Technology has built a set of digital twin system in a large city, which is one-by-one corresponding and mutual mapping with the "whole time, whole range and whole volume" information of urban pavement transportation. In addition, at the "2021 Zhongguancun Forum", Sun Chitian, technical director of Tencent's autopilot simulation, shared the digital twin technology to accelerate the landing of autopilot, and said that it can solve the problems of intelligent Internet connection, autopilot test, efficiency, and safety in the development of intelligent transportation through the characteristics of digital twin.

18.2 Application of Digital Twin in Transportation Equipment Production Line Layer

In the three links of transportation equipment manufacturing, the production line level involves a number of simple and complex processes, which is a very important part. However, due to the limitations of the traditional production line, such as relatively independent process, data transmission delay, and machine accuracy, the production efficiency and cost, product quality cannot be guaranteed continuously and stably. For this reason, digital twin provides managers with multi-perspective status monitoring and visual display of various data, which strongly promotes the construction of Intelligent Manufacturing in China.

18.2.1 *Alike in appearance is not the purpose, but alike in essence is the foundation*

In recent years, with the hot concept of digital twin, in order to strengthen their core competitiveness and seize the digital market, most enterprises in China have carried out digital transformation and actively carried out digital space construction. However, although the external form of the twin built by many enterprises is highly similar to the physical entity, it is "just alike in appearance but not in essence". The twin constructed by these enterprises is only a simple mapping of physical entities and does not fully possess the key features of digital twin, so it can only be called similar in appearance. Similar in essence, on the other hand, is oriented by the business of the production line, can reflect the operation status of the production line in real time according to the information fed back by the sensors, can predict the operation track of the production line through historical data and real-time data, and provide scientific decision-making support to the managers according to the prediction results. For example, we can analyze when we need to replenish materials and timely adjust the movement deviation of the property robot according to the operation speed of the production line.

After achieving similar in essence, digital twin technology is actually a very "simple" technology. Its so-called "simple" lies in the scope of application of digital twin. Digital twin is a "universally applicable theoretical and technical system", which can be applied in many fields, such as logistics, transportation, manufacturing, as well as communication. At the same time, it can also be applied in collaboration with many technologies. For example, it cooperates with Internet of Things, Internet, artificial intelligence, virtual reality, augmented reality, and other technologies.

Of course, digital twin technology is also a very complex technology. Digital twin is a "multi-disciplinary, multi-physical quantity, multi-scale and multi-probability simulation process" which integrates physical model, sensor update, operation history, and other data. The specific development and application of this process require users to go deep into each specific application situation to create global control.

Similar in appearance is not the purpose, but similar in essence is the foundation. Only by mastering the digital twin completely and continuously innovating on the original basis can we better promote the development process of digitalization.

18.2.2 *Three-dimensional is only one kind, decision-making is the soul*

The deep application of digital twin technology in many fields lies not only in the fact that it can conduct three-dimensional simulation of management elements, such as equipment, production lines, and factories, but also in the fact that it can provide comprehensive and accurate decision-making guidance to managers. Therefore, the focus of the digital twin of production line layer is not on the three-dimensional representation, but on the best decision-making evaluated through repeated testing of equipment operation, maintenance, product design, manufacturing, and other aspects of the model.

The core of digital twin lies in decision-making, which is driven by data. Exactly based on massive data can digital twin realize the diagnosis of the past problems of the transportation equipment production line, the analysis and evaluation of the current status of the production line, as well as the prediction of the future operation trend. On this basis, it provides scientific, comprehensive, and accurate decision-making basis for managers.

The decision scheme of the same transportation equipment production line is not unchanged, but needs to be continuously optimized according to the information fed back by the sensors. The rapid development of modern technology is also conducive to the intelligent decision-making of digital twin.

At the 2021 Alibaba Cloud Summit, Alibaba Cloud, and MMI Planning & Engineering Institute IX., Ltd. jointly released the digital factory 1.0 of the automobile industry based on digital twin technology, and the new factory area of FAW Hongqi became the first "crab eating" factory. Here, a car is composed of more than 3,000 categories and 30,000 to 50,000 kinds of components. Digital twin will precisely map every part into digital space, and the tightening force of each screw will be accurately monitored by the digital model.

The core of digital twin is decision-making, and the most important value of decision-making is to interfere with the operation of production lines. For example, the production quality is judged and forecasted in a timely manner to prevent unnecessary parts damage or rework and other problems through the real-time monitoring of stamping, welding, coating, final assembly, and battery in automobile production, so as to ensure the quality of finished products while achieving efficiency increase and cost reduction. Since decision-making can generate such value, it can become the soul of digital twin.

18.3 Application of Digital Twin in Transportation Equipment Factory Layer

Enterprises can further use digital twin technology to build factory level in digital space on the basis of equipment level and production line level, centralize and integrate all control systems of each link of the whole transportation equipment factory, and conduct comprehensive control by a unified platform to ensure the efficient operation of production and manufacturing work in each link.

18.3.1 *Moving factories into virtual space is a future trend*

With the issuance of the strategic document "Made in China 2025", China has gradually moved from a "manufacturing country" to a "manufacturing power". With the rapid development of digital economy, intelligent manufacturing has become the main direction of the construction of intelligent manufacturing power. Therefore, the combination of factory and digital is the development trend in the future.

Through digital twin and other technologies, the details of the factory are constructed in a virtual space, enabling point-to-point information communication between machine parts and components in the factory. Based on the rich data and advanced algorithm model, the factory digital twin can provide whole lifecycle support, optimize the shortcomings between product design and operation and maintenance, make up for the shortcomings, and promote the transformation and upgrading of the manufacturing industry.

At present, China's digital economy is developing rapidly. The scale of China's digital economy will reach RMB 39.2 trillion in 2020, according to "China Internet Development Report 2021". DTT also said that by 2023, the global digital twin market is expected to grow at an annual rate of 38% to reach USD 16 billion (€14.2 billion). However, the development of the digital economy does not mean the elimination of people by machines. On the contrary, the digital economy can effectively promote the transformation and upgrading of the real economy, thus further promoting the economic performance, including the employment rate. In 2020, Tan Jianrong, an academician of the Chinese Academy of Engineering, pointed out that the application of digital twin in the industrial field can achieve resource allocation, intelligent production, and significantly improve production efficiency. While observing the performance of digital twin technology today, we also believe that moving the factory into the virtual space will not reduce the employment capacity of the factory but will create more and more creative jobs

In short, digital twin helps many manufacturing enterprises, including transportation equipment manufacturing. According to Partner, 50% of large industrial companies will use digital twin by 2021, resulting in a 10% increase in efficiency. The factory of the future is a digital factory. The construction practice of smart engineering, such as smart logistics, smart city, and smart water affairs, is also sufficient to prove that human beings have strong control over digital complex systems, and digital chemical plants are bound to move forward from this.

18.3.2 *Digital twin factory, which can not only predict the future but also play back the history*

With the development of modern science and technology, great changes have taken place in the mode of social production. The product lifecycle is getting shorter and shorter, and the demand for private customization is increasing gradually, all of which forces the factory to change to digitalization and automation. In order to enhance their core competitiveness under the premise of cost reduction and efficiency increase, major factories have gradually developed a new approach, i.e. using the ability of digital twin to play back history and predict the future, to carry out intelligent analysis and auxiliary decision-making on the operation of factories, and to escort the design, research and development, manufacturing and testing of products.

The success of digital twin factory needs to go through three links to realize historical revisit and predict the future.

18.3.2.1 *Historical operation data*

Data is the blood of digital twin applications. The digital twin system can obtain massive original data of transportation equipment through the Internet, including the whole lifecycle digital files of physical entities. At the same time, sensors can be used to transmit real-time operational data of physical entities.

18.3.2.2 *Real-time monitoring of data*

The data is integrated and analyzed through artificial intelligence technology, and accurate mapping of digital models is driven to achieve real-time monitoring of physical entities of transportation equipment.

18.3.2.3 *Future simulation data*

On the basis of a large amount of data and real-time monitoring, the simulation model is used to test and evaluate the future trajectory of physical entities of transportation equipment in a repeatable and variable manner, so as to obtain the best test results.

The ability of digital twin to play back the history and predict the future is the result of the collaborative integration of the three aspects of historical operation data, real-time monitoring data, and future simulation data under the support of various technologies.

In the increasingly competitive market environment, it is an inevitable trend for future development to use digital twin technology to digitalize factories. Through the digital model, the personnel, equipment, production lines, workshops, and other aspects are deeply restored in the digital space and controlled by the system. The managers of the factory can monitor the operation of the factory in all aspects in real time through the digital space model and can simulate the product design, production process, and equipment operation through the model to prevent site failures.

Case: understand what digital twin is from the new factory area of FAW Hongqi

In 1958, since the birth of the first Hongqi automobile and becoming a state vehicle for national leaders and major national activities, Hongqi automobile has become a flag of China's automobile industry. A few decades have passed. In the ever-changing technology era, the Hongqi brand has never slowed down, kept up with the development trend of the times, continued to carry out scientific and technological innovation, and insisted on promoting the intelligent and digital construction of the plant area.

In June 2021, the construction of Changchun International Automobile City FAW Hongqi New Energy Automobile Factory was basically completed. At the end of June, the first batch of new energy vehicles were off the production line in the factory, and at the end of 12, the factory officially started mass production.

The reason for the rapid development of FAW Hongqi new energy automobile factory from the completion to the end of the production line is due to the comprehensive control of the intelligent central control system.

On 17 July 2020, Alibaba Cloud announced that it has reached a strategic cooperation with the 9th Institute of Machinery Industry, which is named as the"cradle of China's automobile factory design". The two sides will make use of their respective advantages in research and development, technology, and other aspects to jointly build intelligent factories for automobile enterprises, committed to promoting the digital development of automobile factories and promoting the digital transformation and upgrading of China's automobile industry and manufacturing industry. The intelligent central control system applied in the new factory area of FAW Hongqi was jointly developed by Alibaba Cloud and the 9th Institute of Machinery Industry.

The intelligent central control system is designed and constructed based on digital twin. Under the support of Internet of Things, artificial intelligence, big data, cloud computing, and other technologies, through the self-developed data acquisition monitoring software, the data of millions of points in the five workshops in the plant area are obtained, and all the data are sent to the intelligent central control system platform, which analyzes, tests, evaluates, and improves the business according to the data, so as to achieve efficient and low-cost automatic operation.

The new factory area of FAW Hongqi shows the way of automobile digital factory. If we conclude "automobile digital factory" in one sentence, it is to use the integration of digital twin system and other technologies to help automobile enterprises obtain production data, understand data, and make good use of data, so that the operators of the factory also have the ability of digital innovation.

Chapter Summary

The distinctive feature of the new generation of transportation equipment lies in its digital nature. It is the common development trend of the world's manufacturing industry to move forward from "manufacturing" to "intelligent manufacturing", and the digital twin technology has helped the transformation of the transportation equipment manufacturing mode. Looking forward, China will build an intelligent factory operation system with the characteristics of "comprehensive perception, equipment interconnection, digital integration, intelligent prediction" for the transportation equipment industry by relying on digital twin and other technologies.

Chapter 19

"5G + AI" New Transportation Mode

AI has been used more strongly with the support of 5G as 5G network becomes commercialization, and face recognition, voice recognition, data analysis, unmanned driving, intelligent robot, and other technologies have been improved, providing better solutions for solving transportation problems on urban roads. The integration of "5G + AI" into the construction of intelligent transportation will become the trend of urban development nowadays.

19.1 Digital Twin Transportation Under "5G + AI" Mode

The digital twin transportation under the "5G + AI" mode is the mutual integration and creation of the real world and the virtual space. The digital twin space channel between the physical world and the cloud can be opened up to solve the transportation problems encountered in the real world by collecting the dynamic and real-time transportation data in the real world for analysis and processing and transferring to the virtual space, and then digitalizing and modeling it.

19.1.1 *"5G + AI": new technology piling up and new equipment rising*

Since the concept of intelligent transportation was put forward, countries around the world have accelerated the research and development of technology in the field of transportation, resulting in a variety of innovative technologies that are dizzying. The field of transportation has become a field of innovation and technology.

Many innovative research and development of new technologies are starting the prelude to future transportation. Large-scale automatic driving is no longer just a talk over paper among the many cutting-edge applications and solutions related to intelligent transportation leading to the future, and digital transportation infrastructure will also continue to develop. Among them, virtual simulation, digital twin, and other innovative technologies will play a crucial role.

19.1.1.1 *Virtual simulation is widely used in the field of automatic driving*

It is believed that many people's understanding of virtual simulation technology starts from various large-scale RPG (role-playing) games. They can feel various light and shadow effects and physical effects close to the real nature in the game world supported by data, as well as the AI technology behind the NPC (non-player role) that can make certain simulation behaviors. In fact, in various fields such as industry, aerospace, and transportation, virtual simulation technology has been applied at a deeper level for a long time.

Automatic driving vehicles are tested by virtual simulation technology, which can not only easily build extreme scenarios but also effectively reduce test risks and improve test efficiency. As a matter of fact, the virtual simulation platform applied to automatic driving is like an RPG driving game, and the main character is the vehicle running based on the automatic driving algorithm. Game technology is used to improve the authenticity of autopilot simulation test. Scene modeling, sensor modeling, 3D physical engine, animation engine, etc. are widely used in the field of automatic driving virtual simulation.

At the early stage of the research and development of automatic driving in China, Tencent, which is good at game technology, actively invested in the field of virtual simulation technology, and developed a virtual simulation platform TAD Sim. Combined with cloud computing capabilities, it built a cloud virtual simulation city, allowing test vehicles to test and verify in complex environments.

Compared with virtual simulation, digital twin lays more emphasis on the collection of physical world data and is a "digital parallel world" that is dynamically synchronized with the real world in real time.

19.1.1.2 *Digital twin technology is building intelligent transportation*

Digital twin-related technologies and the continuous development of artificial intelligence, sensors, communication, and other technologies are building the digital infrastructure of intelligent transportation. Gradually, virtual space and physical world communicate with each other based on real data, playing a role in more and more transportation fields. Its specific functions are as follows:

(1) Digital twin technology can be used for scientific deduction of transportation control. The real transportation information is transferred to the digital twin space, and the transportation behavior characteristics and rules are studied by AI, so as to analyze, judge and infer the changes of transportation flow, and then provide scientific basis for transportation management prediction. For example, local places often have large-scale transportation congestion after the concert and major world, national, and provincial events and other large-scale activities, which is a great test for the transportation management department to make predictions and responses in advance according to the actual situation. The deduction on the digital twin platform can provide an efficient and simple solution.
(2) Digital twin technology can be used to improve the authenticity of automatic driving test. The digital twin technology can build a real-time training environment according to the real road data in the cloud test field, in which the automatic driving algorithm can be simulated and tested. In the Changsha intelligent network connection test area, for example, Tencent uses digital twin simulation technology to carry out digital space migration of the geographical overview of the test area, so as to achieve efficient and safe automatic driving experiments under the digital simulation environment.
(3) Digital twin technology will assist in transportation construction planning. Through the design, assembly, and operation of the virtual transportation modeling in the digital twin space, the situation of each stage of the vehicle from design and manufacturing to operation and maintenance can be simulated. In this scenario, the vehicle can be "immersively" put into the whole project and carry out various operations, which can increase the technical means and cognitive field of the employees and save time and cost for the transportation

construction project, and also can improve the quality and efficiency, at the same time, it can also be simulated, so as to timely maintain and respond to emergencies during the construction.

19.1.2 *Upgrading digital transportation management platform according to international first-class standards*

Since the 21st century, data has penetrated into every industry and business function field with the continuous development of computer technology. For the development and utilization of massive data, due to the rapid development of the Internet and big data and other industries, it has attracted more attention. In the future, digital transportation will develop toward an integrated, digital, and intelligent digital transportation management platform based on big data from mobile transportation collection.

Since the gradual commercial use of 5G, the wide application of intelligent terminals, Internet of vehicles sensor collection, Internet of Things, mobile terminal equipment in aviation, water transportation, highway, railway, and other fields has further strengthened the scope, breadth, and depth of transportation data collection, which provides a basis for the development of transportation big data.

19.1.2.1 *Transportation big data development technology*

Efficient data development and machine learning algorithms provide strong support for the development and analysis of transportation big data with multi-source, heterogeneous, massive, and other characteristics, along with the development of AI intelligent learning, cloud computing, parallel computing, and other technologies. The large-scale transportation image recognition and video analysis technology based on AI intelligent learning, the positioning analysis technology based on mobile phone signal, the time-space characteristics analysis of comprehensive transportation data and digital modeling technology, and other big data development technologies provide the possibility for the potential value analysis and further application of urban transportation data.

19.1.2.2 *Transportation big data integration technology*

In the context of social digitalization, economic globalization, and regional integration, all kinds of risks have become more related and cross-border, and no place or department is not in need of coordination. The ability of internal data integration is not enough, so how to talk about external linkage and cooperation? Therefore, the data fusion application within the transportation industry and among various departments is particularly important.

The transportation big data integration technologies such as the transportation cross industry data sharing and exchange technology and the transportation cross industry multi-source and different data processing technology can effectively integrate and share the data information resources, platform resources, network resources and application resources among the multi-source and different data, which not only effectively breaks the information barriers within the transportation industry and among various departments but also provides a basis for the time-space perception and operation state management of large-scale urban integrated transportation system.

Meanwhile, it is pointed out in the "Outline for the Development Plan of Digital Transportation" issued by the Ministry of Transport that China will upgrade the digital transportation management platform according to the international first-class standards. By 2035, China's transport infrastructure will complete all elements, full-cycle digital transportation control network, and the on-demand real-time travel service will be widely used.

19.2 Transportation Incident Management Under the "5G + AI" Mode

From the development trend of digital transportation, it can be seen that the development trend of its technology is to integrate mature information technology such as computers into transportation, and achieve the goal of safe and efficient optimization by strengthening the coordination of road network channels before completely changing the traditional transportation mode. Consequently, the overall development trend of digital transportation is intelligent transportation tools, and integrated transportation intelligent services are carried out based on the "5G + AI" model.

Under the "5G + AI" mode of transportation event management, the redundant loss in identification can be reduced through AI intelligent judgment. The main application technology is the wireless broadband transmission network which transmits large-capacity information. The progress of tool mode is an inevitable condition for realizing digital transportation.

5G wireless transmission network can avoid more geographical restrictions and is the key to the construction of digital transportation system.

The transportation incident management under the "5G + AI" mode is shown in the following forms:

(1) **Transportation information linkage in active mode:** The intelligent service platform App for transportation information is developed and constructed based on the transportation information collection system, which sends information such as congestion status, parking space resources of the destination parking lot, real-time transportation volume, and average speed of transportation flow promotion to the user's mobile phone terminal and the vehicle interconnection system terminal. The user can choose to use the ultra-high data transmission rate of 5G network to view real-time monitoring images according to the needs; thus, first-hand information in the car can be mastered.

(2) **Comprehensive transportation intelligence services based on "5G + AI":** The electronic toll collection system based on the Internet of Things was built to realize the free payment of transportation flow in combination with road identification, mobile payment, vehicle identification, vehicle weighing, and other technical means. Through the vehicle road collaborative service system based on the Internet of vehicles, and through the specific base station provided by the 5G mobile operator, the transportation information is provided to the transportation participants to ensure the safety of the public.

(3) **Accurate perception and intelligent control system in transportation operation:** By means of vehicle identification, magnetic induction, high-definition monitoring, weather monitoring and other multi-source collection, the road abnormal transportation events such as congestion, parking violations, wet rain and fog, as well as fire accidents are monitored dynamically through video analysis technology, real-time warning of transportation conditions is realized, and moreover, vehicles are guided to make early lane change or make a detour. This information will also be sent to the local taxi command system so that the operation efficiency of the taxi will be more efficient and the service will be more user-friendly.

Chapter Summary

"5G + AI" and the development of intelligent transportation promote and support each other. Intelligent transportation can deepen the development of "5G + AI", while the in-depth development of "5G + AI" has greatly promoted the gradual improvement of intelligent transportation, enabling it to have more diversified development situation and more advanced technical indicators in the subsequent development.

Chapter 20

Explore the Digital Space and Construct a Visual Intelligent Transportation Management System

With the continuous development of the information age, the previously envisaged smart city is gradually reflected in real life, which greatly facilitates the daily transportation management and maintenance. Under the background of smart city construction, intelligent transportation supervision visualization system is an important part of it. The management and control of a smart city is extended through a series of monitoring of roads.

20.1 How to Build a Comprehensive Three-Dimensional Transportation Network of "Digital + Intelligent"

With the development of modern science and technology, 5G, Internet of Things, and high-precision map technologies are rendering and filling digital space more refined, real-time, and simulation; at the same time, cross-regional and cross-industry multi-source data are also constantly gathering to form a macro circulation transportation management model. This has brought convenience to the construction of the "digital + intelligent" integrated three-dimensional transportation network.

20.1.1 *Planning objectives of the new integrated three-dimensional transportation network*

In February 2021, the Central Committee of the Communist Party of China and the State Council issued the "Outline for the National Comprehensive Three-dimensional Transportation Network", in which the five development goals of China's new comprehensive three-dimensional transportation network by 2035 are clearly proposed, and the specific contents are as follows.

20.1.1.1 *Convenience and smoothness*

The proportion of the population enjoying the rapid transportation service has increased significantly. Except for some remote areas, the county-level administrative centers in China basically achieve 15 minutes to the national road, 30 minutes to the highway, 60 minutes to the railway, 45 minutes to the high-speed railway, as well as 60 minutes to the airport. Basically, it can be reached between prefecture-level cities on the same day. It only takes half an hour to reach the comprehensive passenger hub from the central city, and the bus transfer time between the comprehensive passenger hub and the central city shall not exceed 1 hour. The barrier-free rate of transportation infrastructure has been greatly improved, and the convenience of the whole travel chain of passengers has been significantly improved, basically realizing the "national 123 travel transportation circle".

20.1.1.2 *Economy and efficiency*

The national comprehensive three-dimensional transportation network facilities are more efficient, the proportion of multi-modal transportation and the efficiency of loading and unloading are significantly improved, the transportation structure is more optimized, and the logistics cost is further reduced. The transportation hub basically has the delivery function, realizing the seamless connection with the delivery hub, and basically realizing the "global 123 fast cargo logistics circle".

20.1.1.3 *Green intensification*

The level of intensive and comprehensive utilization of comprehensive transportation corridor resources has been greatly improved. The whole process and cycle of transportation infrastructure construction is green. The energy consumption per unit of transport turnover continues to decline, the intensity of carbon dioxide emissions is significantly reduced compared with 2020, and the prevention and control of transportation pollution has reached the world's advanced level.

20.1.1.4 *Advanced intelligence*

It has basically realized the digitalization of all elements and full cycle of the national comprehensive three-dimensional transportation network infrastructure. It has basically completed the extensive and advanced transportation information infrastructure to achieve full coverage of Beidou's space-time information services and transportation perception. The technologies of intelligent trains, intelligent Internet-connected vehicles, intelligent general aircraft, intelligent ships, and postal express facilities have reached an advanced level in the world.

20.1.1.5 *Safety and reliability*

The durability and effectiveness of transportation infrastructure have been significantly improved, and the ability to prevent and deal with potential safety hazards of the facilities and the ability to prevent and control them have been greatly improved. The resilience of the transportation network and the ability to cope with various major risks have been significantly improved, the incidence of major and major accidents has been continuously reduced, and the transportation of important materials is efficient and reliable. Basic resume of transportation safety supervision and rescue system with three-dimensional coordination of land, sea, air, and sky. The level of transportation safety has reached the forefront of the world, effectively safeguarding people's lives and property and the overall safety of the country.

20.1.2 *Current situation of China's integrated three-dimensional transportation network*

At present, China's national comprehensive three-dimensional transportation network with railway as the backbone, highway as the basis, and giving full play to the advantages of water transportation and civil aviation is becoming increasingly perfect. The main skeleton space of the national comprehensive three-dimensional transportation network with 6 axes, 7 corridors, and 8 channels has been preliminarily formed, and China's transportation development has made remarkable achievements.

According to the data of the Ministry of Transport, the coverage rate of high-speed railway in cities with a population of more than one million in China has exceeded 95%, that of expressways in cities with a population of more than 200,000 has exceeded 98%, and that of civil transport airports has covered more than 92% of prefecture-level cities. The high-capacity, high-efficiency, and rapid passenger service system dominated by high-speed rail and aviation is accelerating to form. The multiple unit trains bear 70% of the railway passenger volume, and the normal rate of civil aviation flights has exceeded 80% for 3 consecutive years. The transformation and upgrading of road passenger transport is also accelerating. The freight structure has been continuously optimized. From 2016 to 2020, the proportion of railway in freight volume increased from 7.6% to 9.8%, and the proportion of water transportation increased from 14.5% to 16.4%.

It has basically realized access to roads, buses, and postal services in towns and villages with conditions. The problem of bad road conditions in poverty-stricken areas has been solved historically. China has also launched express delivery to the countryside, slow trains, civil aviation short-distance routes, etc. to provide better services for the masses. A large number of key projects such as the Hong Kong–Zhuhai–Macao Bridge, Beijing–Zhangjiakou Railway, Beijing Daxing International Airport, and Shanghai Yangshan Port Automation Terminal have been put into use. China road, China bridge, China port, and China high-speed railway have become beautiful Chinese name cards.

With the gradual improvement of China's comprehensive three-dimensional transportation network, the income scale of China's postal express industry has expanded rapidly, from RMB 198 billion in 2012 to 11038 billion in 2020, with a compound annual growth rate of 24%. The volume of express business has increased from 5.7 billion units to 83.5 billion units, with a net increase of 14 times and an annual growth of 40%.

Since the 18th National Congress of the Communist Party of China, the operating mileage of China's newly put into operation railway of 53,000 kilometres is the world's first, and major railway transportation economic indicators such as railway passenger turnover, freight delivery volume, as well as freight turnover volume have steadily ranked first in the world.

In 2020, China's aviation services covered 92% of the country's prefecture-level administrative regions, 88% of the population, and 93% of the total economy. The number of routes reached 5581, an increase of 3124 or 120% over 2012. Aviation services covered 82.6% of the population in poverty alleviation areas.

The current situation of China's comprehensive three-dimensional transportation network has also been publicized in the "Outline for the National Comprehensive Three-dimensional Transportation Network". It is expected that the annual growth rate of passenger transportation (including household car transportation) in China from 2021 to 2035 will be about 3.2%; the annual average growth rate of freight volume of the whole society is about 2%, and the annual average growth rate of postal express business is about 6.3%; a modern high-quality national comprehensive three-dimensional transportation network with convenient and smooth operation, economic efficiency, green intensification, advanced intelligence, safety, and reliability has been basically established.

20.1.3 *Hierarchical coverage of "digital + intelligent" technology*

The initial manifestation of technology changing life is not limited by cash, bank cards or even mobile phones, only one face can realize the subway ride; nowadays, with the help of AI and big data technology, Shenzhen Airport has reduced the allocation of seats for more than one thousand flights from the original 4 hours to 1 minute, and 2.6 million passengers are exempted from shuttle bus every year; with the help of "5G + AI" technology, Port of Xiamen has realized "unmanned tallying" and remote control of port cranes.

20.1.3.1 *Transportation "new infrastructure", from the traditional "transportation" era to the innovative "data" era*

The promotion and application of new technologies have significantly improved people's logistics efficiency and travel experience under the background of the information age. However, with the acceleration of urbanization, due to the limitation of limited resources such as roads, routes, and ports, the efficiency contradiction of vehicles is still prominent. Firstly, the efficiency of existing vehicles has not yet been fully released, and secondly, travel and logistics demand is still growing with the development of economy. In the case of limited transportation supporting resources, improving travel experience and transportation efficiency has become an urgent and urgent problem to be solved.

With the deepening of digitalization of transportation business, in the foreseeable future, transportation is no longer only independent passenger and cargo transportation in railway, highway, waterway, and aviation vertical fields, but an integrated three-dimensional transportation network formed around the three business flows of public travel, cargo transportation, and transportation. The key to solving the problem is to deeply integrate the "data" provided by new information and communication technologies, such as cloud computing, 5G, AI, and big data, with the transportation business, so as to create a safe, efficient, and experienced intelligent transportation. The transportation industry will change from the traditional "transportation" era to the innovative "data" era.

20.1.3.2 *Hierarchical coverage of "digital + intelligent" in transportation agents*

HUAWEI has proposed a reference architecture for intelligent upgrading of the transportation industry, namely "transportation agent". An open, accurate judgment, multi-domain collaborative, three-dimensional perception, and continuous self-evolution intelligent system is gradually built to achieve efficient collaboration through the effective integration of cloud, AI, computing, and other technologies. With a "stable and efficient" structure, it responds to the diversified innovative needs and challenges of intelligent transportation in the future and supports the intelligent upgrading of the transportation industry.

According to this concept, transportation agents will become an open technology architecture shared by the industry. Through the hierarchical coverage of the four-layer technical structure of intelligent connection, intelligent hub, intelligent interaction, and intelligent application, it supports the construction of the digital collection system, network transmission system, and intelligent application system of the transportation industry, and speeds up the development of transportation information in the direction of digitalization, networking, and intelligence. Details are as follows:

(1) **Intelligent connection:** Including 5G, Wi-Fi, IP, and other key information technologies, it can realize the internal connection of the intelligent center, the connection from the intelligent center to the intelligent interactive equipment, and the mutual connection between the intelligent interactive equipment. Through the seamless coverage and the Internet of Everything, application collaboration, data collaboration, and organization collaboration can be realized.
(2) **Intelligent hub:** The intelligent center is the decision-making system of transportation agents, just like the human brain. Relying on the cloud infrastructure, it orderly integrates key digital technologies and capabilities, and provides and calls them through API interface to enable applications and make data inclusive of AI.
(3) **Intelligent interaction:** Intelligent interaction is the basis of connecting the physical world and digital space of transportation, which includes various perception devices distributed in the field stations, hubs, and road networks, such as smart cameras, radars, as well as sensors.
(4) **Intelligent application:** Intelligent application is the value presentation of transportation agents. It can optimize the experience and restructure the process through collaborative innovation with customers and partners, so as to help transportation customers achieve a comprehensive improvement in safety, efficiency, and experience in the business layer.

20.1.4 *Full replication of virtual space in integrated transportation network*

Any country in the process of urbanization will encounter the symptoms of "urban disease". For large and medium-sized cities, the transportation problem of difficulty in traveling and parking has become a common problem. Through the full replication of the virtual space of the real integrated transportation network, various transportation problems can be effectively managed. "Digital + intelligent" is used to empower transportation and help to realize the smart city and a better life in an all-round way.

20.1.4.1 *Real-time "high precision data service"*

The higher precision and more real-time data service is the last tile for the construction of transportation digital space and also the base and foundation of the "digital + intelligent" integrated three-dimensional transportation network.

If the world we see is ambiguous and has delayed refresh, we cannot make accurate judgment and behavior feedback on it; this is especially true in the transportation world, because the essence of transportation is fast, dynamic, and interactive.

When the space construction and positioning with higher resolution significantly increase the amount of space information, intelligent transportation (assisted driving/automatic driving) can only approach or even exceed people's perception of the road and road conditions, which is beyond the requirements of transportation scenarios that cannot be achieved in the past with low resolution: "know" from mechanical information and "perception" from dynamic biomimetic, the "super sense" qualitative change from the perspective of God is a space-time cognitive process jointly completed by high-precision map, high-precision remote sensing, high-precision positioning, and the Internet of Things.

Taking high-precision map as an example, its collection principle and data update process are similar to the perception process. For example, Emapgo, a domestic high-precision map merchant, adopts the combination of field collection and the new "crowdsourcing + deep learning" mode. While collecting data with high-precision field equipment, it also regularly pushes crowdsourcing tasks or UGC to Internet user groups to collect road and POI data; through AI image recognition, the information of road signs is updated automatically.

20.1.4.2 *Data graph conversion technology and time-space index technology*

If the multi-source and massive time-space data is not organically integrated, managed, organized, distributed, and shared, the data will still be randomly stacked construction materials, unable to build the overall digital space, and even unable to form digital twin.

The digital space of transportation is not a comprehensive multi-source data, but more like a transportation network system. The data is arranged and combined in a structured and orderly manner, which can not only quickly and deeply see the micro world of precise positioning of a certain sensor beside the road but also make a comprehensive view of the overall three-dimensional transportation structure of multiple and cross regions.

The multi-source data is automatically organized and released in accordance with the time-space order, and rapid query and real-time information extraction are realized through the time-space index technology. While improving the management efficiency of multi-source data, it can also effectively realize the integration of spatial data, video stream, Internet of Things, and other data at the bottom level, and then interactive analysis; at the same time, it can maintain the integrity of the client business application data and the server data, and the spatial relationship remains unchanged. It can be accurately and rapidly presented at the front end, which can solve the problems of efficient management of multi-source and massive data convergence of transportation departments, accurate time and space matching, etc.

20.1.4.3 *Real-time spatial operation based on high-precision positioning of Beidou*

The three powerful tools, namely high-precision spatial data service, Beidou system and real-time spatial computing, are integrated and applied to realize the overall digital base formed after the end-to-end real-time perception and early warning feedback, which will release more powerful energy in the field of intelligent transportation: It is not only essential in emerging fields such as intelligent driving, vehicle road collaboration, and Internet of vehicles; even at present, it can also benefit the public in a variety of travel scenarios, such as high-speed, night, rain, and fog weather, and other risk driving environment, real-time warning of abnormal conditions in front of the lane, and smart parking on the urban side.

At present, the digital base has been put into operation in the field of smart high-speed, which effectively helps the transportation management department to improve the safety guarantee and emergency rescue capabilities of high-speed drivers and passengers, reduce the subsequent chain accidents caused by abnormal parking and accidents in high-speed scenarios, as well as the driving risks caused by high-speed bad weather to drivers and passengers.

20.2 Digital Twin: Intelligent Transportation Visual Decision System

The intelligent transportation visual decision-making system is a real-time, accurate, and efficient comprehensive transportation management system which integrates advanced information technology, data communication and transmission technology, electronic sensor technology, control technology, and computer technology into the whole ground transportation management system, and it is an important part of intelligent transportation in the future.

20.2.1 *Construction process of visual transportation monitoring system*

In the construction of visual transportation monitoring system, real-time data, equipment status, and video monitoring are indispensable. Among them, video surveillance has always been a part of the main body. "China Skynet" emerged under the influence of the Internet of Things. This is actually a surveillance system covering all the streets and lanes of the city, with 170 million surveillance cameras. In the next three years, another 400 million cameras will be installed to become a "public security video surveillance system", and it is related to people's daily safety and security management.

"To be rich, build roads first", transportation is the lifeblood of the economic development of a city, which is closely related to all aspects of people's daily life. Road, bus, rail transit and other facilities are the main embodiment of urban transportation. With the accelerated urbanization process brought by the promotion of economic and social development, the number of domestic vehicles in China has increased rapidly, and urban transportation problems have become increasingly severe. In order to alleviate various problems faced by urban transportation, transportation management departments have adopted a variety of solutions, such as the construction of a series of intersection checkpoint monitoring, video monitoring, signal light control and other monitoring and control equipment, which has played an effective role to a certain extent, but each equipment has independently solved its corresponding problems, which is unable to form an overall system and overall control over transportation forms. The construction of visual transportation monitoring system can deal with this problem well. The specific construction process is described in the following.

20.2.1.1 *Transportation situation monitoring*

It supports the monitoring and visual management of the quantity, spatial location distribution, real-time status, and other information of the basic transportation resources such as cameras, bayonet equipment, flow detection equipment, as well as transportation lights, supports the query of detailed information of equipment, alerts the equipment that does not work normally, strengthens the monitoring and perception of equipment status by the managers, and improves the operation and maintenance management efficiency of transportation infrastructure.

20.2.1.2 *Video inspection and monitoring*

Through the integration of the front-end video inspection system and the effective combination of video intelligent analysis, intelligent positioning, and intelligent research and judgment technology, visual monitoring is carried out for the sections with safety hazards, sections that are prone to sudden accidents and sections that are often congested, so as to realize real-time alarm and rapid display of abnormal events, and the peripheral monitoring video of abnormal points can be accessed intelligently, so as to effectively improve the efficiency of receiving and handling police.

20.2.1.3 *Intersection signal light monitoring*

The real-time monitoring is conducted through the collection of transportation flow, flow rate, vehicle and road abnormal events, signal light status, and other information, and moreover, AI modeling intelligent algorithm can be combined to compare the historical best transportation speed and the best transportation volume, at the same time, visual analysis is conducted to study and judge the transportation situation at the intersection, so as to provide scientific decision-making basis for the optimization of transportation organization and signal timing at the intersection, and fully improve the transportation operation efficiency.

20.2.1.4 *Analysis of illegal cases*

The existing data resources of the transportation management department are efficiently integrated. Through a variety of visual interaction and analysis and judgment technologies, the massive historical violations of laws and regulations cases data are visually analyzed in series, and the time-space distribution rules of cases are deeply explored, providing support for the transportation management department to carry out cause analysis, active prevention, and other business applications.

20.2.2 *Construction method of immediate emergency command and dispatching system*

In the process of transportation, materials and equipment cannot be transported in time once the road transportation is blocked and the personnel, it will bring incalculable losses to people's lives and property safety. Therefore, ensuring the safety and smooth of road transportation is an important support for the safety and convenience of transportation. The purpose of the urban transportation emergency command and dispatching system is to establish a scientific and standardized transportation emergency management information system, so as to efficiently conduct the emergency command and dispatching in case of emergency transportation accidents in road transportation. Its main construction methods are described in the following:

(1) The on-duty emergency communication means adopts the wired communication system and is equipped with wireless communication equipment (such as mobile police terminals), telephone scheduling, multi-channel fax, digital record, and other systems to ensure safe and smooth communication between the emergency management department and the relevant departments at the upper and lower levels. At the same time, the image, voice, data, and other information transmission between the emergency scene and the emergency platform is realized by means of Beidou satellite, cellular mobile, and cluster communication.
(2) Based on the database server, message server, application server, GIS server, video monitoring disk array, etc., the construction of database management system, application middleware, GIS middleware, PGIS call interface, as well as message service mechanism further

strengthens and improves the construction of technical support platforms such as database and geographic information services.

(3) Sort out the data sources of the existing business system and other industry information resources, establish a data exchange and sharing information platform through the integration of information resources, and form a special data resource database of urban transportation immediate emergency command and dispatching, including road facilities database, transportation information database, vehicle management database, emergency resources database, emergency plan database, as well as event information database.

(4) In view of the actual needs of the real-time emergency command and dispatching management, the mature, advanced, and applicable information network technology is fully used to realize the functional design and development of various systems, and further enrich the information display and release means.

(5) Fully consider the development of data interface for data exchange and sharing with the relevant systems of transportation management departments, relevant government departments, and other units, so as to realize the co-construction and sharing of emergency resources and the coordinated command of emergency management.

20.2.3 *Core module of efficient and accurate data research and judgment system*

Aiming at the transportation data collected by various monitoring means, an efficient and accurate transportation data research and judgment model is established to analyze and process the transportation data. At the same time, under the guidance of the overall construction idea of "mastering the current status, finding out the rules, and scientifically guiding", the transportation parameters within 15 minutes are predicted according to a large number of real-time detection data and historical data, which can make the transportation managers timely understand the changes in the road network operation, take countermeasures in a timely manner, and provide reference information for travelers.

20.2.3.1 *Analysis of road transportation operation*

A comprehensive analysis of the road travel time is made according to the vehicle number information collected by the transportation flow information collection system, road vehicle intelligent and detection record system, then the travel time status of different road sections is obtained, and displayed based on GIS map. Travel time analysis can be divided into three steps:

(1) Monitoring is set up at the upstream and downstream intersections of the road, and the average time and the minimum time from the upstream intersection to the downstream intersection are calculated based on experience.
(2) For the transportation record of the downstream intersection, the system will query the transportation record of the same vehicle in the transportation record buffer of the upstream intersection. If it is found, the transportation time will be judged first, and if it is less than or greater than the preset time, it will be eliminated. For vehicles with "appropriate" driving time, the system will establish a new driving record based on the driving records of upstream or downstream intersections, and add an automatic driving time calculation system.
(3) After collecting a certain number of driving records, the calculation system uses some variable parameters to process all the records with a de-boundary value, i.e. removing the records with too long or too short travel time and taking a small amount of more appropriate average travel time as the estimated travel time.

On the basis of the transportation parameters generated in different time and space, the change rules of transportation parameters of upstream and downstream sections are excavated, and the transportation parameter prediction model is established, so as to predict the transportation parameters in the future in advance.

20.2.3.2 *Prediction of transportation parameters at intersections*

For a specific time (such as rush hour), the transportation volume and speed of the adjacent sections of an intersection are predicted. The middle level indicators (average formation speed ratio of the road, transportation saturation of the road, etc.) and the bottom level indicators (queue length ratio) are selected to analyze and determine the operation of the urban road sections, and then the road operation of the intersection is determined, from the perspective of time, generate early warning indicators of whether there is congestion at intersections, and set critical points for the formation and dissipation of short congestion.

At the same time, the travel time of a certain section is forecasted, and the travel time within 15 minutes of the key section is calculated. The final forecast results can be published in the whole city through the public media, such as transportation indicator panel and transportation radio, and can also be provided to the intelligent transportation service platform to meet the needs of the citizens for the early perception of intelligent transportation city intersections.

20.2.4 *Effective combination of multi-dimensional data perception system*

In the information age, data is the basic support for all algorithms and artificial intelligence. In intelligent transportation, the effective combination of multi-dimensional data perception system can effectively alleviate transportation congestion, improve transportation efficiency, improve transportation services, and maintain the optimal matching and dynamic balance of transportation supply and demand.

20.2.4.1 *Big data of Internet of Things: basic support of intelligent transportation*

Data is divided into collection, transmission, and storage. Taking the most simple GPS navigation software as an example, the data required by the system is the speed limit, number of lanes, number, length, congestion status, and pre-recorded navigation voice of each road, etc. These data do not need to be transmitted in real time except for the road congestion status. The data required by smart city needs to be collected by installing different sensors in various corners of the city and vehicles, and through 5G real-time transmission, the Internet of Things, vehicle network, and urban brain are connected, and stored in the data center of urban brain.

The video monitoring records the number of vehicles on the road, road conditions, driving speed, and the driving route of each vehicle. The bus card swiping records how many people have taken the bus at all time periods. At the same time, personal travel path can also be recorded through mobile App positioning to establish a large database of transportation Internet of Things. Most of the data will be processed in real time at the sensor terminals, and then uploaded to the cloud to connect the city's smart brain.

Some data are directly transmitted to the cloud through 5G and processed by the urban brain in real time to help urban managers make the best judgment and decision. However, not all data can do this. It requires data development and other technologies to screen out and store useful data for the urban smart brain to make long-term planning.

20.2.4.2 *Intelligent transportation lights*

Unscientific waiting for a red light is now considered to be the main cause of transportation jams. Therefore, major cities around the world have developed "intelligent" transportation light management system to improve the load-bearing and operation efficiency of key congested sections.

This intelligent transportation light system uses the computer vision chip built in the cameras that are commonly used in cities to carry out edge calculation and collect real-time transportation flow data, such as the number of vehicles, driving speed, and road congestion in a certain section of the road during the rush hours of the city. After collecting these data, it should conduct analysis and judgment, optimize the transportation light timing at intersections, improve road transportation efficiency, and effectively reduce transportation jams.

Relying on the selected big data of transportation flow in various weather and various periods, and through artificial intelligence to predict the transportation flow in a certain period of time and weather in the future, the transportation light timing planning at all intersections of the city is realized in advance.

At the same time, the intelligent transportation light system will also realize the priority of public transport signals, namely the "road terminal" will obtain real-time data such as the location, driving status, speed, and number of vehicles of the intelligent Internet connected public transport vehicles, and carry out real-time linkage with the transportation light control system with the support of the vehicle road collaboration technology.

The bus interacts with the intelligent transportation light control system when it approaches the intersection, adjusts the configuration of the time length of each phase of the signal light immediately under the scientific multi-dimensional control mechanism, and achieves the priority of public transportation by shortening the red light, extending and other ways.

Chapter Summary

The national integrated three-dimensional transportation network will be the space network of top-level design of China's transportation infrastructure and the foundation of the integrated transportation system in the coming decades. It includes the main channels and nodes of various modes of transportation, such as railways, highways, water transportation, civil aviation, and pipelines. It is a sea, land, air, and space skeleton network with complete layout, reasonable scale, optimized structure, resource-intensive, highly connected, and interconnected.

Chapter 21

Digital Twin: Creating A New Pattern of Transportation Infrastructure

With the research and development and application of artificial intelligence + 3D digital automatic modeling and other technologies, the implementation efficiency of digital twin will be greatly improved, and the visual operating system of intelligent transportation will be realized. Moreover, it will also enable the industrial Internet to upgrade from 2D content to 3D, or even the promotion of 4D in the future, and give human vision the leap from 2D to 4D digital everything.

21.1 New Infrastructure Model of "Digital Twin + Cloud Computing"

The next 10 years will be the construction and installation period of new infrastructure, according to the current situation. Nowadays, the most important theme of the global digital economy is how to restructure and migrate the digital infrastructure, as well as the ecosphere reconstruction based on the new digital infrastructure. The advanced digital technology represented by digital twin provides an economic, applicable, and reliable technology base for the future development of digital economy in the process of continuous overlap, integration, and upgrading. Human society will be promoted into an era of Intelligence of Everything with comprehensive perception, reliable transmission, intelligent processing, and accurate decision-making, with the development of the new generation of information technology.

21.1.1 *Endogenous development of the transportation industry in the digital era*

The development process of urban transportation, in general, involves more and more extensive areas. Since the reform, transportation management has always been the responsibility of a few relevant departments, such as transportation management departments. Transportation planning is part-time, dominated by urban planning departments, without specialized transportation planning units. As time goes by, urban transportation nowadays is no longer just the work of a certain management department, but the core point to be concerned with by the government, enterprises, and all walks of life.

21.1.1.1 *Intelligent transportation mode in the digital era*

At the beginning of the 21st century, the development of information technology has entered a fast lane, and at the same time, it has accelerated the upgrading and transformation of traditional transportation to intelligent transportation. In 2000, the Ministry of Science and Technology of China, together with the relevant departments such as the State Planning Commission, the Ministry of Public Security and the Ministry of Transport, set up a national intelligent transportation system coordination and guidance group and office to officially organize research on the development of China's intelligent transportation system, from which China began to paint the future of intelligent transportation.

Intelligent transportation uses advanced science and technology, i.e. AI technology, sensor technology, 5G as well as communication technology, to be applied in transportation, service control and vehicle manufacturing, strengthens the connection among people, vehicles, roads, environment, and other transportation participating parts, so as to form a comprehensive transportation system to ensure safety, improve efficiency, improve the environment and save energy.

In short, it is to make use of high-tech to transform the traditional transportation mode into an intelligent one, so as to achieve the purpose of safety, energy saving, and high efficiency.

21.1.1.2 *Intelligent transportation construction in the digital era*

IBM put forward the concept of intelligent transportation in 2009. Its main purpose is to make full use of new generation information technology, such as space perception and Internet of Things, on the basis of intelligent transportation, integrate transportation science and AI innovation, and achieve the objectives of active service, scientific decision-making, and global perception. The relevant data of transportation are grasped and transformed through the construction of real-time dynamic information service system, thus forming a problem analysis model, and promoting the safe, efficient, convenient, economic, environmental protection, and comfortable operation and development of transportation.

The key point of intelligent transportation lies in "wisdom", which generally refers to a kind of advanced creative thinking ability based on physical and mental organs possessed by a living body, namely understanding, analysis, judgment, perception, memory, sublimation, etc. of events. It belongs to the concept of biological level, and its core lies in spirituality and perception. In short, "wisdom" emphasizes a state of unconsciousness or subconsciousness, which can exert its own subjective initiative without much external intervention.

The implementation model of intelligent transportation is to install a "brain" for urban transportation so that it can see, hear, and respond to relevant information in a timely manner, and fundamentally solve the problems of urban transportation road congestion, resource waste, frequent accidents, and difficulty in real-time control of the situation, so that the development of urban transportation can embark on the track of sound development.

China is already a big transportation country worthy of the name, but it is still not a country with strong transportation network. At the end of the day, the gap is in services. Service is the essence of transportation. In terms of transportation services, there are two focuses: One is safety, and the second is convenience. This has always been a good vision for people to travel and live and is also the main content of intelligent transportation construction.

21.1.2 *New transportation infrastructure, production of "physical products" in "virtual design"*

"New infrastructure for transportation" emphasizes that the transportation system should have all the ability to connect and be connected. Its implementation method is to transfer the transportation elements such as road network, vehicles, travelers, and managers from the physical real world to the digital virtual space, so that each digitalized individual becomes a CPS. The production of "physical products" in "virtual design" maximizes the enrichment of transportation data, and the data flow tends to be diversified and networked, ultimately realizing the digitalization and networking in the transformation of transportation digitalization.

21.1.2.1 *Virtual design: establishment of "data organization structure"*

In the era of digital economy, as a "key factor of production", data has become a consensus of all walks of life. In 2019, China's Ministry of Transport issued the "Outline for the Development Plan of Digital Transportation", which defined the role of data in the transportation level: Digital transportation is an important area of digital economic development, and is a modern transportation system that takes data as the key element and core driver, and promotes the continuous integration and interaction of physical and virtual space transportation activities. The development of digital transportation should take data as the key element, empower the transportation and related industries, promote the linkage innovation of modes, business types, products, and services, improve the quality of travel and logistics services, and make the digital dividend benefit the people and enhance their sense of gain.

An organization responsible for data management needs to be constructed as data (as a survival factor in the information age) should be managed as real assets in order to become assets capable of producing business and economic value. This organization is responsible for defining data strategic objectives, short-term and long-term data tasks, promoting and operating the construction of the entire data system, defining governance-related system specifications and safeguard measures, and, as a shared service entity, empowering business management with data.

21.1.2.2 *Physical production: grasp the three main lines of "data tools, data experience, and data operation"*

After establishing the data organization structure, we need to promote the key work of producing "physical products" in "virtual design" based on the three main lines:

(1) **Data tools:** With the increase in data size and types of data sources, the higher the complexity of data, the higher the management cost. If the data type scale is in the thousand level (1,000 + data type), it is necessary to consider the use of data tools, because if the system has 1,000 types of data, the new access to 1 type of data is not a simple "1,000 + 1" arithmetic operation, but a product operation, it needs to establish a data and other 1,000 types of "network relations", including information relations, index dependence, as well as application sharing, etc.

(2) **Data experience:** The construction of transportation data management system should not only integrate into the industry and understand the actual business needs; in addition, it should also step out of the industry and establish the industry data information domain, indicator information system, data asset system, and shared service system from the overall situation in combination with national standards. This part is the focus of data construction.

(3) **Data operation:** As a key factor of production, data is the actual "asset" of the industry departments and should be operated and managed according to the requirements of assets. After data capitalization, the core purpose is to provide business services, empower management, and share business. In the end, the ownership of the data is separated from the trading right, and the transaction can be carried out and the data is capitalized. The premise of all these needs to have a unified "legislation and law enforcement", carry out system design and asset management, and form a "skeleton" of data governance and operation.

21.2 Virtual Simulation Becomes Rigid Demand for Transportation Infrastructure

Transportation virtual simulation refers to the three-dimensional virtual simulation effect based on the real urban transportation road and combined with the street, greening, vehicles, and other transportation facilities, so as to enable users to immerse in the transportation of the virtual city. Or through virtual driving, it improves driving skills and road psychological quality, or in the "simulation" world, it reviews and optimizes the overall layout of the transportation system. The advantages of virtual simulation technology make it a rigid demand for transportation infrastructure.

21.2.1 *Digital integration of future road facilities and environment*

On the basis of the rapid and vigorous development of new technologies such as big data, artificial intelligence and Internet of Things, and the promulgation of the "new infrastructure" policy, we can predict that the development of the transportation industry, especially the direction of intelligent transportation, will be carried out around the digital integration of infrastructure transportation facilities such as road facilities and the environment in the next 15 years.

By using computer system to build a digital road space with global perception, comprehensive connection, real scene and AI intelligence, it is of great significance to improve the operation level of the whole road transportation system, reshape and rebuild the operation management mode, service mode and business mode of the traditional road transportation system, and realize the transformation and upgrading of the industry. According to the current development trend, there are mainly four aspects in the digital integration of road facilities and environment in the future:

(1) The deep integration of transportation infrastructure and information technology refers to the use of information technology to change the functions of the existing transportation infrastructure, so as to improve the operational efficiency of the facilities, and integrate into the daily business to better serve transportation.
(2) Digital construction is carried out for attribute information of different road facilities. The location, function, status, and other trivial and changing information of road facilities are transformed into measurable feature data.
(3) For highway transportation facilities, including roads, bridges, tunnels, and some ancillary facilities, corresponding digital models are established according to their characteristics and application scenarios, and the models are used for calculation and processing in reality, and corresponding measures are taken for classified management.
(4) The digital integration of road facilities and the environment, that is, to promote the transportation infrastructure from passive perception, others' perception, to self-awareness, visual perception, so as to obtain the front-end data conveniently and accurately, and then output the terminal through the digital platform and vehicle road collaborative system, so that road users can know "what's on the road", and make the management decision-maker know "what the car is doing".

At present, there are still many defects in the field of information collection and information release. For travelers, when they are on the road, they do not know the road conditions in front of them or the weather in front of them. It is difficult to obtain relevant information in real time; as for managers, although basic transportation flow statistics are carried out every day, more detailed data can be obtained, for example, how many people are in the car? What's in the truck? What is the driver's level of fatigue? How is the condition of the vehicle? This information is not collected. Therefore, the goal of intelligent transportation nowadays is to establish a complete set of digital integration system and supporting platform for road facilities and environment. With this platform, it will be easy to collect these data through roadside equipment for analysis in the future, so that transportation participants can travel more safely and conveniently, and transportation managers can manage it more economically and efficiently.

21.2.2 *Dynamic scene simulation in virtual space*

With the rapid development of science and technology, the functions of computers are becoming increasingly powerful, and the content that can serve the people is also increasing. In reality, transportation can be dynamically simulated by computer in real time, and the solution to the problem can also be realized by its simulation. Taking large-scale event venues as an example, the huge volume of tens of thousands of people makes the organization of on-site pedestrian and vehicle transportation difficult to distinguish. For example, the time for the audience to enter the venue is different, but tens of thousands of people leave the venue at the same time after the event and the program. For this huge flow of people, how to achieve timely, reasonable, and effective crowd evacuation and avoid overcrowded walkways resulting in safety accidents requires transportation engineers to change the traditional design methods based on experience and organizational norms.

With the development of economy and society, people have more needs for transportation. The transportation flow in time period and large volume requires the transportation managers to use "scientific quantification" as the support of detailed design, and then use the method of transportation forecast to achieve the demand of "scientific quantification". The dynamic scene simulation of virtual space is used to test the transportation rationality of the construction scheme of large-scale sports venues, predict the flow of people, simulate the state of people in the building flow in the virtual space, and realize the "transportation analysis under dynamic scene simulation", which can truly reflect the needs of space.

For example, in 2019, Beijing Wukesong Stadium held a grand Basketball World Cup, with a large and complex audience. How to evaluate the security of a building that has been built for many years is a very important and complex project. The transportation center of the Chinese Academy of Sciences has easily solved this technical difficulty by using the dynamic scene simulation and analysis of virtual space. The most intuitive and real simulation technology reflects the potential congestion risk and effectively realizes the security assessment.

From this, it can be concluded that in future urban planning, from the design and construction of a certain urban road to the planning of the entire urban road network, the most intuitive presentation and implementation can be verified in the simulation scenarios. At that time, the problem of urban road congestion will be effectively solved.

21.2.3 *Improvement method of transportation infrastructure simulation*

Transportation infrastructure simulation system is to develop urban, national and even global road terrain simulation scenarios based on real geographic data. At the same time, the integration of transportation facilities systems such as long-distance bus, highway, road, parking lot, aviation, shipping, ferry, transportation signs, as well as railway makes a certain amount of transportation in the designated road area with a reasonable transportation density for motion presentation, thus effectively improving the visual effect of urban transportation simulation planning. The methods to improve the transportation infrastructure simulation mainly include the following contents:

(1) Promote the integration of transportation network and information network. It is planned to build an intelligent transportation communication network, orderly promote the construction of a new generation of information communication base station integrated with 5G of expressways, high-speed railways, urban roads, and comprehensive hubs, and gradually realize a large area of continuous coverage.

(2) Through cutting-edge communication technology, the transportation system and Beidou satellite data are integrated to create Beidou space-time information service network. Relying on China's Beidou satellite navigation and positioning to create a reference station network, and along the urban roads, railways, airports, ports, and other transportation infrastructure, it supplements the construction of a navigation integrated reference station network and promotes the comprehensive and continuous coverage of Beidou navigation information service network.

(3) Build a full coverage network of transportation perception. The construction of transportation perception network will be promoted by the separate implementation and separate governance, so as to create a comprehensive perception coverage network for all road and waterway transportation conditions. Strengthen the deployment of active early warning facilities for key parts, such as key tunnels, bridges, and accident-prone sections and shallow danger sections.

(4) Fully promote the intelligent sorting system. It will further improve the efficiency and safety level of parcel transportation in the postal express sorting center and the passenger hub by using automatic sorting, intelligent security detection, mechanical loading and unloading, and other technical equipment.

(5) Build transportation intelligent operation control network. Actively promote high-precision map services at the component level and lane level of transportation infrastructure. It is planned to build the edge cloud and cloud control center of transportation intelligent operation along the transportation line, study and promote the development of intelligent transportation, establish AI transportation control network, and achieve the overall optimal transportation operation through cloud control. Strengthen the dynamic simulation of transportation emergencies, improve the transportation emergency command ability, and ensure transportation safety and smoothness.

21.2.4 *Construction mode of big data and vehicle road collaboration*

In the 5G era, as the connection of city and vehicle, road is gradually becoming the intersection of communication network and intelligent sensor integration innovation. How to use the construction of big data and vehicle road collaboration to improve the level of urban intelligence and improve the efficiency and security of urban road network and vehicle coordination has become a new opportunity and challenge for technology to change life.

21.2.4.1 *Artificial intelligence*

With the popularization of artificial intelligence technology, the ability of image recognition and voice interaction has been greatly improved, which greatly improves the degree of intelligence of travel business, helping vehicle road collaboration to establish efficient and stable intelligent operation system.

Intelligent voice interaction is widely used in vehicle road collaboration, but to quickly and accurately identify and execute the voice of drivers and passengers requires vehicle AI combined with travel scenarios for targeted research and analysis.

21.2.4.2 *Cloud computing*

The simulation scenarios of vehicle road collaboration will generate a large number of data and data processing businesses, which require efficient computing capabilities, while cloud computing is a solution. For example, the gradual rise of automobile cloud services in China, based on the accumulation of cloud computing technology, is generally divided into automobile cloud platform and automatic driving cloud platform. Among them, the automobile cloud platform has the industry advantages of public cloud, private cloud, and hybrid cloud, and is an effective technical support for data cleaning, management, and analysis.

21.2.4.3 *High precision map*

Only by mastering the latest, real-time, and dynamic information of the surrounding environment of vehicles can the vehicle road collaboration ensure the safety and efficiency of vehicles and drivers, so as to make timely response through other driving assistance systems. Therefore, an important basis for building vehicle road collaboration through big data is high-precision map. First, in order to describe the whole road more accurately, clearly, and comprehensively, high-precision map needs to visualize all objects and all visible features that affect transportation driving behavior. Second, the data generated by the high-precision map will be more abundant and larger, and also needs dynamic and real-time updates, including static data such as lane center line, boundary point, reference point, virtual connection line, as well as dynamic data such as weather, geographical environment, road transportation, and self-driving status.

Chapter Summary

Digital twin infrastructure requires not only huge capital investment but also profound technical precipitation, as well as management level and staff skills to reach the corresponding level. At present, most of the domestic enterprises do not meet these conditions. Therefore, if we follow the digital twin technology in a blind way, we may lose all our money. Only after careful research, objective assessment, and careful investment can we open the right path to the space of digital twin.

Chapter 22

Digital Twin Activating China's Comprehensive Three-Dimensional Transportation Network

New Structure

In Guizhou, China has piloted the digital twin transportation system in advance. By using video monitoring, integrating MMW radar, and hologram perception of transportation elements such as motor vehicles, non-motor vehicles and pedestrians, the mapping model of the real transportation system in digital space has been established to conduct real-time analysis and tracking, which can effectively solve transportation problems such as the waste of transportation resources, the rigid function of the signal system, unpredictable transportation events, and unable to rapidly response. It is just around the corner to realize the digital twin and fully activate the new structure of China's comprehensive three-dimensional transportation network.

22.1 "Three-in-One" Definition of Digital Twin Three-Dimensional Transportation

Despite the rapid development of information technology, up to now, the information system is still unable to drive the automatic operation of the transportation system after the completion of the hardware construction as the factory pipeline. As a matter of fact, the information system can neither drive the visible automation of the physical world (the real bottleneck) nor drive the invisible automation of the digital space (the data bottleneck).

In view of this reality, R&D institutions need to adopt the three-dimensional mode of governance thinking, reapply IT technology to manage various problems in transportation operation, and reconstruct digital twin transportation, focusing on the reconstruction of main processes and laws. This requires three dimensions of thinking.

From the physical dimension, we can thoroughly sort out the road network, right of way, road setting, roadblocks, and road efficiency, and excavate the spatial value of roads with the help of information technology. From the social dimension, it can be sorted out from the perspectives of people's livelihood, people's trust, and public opinion, and tap the value of people's trust and public opinion by means of information technology. From the information dimension, nothing is more than data, models, algorithms, and platforms. Of course, these may be representations, and what really needs to be governed is the functional aspect of the data.

The following is the "three-in-one" governance mode of digital twin three-dimensional transportation.

22.1.1 *Data is soil, AI is perspective*

Nowadays, many IT companies are inculcating the idea that data is the means of production, water, and electricity. However, this description is not accurate. If it is the production materials, it means that we can use them directly. But is this the reality? Practitioners who actually use transportation data will understand that most of these data cannot be used immediately. Therefore, we must make a comparison to the data. The data is more like "soil". What is planted on the "soil" must be designed by ourselves, and at the same time, the "soil" itself must be managed and maintained.

From the perspective of China's national conditions, in the digital twin three-dimensional transportation system, the value of AI in the short-term reality should be perspective, rather than the expectation that AI can achieve unattended automation, which is different from the traditional AI algorithm based on mathematical derivatives of AI.

22.1.2 *Urban transportation perspective by digital twin*

The so-called perspective means that the problems that were not noticed before can be clearly identified, so as to be visible, identifiable, manageable, controllable, manageable, and convincing. In other words, it can be seen from the status of city-level transportation operation, identify frequent and occasional problems, manage daily business processes, control transportation congestion and abnormal situation, treat comprehensive and difficult problems, and provide quality services to managers and the vast number of travelers.

22.2 Digital Twin Built on Intelligent Transportation

Digital twin refers to the "clonal" of physical objects based on a device or system, and this "clonal" is called "digital twin". It is created on an information platform and is virtual. The biggest characteristic of digital twin is that it is a dynamic simulation of physical objects. Nowadays, with the complexity of domestic driving scenarios, people are increasingly calling for the stability of vehicle driving, the safety of driving routes, and the convenience of daily travel. It is bound to be a new direction for future travel to enable digital twin to enable intelligent transportation.

22.2.1 *Global vision and overall planning to accelerate the development of intelligent transportation digital space*

In 1998, Seoul (South Korea) began to construct the intelligent transportation system. In 2005, the system was officially named "TOPIS". With social changes and continuous iteration of technology, TOPIS system has now achieved linkage with the transportation police, Meteorological Bureau, road transport management, and other relevant departments, and has evolved into a set of comprehensive management platform for real-time collection, feedback, analysis, and processing of urban transportation data.

The big data collection terminal of TOPIS system includes GPS equipment carried by more than 7 thousand taxis in the city, nearly 10,000 buses, and bus card reading equipment installed in metro stations, and nearly 800 surveillance cameras. Through these devices, Seoul has achieved real-time monitoring of approximately 1,436 kilometres of roads in the city, collecting approximately 26 million real-time transportation data and 85 million bus cards reading information every day. Citizens can check the mobile map application or the website designated by the Seoul municipal government to understand the real-time road condition and congestion forecast. Seoul has also installed more than 300 large real-time transportation displays on its main urban roads to help people avoid congested roads.

In China's "14th Five-Year Plan" and the long-term goal of 2035, it is proposed for the first time that "strengthening itself through science and technology as a strategic support for national development". At the same time, the strategy of "strengthening country with strong transportation network" related to industrial upgrading has been proposed for many times, which has also been given more missions and tasks.

The "14th Five-Year Plan" is the first five-year plan for China to embark on a new journey of modernization after building a well-off society in an all-round way, with great significance. Therefore, intelligent transportation, which is closely related to strengthening country with strong transportation network, has gained a huge development opportunity. In terms of industrial upgrading, the "14th Five-Year Plan" pointed out that the focus of economic development should be placed on the real economy. Under the guidance of this goal, the main tasks are to build a manufacturing power, a quality power, a network power, and a digital China, promote the upgrading of the industrial base, the modernization of the industrial chain, and improve the economic quality benefits and core competitiveness.

It is the top priority to develop strategic emerging industries, accelerate the development of modern service industry, promote infrastructure construction as a whole, accelerate the construction of a transportation power, promote the energy revolution, and accelerate the digital development of intelligent transportation during the process of improving the modernization level of the industrial chain supply chain.

At present, China's Guizhou, Shanghai, Shenzhen, Beijing, and other places have gradually begun to use digital twin technology to build intelligent transportation, focusing on global vision and making overall planning. In the foreseeable future, the digital space of intelligent transportation will be accelerated.

22.2.2 *Digital space migration of the four transportation elements of "people, vehicles, roads, and environment"*

Transportation is the key to the rejuvenation of the country and the foundation of a strong country. In terms of the development strategy of automatic driving technology, there are two technical routes around the world: one is the "bicycle perception" automatic driving program led by the United States; the other is China-led "Internet awareness" automatic driving technology.

The major difference between the two is that the single vehicle perception does not rely on the deployment of additional sensor units in the external environment, but mainly relies on the vehicle's own assembly perception system to obtain the information of the surrounding environment of the vehicle body; in order to reduce the technical and cost requirements of vehicle body perception and computing capacity, the network connection perception needs to install additional sensors and communication units on the road side to expand the perception range and accuracy of vehicles. It can be seen that the construction and development of digital transportation infrastructure is the key technology to support automatic driving and vehicle road collaboration. It is imperative and of great significance to carry out research in this field.

Under the intelligent transportation mode, the experience of transportation trip will be reshaped, and the smart travel service Maas, the smart Internet car V2X, the smart Internet facility I2X, and the smart car road collaboration will be realized through the interconnection of the data of the car connection and the road connection among people, vehicles, roads, and clouds.

Based on the standardized communication protocol, the integrated perception, communication, calculation, control, and other technologies of intelligent vehicle road collaborative management and control can realize the digital space migration of physical space including the four elements of "people, vehicles, roads, and environment", standardize interaction and efficient collaboration, utilize the big data capabilities of edge, end and cloud computing, and solve the systematic resource optimization and configuration problems. The four core elements of the intelligent transportation system supported by the intelligent vehicle road collaborative system mainly include the following contents.

(1) **Human-intelligent travel service maas:** behavior management and guidance (pedestrian, driver, automatic driving algorithm) system.
(2) **Vehicle-intelligent Internet-connected vehicle V2X:** intelligent vehicle terminal system.
(3) **Road-intelligent network connection facility I2X:** intelligent road measurement facility system.
(4) **Environment-intelligent vehicle road collaboration platform:** regional transportation and urban transportation management and control decision-making visual deduction system.

22.2.3 *The era change of new ideas, new approaches, and new concepts of intelligent transportation*

Under the traditional transportation mode, from transportation planning to construction, to management and operation and maintenance, they are passive subordination to a large extent and have strong dependence on urban land planning. The use and layout of urban land limit or even determine how urban transportation will occur and be distributed to a large extent. In this way, the traditional transportation system for urban management is more shown as the nature of supporting equipment, just to meet the basic transportation functions of various regions of the city.

Due to this subordination, the transportation development of cities under the traditional transportation mode is always behind the growth of transportation demand caused by land use.

The transportation management mode of "treat head for headache and treat foot for foot pain" has been widely accepted by local planning departments in the past period, and most of the transportation improvement methods sought are to enhance the transportation capacity by continuously building new roads. However, there is a well-known "Braess's paradox" in transportation engineering, that is, adding a road section to a transportation network will increase the travel time on the network and reduce the service level of the entire transportation network. In other words, the more new roads are built, the more crowded the city may be.

Intelligent transportation, on the other hand, takes demand as the core and promotes application services. In response to complex and random demand, it provides dynamic generation services, dynamic matching services, and dynamic derivative new services to achieve accurate supply of transportation information. The Internet will deeply penetrate and integrate with the transportation industry, and bring profound changes to relevant links. It will become an upgrading technology and important idea for the construction of intelligent transportation. Based on big data thinking, urban transportation data is conditionally open, and data fusion and in-depth mining are conducted based on open data to provide more intelligent and convenient transportation information services for transportation travelers and managers. Based on the user's thinking and using the Internet interactive experience, it conducts public demand survey to understand the most urgent problems that the public wishes to solve, and provides personalized and diversified information services at any time and anywhere.

22.3 Digital Twin: Opportunities for China's New Infrastructure

In recent years, the concept of digital twin has gradually come into everyone's view. Its core technology is to use the latest three-dimensional visual graphic engine to reflect the whole lifecycle of the real world. In comparison with the traditional two-dimensional management interface, it is a revolutionary change, especially excelling in the display of high-tech and extremely complex management objects, such as new infrastructure. After combining with the new infrastructure, it can solve the problem of insufficient demonstration ability and poor actual effect when the new infrastructure projects are implemented, which is an opportunity for the successful implementation of the new infrastructure in China.

22.3.1 *In the new era of China's transportation, digital twin is launched at the right time*

Digital twin emphasizes the real-time, comprehensive, and interactive collection of physical world data, which is a "parallel world" in dynamic synchronization with the real world. The synchronization with the real system is ensured through all kinds of simulation, analysis, data accumulation, mining, and even the application of artificial intelligence.

The realization of digital twin in intelligent transportation can rapidly construct the three-dimensional scene of transportation in large areas and present massive dynamic transportation information in real time:

(1) Rapid construction of three-dimensional scene for transportation in large areas. "Digital twin" should be able to quickly build a complete set of three-dimensional transportation scenarios. Based on the two-dimensional vector data of the road, the parameterized modeling technology makes full use of the spatial model algorithm to automatically generate the three-dimensional road model according to the information such as the distribution of road facilities and design parameters and combined with the terrain where the road is located, which is basically consistent with the actual shape of the road. Users can make full use of simple data input and rely on software to achieve rapid modeling of large-scale transportation facilities.

(2) Massive dynamic transportation information is presented in real time. With the support of 5G technology and Beidou technology, the dynamic data in transportation facilities management and control and operation supervision has become more and more time-efficient, and the amount of data has become larger increasingly. It is the second key technology for the construction of "digital twin" scenarios to load and present massive dynamic data in three-dimensional scenarios, and conduct management, statistics, and analysis.

Digital twin in intelligent transportation supports the overlapping presentation, statistics, and analysis of multi-source and massive dynamic information in the three-dimensional scene. It can not only present the high-precision positioning information of Beidou in the three-dimensional scene but also connect the transportation flow data services, and present the sensor information in the transportation facilities in the three-dimensional scene in a visual manner. The back office manages and constructs analysis services such as electronic fences, analyzes and supervises information such as tracks and deformation, and makcs statistics of their operational losses, so as to achieve real-time control of transportation facilities management and control.

Future transportation is the intersection and integration of the real world and the virtual space. By acquiring real-time transportation data in the real world and digitalizing and modeling it in the virtual space, we can open up the digital twin space of the physical world and the cloud to extrapolate and solve the transportation problems encountered in the real world.

22.3.2 *Top ten glorious awards of China's transportation highlight new heights of digital achievements*

On 23 February 2021, Bentley China released the 2020 outstanding achievement award for infrastructure digitalization in China, which was awarded by 10 individuals and 6 institutions, two lifetime achievement awards for infrastructure digitalization, three awards for infrastructure industry masters and five pioneer awards for infrastructure digitalization innovation, three infrastructure technology promotion best institutions awards and three infrastructure best innovative enterprise awards, and more than ten institutions or individuals related to transportation digitalization won the awards.

Users have always been the source of digital innovation, and China's infrastructure industry is a strong innovation force, achieving the best practices drawing worldwide attention, and also a star on the stage of the Glorious Grand Prix of the infrastructure conference.

Now, China transportation is fully promoting digital and digital twin to improve the quality of life of the people, and plans to promote the digital twin technology to take root and bear fruit in China within two to three years, so as to fully assist in the construction of "new infrastructure" and smart China.

And those winners who represent China's transportation have achieved many China's first and the world's first in the process of promoting digitalization, highlighting the new heights of China's transportation digitalization achievements.

Standing at a new historical starting point, China's transportation is closer to the goal of strengthening the country with strong transportation network in the world than at any time in history. Looking back, we are proud of our achievements; based on the current situation, the task is full and substantial; looking forward to the future, the dream urges people to work hard. Only by doing things can we achieve great success, and only by going can we open up a road. China's transportation must firmly grasp the development opportunities given by the digitization and the digital twin, always keep in mind the historical mission entrusted by the state and the people, and strive to start a new journey of strengthening the country with strong transportation network.

22.4 Development Trend of Intelligent Transportation in the Context of Strengthening Country with Strong Transportation Network

According to the planning objectives in the "Outline for the Construction of Strengthening Country with Strong Transportation Network", China will become a country with strong transportation network in an all-round way by the middle of the 21st century. The level of digitalization, networking, and intellectualization of transportation information is in the forefront of the world. Meanwhile, the rapid development of computer, Internet, big data, artificial intelligence, digital twin, and other technologies will provide strong technical support for the construction of intelligent transportation.

22.4.1 *New mode of "extensive Internet of Things, comprehensive perception" intelligent transportation*

At the beginning of 2018, the new infrastructure wave guided by "accelerating the pace of 5G business and strengthening the construction of new infrastructure such as artificial intelligence, industrial Internet and Internet of Things", especially the characteristics of digital infrastructure, requires top-level design and targeted layout across the country. This is also an important issue faced by intelligent transportation.

It is an important challenge for transportation management departments to make use of Internet high and new technologies such as machine intelligence technology, cloud computing technology, and big data integration technology to provide new means and new thinking for urban intelligent transportation governance and meet the new needs arising from the continuous progress of technology and the continuous improvement of people's quality of life. Starting from the transportation integrated command and dispatching platform, transportation signal control system, video monitoring resources, transportation guidance system, transportation flow information collection, and other practical business systems, the introduction of mainstream Internet information technology and the construction of precision business service scenarios for intelligent transportation management can effectively improve the data and information service capabilities in terms of application of intelligent transportation in data integration, video intelligent analysis algorithm, business innovation support, efficiency enhancement of transportation management business, and services for public travel.

The new mode of intelligent transportation under the "ubiquitous interconnection and comprehensive perception" adopts the Internet + intelligent transportation governance framework. Through the establishment of a perception "ubiquitous network" and the coordination of people, vehicles, ships, roads, and ports related perception equipment, the integrated transportation operation status can be seen, measured, and controlled; the perception network is a collection of information from various departments, users, processes, equipment, and systems. Meanwhile, the perception network serves for different applications, which greatly expands the range of information perceived by various applications.

Intelligent transportation perception and information collection are completed from the perception layer system and facilities, and various transportation signals, data, and other information collected are transmitted to various business subsystems, which realize various basic transportation business functions respectively.

This layer is mainly responsible for the business management system, including the transportation police department, the transportation committee department, and the urban management department. These systems include both the new business system of this period and the business system completed by each department in the early stage. It can also connect with the railway, civil aviation, rail transit, and other related transportation department sub-systems through the access layer in accordance with the development of intelligent transportation in China in the future.

22.4.2 *"Real replication, virtual simulation" transportation mirror development*

Transportation mirror is transportation simulation technology. From the overall trend, the United States is the country with the rapid development and earlier development of transportation simulation, and the simulation software of other countries in the world is all developed on the basis of the transportation simulation technology of the United States. From the perspective of the whole development process of transportation simulation software, transportation simulation software has experienced the initial stage, rapid development stage, and commercial stage.

22.4.2.1 *Preliminary development stage*

At the initial stage, the main goal of transportation simulation development is to achieve a reasonable design of transportation signals. At this stage, the design model is mainly the macro design model used. This model has certain limitations, and its mobility and expression are not ideal. The conclusions obtained through this model naturally do not have the true meaning.

22.4.2.2 *Rapid development stage*

Transportation simulation software ushered in a high-speed period of development from 1970 to 1980. The rapid development of computer technology has promoted the synchronous and rapid development of computer-related industries, of which transportation simulation software is a classic example. Based on the rapid development of electronic computers, the simulation accuracy of simulation software has been greatly improved, and the functions of software also tend to be diversified. With the rapid development of macro software, micro transportation simulation has also stepped on the high-speed train of the times. The two most prominent software programs are NESTSIM and AIMSUM2 transportation simulation software. Among the two software programs, NETSIM simulation software developed and designed by the U.S. government is the substitute. This model is a network micro transportation simulation for the motion state of a single vehicle. The emergence of NETSIM has pushed the description of the transportation phenomenon of urban roads to a new level.

22.4.2.3 *Commercialization stage*

At the mature stage of the development of transportation simulation software, a new development and design direction of simulation software has emerged, which has shifted from the traditional development model to the software development and design oriented to the software use target. There are more and better choices for the methods and ways used in software development. The development mode has shown the phenomenon of hundreds of flowers in bloom and contention of a hundred schools of thought. People have abandoned the development and design of traditional FORTRAN language or special simulation language, which has greatly improved the performance of simulation software.

At present, the development and design of domestic transportation simulation software is still in a relatively basic stage of application and design, and there is a large gap with foreign simulation software. China has a weak foundation in terms of both technology and human resources. In the future, in order to achieve a larger scale of commercialization of simulation software in China, more capital and policy inputs are needed. However, real replication and virtual simulation are the main development direction of this technology.

22.4.3 *Achieving "strengthening country with strong transportation network" through digital power strategy*

In 2019, the Ministry of Transport of China formulated and issued the "Outline for the Development Plan of Digital Transportation", which proposed for the first time to use the strategy of digital power to achieve "strengthening country with strong transportation network". It also points out the specific direction for the comprehensive integration of intelligent transportation and digital twin.

By 2025, a new step will be made in the digital upgrading of all elements and full cycle of transportation infrastructure and transportation equipment, and the digital acquisition system and network transmission system will be basically formed. The public network, such as the fifth generation of mobile communication (5G) and the new generation of satellite communication system, has been initially applied in the industry. The application level of transportation big data has been greatly improved, travel information services have been fully covered, and the logistics service platform and integration have entered a new stage. The transportation infrastructure will be digitalized in all elements and cycles, and the transportation control network integrating heaven and earth will be basically formed by 2035. The on-demand instant travel service will be widely used. "Moreover, China will become a major international standard setter or participant in the field of digital transportation, and the overall competitiveness of the digital transportation industry will be leading in the world".

In order to achieve this wonderful prospect, innovation must become the first driving force for the future development of digital transportation, and the new concept of "travel as a service" will be fully popularized. Data is used to link travel demand and service resources, so that travel becomes an on-demand real-time service and makes travel easier. It is emphasized to build a digital collection system, a network transmission system, and an intelligent application system with the "data chain" as the main line, so as to accelerate the development of transportation informatization to digitalization, networking, and intelligence, and provide support for the construction of "strengthening country with strong transportation network".

In the future, digital transportation mainly consists of seven systems, including digital collection system, network transmission system, intelligent application system, industrial ecosystem, network and data security system, standard system, support and guarantee system, etc.

It can be predicted that in the near future, China will take data as the key element, fully empower transportation and related industries, improve the quality of travel and logistics services, and create a more happy and harmonious social living environment.

Chapter Summary

There is still a certain gap from the real comprehensive coverage, synchronous visibility, and virtual real interaction even though digital twin is the cutting-edge trend of intelligent transportation. However, driven by the 5G technological transformation and demand upgrading, digital twin technology provides new ideas, new approaches, and new concepts for intelligent transportation, which will continue to develop in the future and eventually form a complete set of technical operation system so that the “real world is moved into virtual and the virtual world is gone into the reality” can be realized.

Postscript

1. Why Write a Book

Internet, big data, cloud computing, artificial intelligence, blockchain, and other technologies have accelerated innovation and are increasingly integrated into the whole process of economic and social development in various fields. The rapid development of digital economy, the wide scope of radiation, and the depth of influence are unprecedented and are becoming the key force to restructure the global factor resources, reshape the global economic structure, and change the global competition pattern. In order to strengthen the theoretical research on the development of digital economy, this book puts forward countermeasures and suggestions on the problems related to the development of digital technology and digital economy. In order to promote digital upgrading and promote the deep integration of emerging technology and transportation, the Ministry of Transport has successively issued policies such as the "Outline for the Development Plan of Digital Transportation". Digital technology, including digital twin, is widely used in all aspects of transportation. China's transportation development has begun to show the characteristics of dualization, quantification, and coding. As a digital technology with high relevance and outstanding value to the transportation industry, digital twin has emerged and promoted the strategic process of China's "strengthening country with strong transportation network" in depth.

The digital twin is in line with key policies such as the "Outline for the Construction of Strengthening Country with Strong Transportation

Network", the "Outline for the Development Plan of Digital Transportation", and the "Outline for the National Comprehensive Three-dimensional Transportation Network". It puts forward practical suggestions on the application of digital twin in the field of transportation in the process of China's intelligent transportation upgrade, including strengthening the application depth of digital twin and building a comprehensive three-dimensional transportation network with outstanding intelligence. The experience and methods used for reference will promote the transformation of China's transportation development from traditional factor-driven to innovative technology-driven, and rely on digital power to support and accelerate the construction of strengthening country with strong transportation network.

2. Digital Twin Transportation Leads the Modern Development of Transportation

Digital twin will become a powerful tool in China's strategy of becoming a strong country with a strong transportation network. The digital road cast by it indicates the direction and trend of China's transportation development in the future. In the digital era where everything can be digitalized, the development of digital twin constantly changes the quality of life of the public. It not only realizes the real replication and evolution of the real world in the digital space but also completes the twin connection of the real things in the virtual space under the background of the Internet of Things. This technology greatly promotes the improvement and development, upgrading, and iteration of things since it can provide dynamic demonstration and corresponding changes of the state information of real things. After the digital twin landing in China's transportation field, it has gradually become an important technology for China's transportation development. Digital twin is not only the technical support for China's transportation development but also the most common application field in the global transportation industry. For example, digital twin upgraded the smart system of high-speed rail in Germany, helped Japan complete the integration of ETC 2.0 system, vehicle navigation system, and VICS, and realized the collaborative management of vehicles and roads. Digital twin has not only become an important sign of intelligent development in the transportation developed countries such as the United States, Japan, and Germany but also driven the economic progress of these countries. China's transportation has undergone many changes driven by the digital twin, and the overall wisdom, convenience, and effectiveness have been greatly improved since 2019.

Digital twin will become a powerful tool in China's strategy of becoming a strong country with strong transportation network. The digital road cast by it indicates the direction and trend of China's transportation development in the future. The "Outline for the Development Plan of Digital Transportation" issued by China clearly pointed out that "the future development of digital transportation includes building a digital collection system, building a network transmission system, building an intelligent application system, cultivating an industrial ecosystem, improving the network and data security system, improving the standard system, and improving the support and guarantee system". Each development focus reflects the technical core of digital twin, and the correspondence between digital twin and the development of digital transportation in China is very significant.

The digital development of transportation has become a tangible development trend. The original intention of the digital development of transportation is to improve the development effect of transportation and ensure that the construction of transportation network meets the needs of social development. The application of digital twin technology is constantly deepening in the process of the application of digital technology in transportation, which brings more prominent help to the transportation industry and creates greater value in the field of transportation. In March 2020, the Ministry of Transport officially issued the "Outline of Smart Urban Rail Development of China's Urban Rail Transit", which proposed six connotations of smart rail transit, namely "smart city rail connotation, intelligence and smart, smart system and smart city rail, smart city rail and urban rail information, smart city rail and transportation power, smart city rail standards and urban rail information standards". In the subsequent construction of smart rail transit, digital twin technology has fully highlighted the wisdom of rail transit and has become a key guarantee for the deployment of China's rail transit so that the six connotations of smart rail transit are fully highlighted in various major initiatives. For example, Changsha Metro in China has completed intelligent upgrading through digital twin and successfully solved the transportation and turnover problems of 15.2113 million passengers and 2.5352 million passengers per day during the May Day period of 2021 based on the outstanding decision-making basis of digital modeling. In April 2020,

China's National Development and Reform Commission and the Cyberspace Administration of the CPC Central Committee jointly issued the "Implementation Plan on Promoting the Action of "Using Data to Empower Intelligence in The Cloud". Fostering New Economic Development", which officially identified digital twin, big data, artificial intelligence, and 5G as the core technology of China's economic development and specifically discussed the "digital twin innovation plan" in a chapter, which requires "involving all parties to propose digital twin solutions". After the release of the implementation plan for economic development, the speed of the combination of various industries and digital twin in China has increased again, and the simulation ability of the combination of virtual and real digital twin has begun to play a greater role in the field of transportation in China. In the following more than 1 year, the digital infrastructure of transportation began to develop vigorously, and the relevant policies of digital twin have been launched in various places, and the digital twin has gradually been popularized in the field of transportation in China. In 2021, the "Digital China Development Report (2020)" issued by the Internet Information Office of China also clearly stated that digital transportation has spawned a lot of new technologies, demonstrated new business types and new hotspots, focused on practical application in development which has accelerated the development of the transportation industry, focused on policies to promote the livelihood industry and poverty alleviation projects in China, and focused on international standards, so as to lead China's transportation to a high-quality and international high-speed.

Digital transportation has become an important growth pole of China's digital economy and a representative industry in China's digital field. Digital technology has brought great changes to China's economic development that are hard to meet in a hundred years. After clarifying the complex structure of various fields, it has brought great changes to China's scientific and technological development and industrial development. In this era, all things can be connected and all things can be programmed. Digitalization and networking have become the main way to drive the progress of the world. In the process of improving the quality of China's "strengthening country with strong transportation network" strategy, digital twin has played an important role in three aspects: improving the quality of public intelligent travel, creating a new era of transportation, and boosting the intelligent development of transportation. At the same time, it benefits the economy and promotes the continuous improvement of people's living standards, which has become the key driving force of China's transportation development.

However, after various fields such as energy, finance, and transportation are linked to the digital, although the development speed has doubled and improved, security has also been accompanied by a number of hidden dangers. As Zhou Hongyi, CEO of 360 Group, said at the Eighth Internet Security Conference, "entering the digital twin era, cyber attack has more influence than nuclear weapons".

China's transportation departments have carefully analyzed the characteristics of digital transportation safety systems in developed countries and summarized seven new elements of transportation networks, based on which a set of intelligent systems to meet the needs of transportation development has been built, and a perfect safety protection system has been established on the basis of safety data, digital twin technology, and professional teams. Currently, China's transportation network can effectively prevent APT attacks and ensure the efficient operation of the transportation network under the protection of this security system. On the premise of safety and stability, it is predicted that China will enter a digital and intelligent development period in the next 5~10 years. As pointed out in the "Outline for the Construction of Strengthening Country with Strong Transportation Network", "5G, cloud computing, big data, Internet of Things, artificial intelligence and blockchain are the core of digital technology to drive China's development". In this change of society, transportation will remain the main development force, and transportation construction will also focus on digitalization and intelligence for upgrade. Compared with the previous vertical development of land, sea, and air, the future transportation will focus on the coordinated development of travel transportation, and China's comprehensive three-dimensional transportation network will show a strong leading role.

The combination of transportation and digital can be regarded as the inevitable law of the development of the times and the inevitable trend of China's transportation development. The digital is the way to connect the virtual space since transportation is the way to connect the real world, and moreover, the integration of transportation and digital has become inevitable with the combination of the virtual and the real world. In order to promote the orderly development of China's digital transportation and intelligent transportation, the digital new infrastructure has established the primary goal of overall planning, industrial layout, and competitiveness improvement of China's transportation departments. The effective integration of digital twin and new transportation infrastructure can not only fully meet the needs of transportation development but also ensure the synchronization of China's core transportation technology with the world.

Therefore, under the premise of comprehensively improving the transportation network security, China's transportation management departments have begun to vigorously develop the construction of transportation digital infrastructure and, at the same time, upgrade the digital system of the storage and logistics system, qualitatively change the development mode from various basic aspects of transportation, and strive to create a new pattern of intelligent transportation development. First of all, China's transportation management departments take the safety of infrastructure, transportation tools, and operating environment as the core to further improve the transportation operation efficiency according to the construction requirements of China's "Outline for the Construction of Strengthening Country with Strong Transportation Network". Second, we have strengthened the construction of digital infrastructure and also increased the construction of public security. After that, we will focus on the combination of network application and digital application to accelerate the digital upgrading of the overall transportation network. Finally, we will strive to carry out digital technology upgrading and innovation and strive to maximize the performance of the maximum value of digital upgrading of transportation infrastructure.

After the digital characteristics of China's transportation infrastructure, the warehousing and logistics industry in China has also carried out matching development. China's transportation departments have strengthened the intelligent development of the warehousing and logistics industry according to the direction guidance of the "Outline for the Construction of Strengthening Country with Strong Transportation Network", in the past two years. They have promoted the deep integration of the warehousing and logistics industry with emerging technologies such as digital, 5G, and artificial intelligence in the process of striving to create good policy conditions and high-quality development environment. Digital twin plays an important role in this process. It not only establishes a fully intelligent logistics and warehousing system but also fully improves the intelligence and efficiency of the warehousing and logistics industry through digital model, providing strong support for the industry to reduce costs and increase efficiency and high-quality development.

At the same time, many warehousing and logistics markets have begun to carry out digital transformation with the support of local policies, and major logistics brands have established intelligent logistics centers and built digital models that can be simulated and tested. This process of digital twin enables personnel engaging in warehousing and logistics industry to see the potential problems of the logistics system, make up for them, and stimulate new thinking and new path for the development of the industry, which makes the storage and logistics industry improve its own value and accelerates the pace of development.

3. Global Trend and Situation of Digital Twin Transportation Development

The wisdom of the transportation industry in the world is becoming more and more prominent with the increasing value of digital twin in the global transportation industry. Some developed countries have issued a number of policies to strengthen the application of digital twin to improve the development speed of intelligent transportation. For example, the United States has promulgated and repeatedly upgraded the United transportation law in order to promote the development of the transportation industry, which has also been implemented in the development process of intelligent transportation and deepened the integration of intelligent transportation and the transportation industry; Japan has successively issued the "Declaration on Creating the World's Most Advanced Digital Country" in 2019 and the "Land and Transportation White Paper 2020" in 2020. The two laws and regulations have clarified the enhancement of digital application in the field of transportation and striven to improve the construction results of digital transportation.

From the new wave of global transportation development, it can be seen that the deep application of digital twin is a major direction of national transportation development. Digital twin has shown its unique advantages in depth, width, and matching of transportation applications. China's transportation industry has long been studying the application of similar digital twin simulation technology. These simulation technologies carry out the simulation of transportation development through the correlation analysis of individual transportation location data and transportation network passenger flow data. The matching degree and integration degree of the transportation industry are significantly higher than those of the traditional simulation technology after the birth of the digital twin. The digital twin realizes the virtual and real synchronization, real-time tracking, and scenario online of the transportation network, enabling the full element simulation to show great value in various fields of land, sea, and air in China.

For example, the digital twin application of Beijing Daxing International Airport in China has not only improved the airport operation efficiency, passenger travel experience, and the safety guarantee level of aircraft operation but also promoted the development of China's aviation economy. Beijing Daxing International Airport, as a representative transportation hub of digital twin for the digital upgrade of China's air transportation, has a direct impact on its upgrade effect and has far-reaching significance for the overall simulation and upgrade of China's intelligent transportation.

From the application effect of digital twin, it can be seen that the application of digital technology meets the needs of China's transportation development from multiple levels, and China has also issued a number of related policies according to the application effect of digital twin, such as the "Integrated Development Plan for Higher Quality Transportation in the Yangtze River Delta region" issued by China's National Development and Reform Commission in 2020; the effective application of digital technology is mentioned in the "Planning for Integrated Development of Transportation in Zhengzhou Metropolitan Area (2020–2035)" issued by Henan Province in 2021, and the digital upgrade of transportation has become the main feature of digital twin.

In future development, digital twin will also play a greater role in various fields of land, sea, and air with the deepening of China's various transportation policies.

For example, in terms of ground transportation, the largest role of digital twin lies in the three-dimensional display of transportation construction and transportation facilities operation in various complex environments, as well as the prediction of various accident conditions and special conditions. It can be said that the application of digital twin has led the ground transportation data to achieve a breakthrough from two-dimensional to three-dimensional, and fully combined with external data to make the real and all-round simulation of the development of ground transportation; in the aviation field, digital twin will play an important role in three aspects: the visualization of aviation operation and maintenance management, the visualization of aviation operation, and the enhancement of airport emergency response capability; in terms of water transportation, the simulation application of digital twin solves the existing blocking points of China's water transportation development, makes up for the existing deficiencies of China's water network, makes clear the development direction of China's water transportation more intuitively under the simulation of digital twin, and learns from the experience of digital transformation of water transportation in developed countries, thus the breakthrough of China's water transportation is expected to be made in the near future.

After summarizing the effects of digital twin in China's transportation development, the author finds that digital twin brings more possibilities for transportation development, including the possibility of new transportation control, the possibility of transportation operation improvement, and the broad prospect of transportation development, which greatly expands the space for transportation development.

After the digital twin helped China's intelligent transportation to complete the simulation upgrade, a variety of new business types in the field of transportation emerged frequently. The intelligent characteristics of "people are happy with their behavior, and materials are excellent with their flow" have become increasingly prominent. From the perspective of the development of transportation technology, the importance of digital twin is becoming increasingly prominent. It has built a transportation network with vehicle road collaboration, helped the transportation industry to achieve digital transformation, and has been supported by a number of policies. However, from the perspective of transportation globalization, it can be seen that the application of digital twin in China integrates the advanced experience of many developed countries.

This is because the digital twin is not only an important scientific and technological force in China's transportation development but also has been highly valued and achieved remarkable results in the world. The rapid development of China's transportation is promoting China to transform from a transportation country into a transportation power. In this process, the road of intelligent transformation of the developed countries in Europe and the United States has become the main object for us to learn from. Although China's level of transportation development is close to the international level, many European and American countries started their intelligent transportation earlier, and their transformation experience and upgrading mode are still worthy of our reference and analysis.

Taking the application status of digital twin in Europe and the United States as an example, this technology has taken root and thrived in various transportation fields around the world since the digital twin was applied by the United States to the aerospace field. In more than 10 years of development, digital twin has led a number of European and American countries to take the lead in realizing the upgrade of intelligent transportation and comprehensively promoted domestic economic development. For example, the 27 countries in Europe jointly proposed the concept of "fifth generation transportation" in 2015, which aimed at the network management of the road network of more than 30,000 kilometers in Europe, and the identification of key sections of the overall network according to the unified rules. Under the support of digital twin technology, the "fifth generation transportation" in Europe has developed smoothly. With the improvement of the effectiveness of road network management, congestion, transportation casualties, and carbon emissions in 27 European countries have been significantly reduced.

Currently, the transportation of Germany, France, the United Kingdom, Italy, and other countries has achieved outstanding development results through digital twin technology, and these results provide a reference for China's intelligent transportation construction, indicating the direction.

Compared with the development status of China's intelligent transportation, many European and American countries have obvious advantages in building intelligent transportation systems, and the reason is mainly the early use of digital twin in these countries. In the case of the United States, the United States can establish the intelligent transportation system to ensure safety, improve the environment, and improve efficiency earlier and more comprehensively since the United States has used the new generation of information technology, digital communication technology, sensor technology, network control technology, and artificial intelligence technology earlier than China.

As of 2021, the construction and development of China's intelligent transportation has shortened the gap with the transportation powers in Europe and the United States and even exceeded in urban transportation intelligent scheduling, highway intelligent operation and maintenance, etc.; however, the overall combination of people, vehicles, roads, and environment still has obvious deficiencies, which is mainly reflected in the maturity of intelligent transportation system (ITS).

At present, the U.S. ITS has reached the international first-class level, and its intelligence is very prominent. Overall, the U.S. ITS includes seven systems, namely transportation management system, transportation demand management system, public transport operation management system, interstate transportation management system, electronic toll collection system, emergency management system, vehicle control, and safety system. It promotes the development of intelligent transportation in the United States from multiple perspectives; meanwhile, ITS has also realized the vehicle road collaboration in Japan and the comprehensive upgrade of ETC 2.0 in Japan; in addition, it improves the overall transportation intelligence of other European countries in terms of transportation management, travel planning, travel monitoring, vehicle control, as well as automatic toll collection.

In general, digital twin has explored transportation development strategies suitable for the national conditions of these countries with developed transportation, which has shown great value in transportation construction, transportation development, and transportation operation and maintenance management, which is also the main reason for these countries to develop into a transportation power.

In recent years, the construction of China's overall three-dimensional transportation network has achieved remarkable results under the promotion of the strategy of a strong transportation country. According to statistics of the transportation department, China has ranked first in the world in terms of mileage of transportation route construction and transportation technology upgrading level in recent years. The development strategy of "giving priority to transportation" in China is promoting the realization of the goal of a strong transportation country.

Although the overall situation of China's transportation development is very good, and advanced concepts have been used for reference, the intelligence gap still exists. This gap exists in not only the development of transportation construction but also the field of transportation to serve the people's livelihood and stimulate the national economy. It is more and more obvious that the transportation developed countries use digital power to drive the overall development of the country, especially after the arrival of the digital era. These development experiences and achievements are worth thinking and learning from the people in the transportation industry in China in exploring the development path suitable for the national conditions:

(1) From the current situation of intelligent transportation construction in developed countries, we can see that their definition of intelligent transportation is very comprehensive. These countries believe that intelligent transportation is not only the intelligent transportation itself but also the control and support of transportation management, transportation communication, public travel, and transportation construction and development by fully using emerging transportation technologies, making the transportation network an important force for national perception, connection, analysis, control, and driving national development. In the meantime, the new transportation technology also needs to ensure the development of transportation safety, give full play to the efficiency of transportation infrastructure, improve the level of transportation operation and maintenance management, and provide a steady stream of power for the sustainable development of the country.
(2) Be good at promoting the overall progress of society through the development of science and technology in various fields. Just as each of the developed countries is using digital twin technology to solve the uncoordinated relationship among people, vehicles, roads, and links, and has obtained many representative development results in the monitoring and coordination management of intelligent transportation, toll and parking management, transportation assistant driving function, logistics and transportation management, and other aspects.
(3) Note the construction of new infrastructure, transportation resource allocation, intelligent transportation management, transportation development policies, urban transportation management, and national transportation coordination management, which have achieved today's development results.

From the development of intelligent transportation in transportation-developed countries, people in China's transportation industry can feel that emerging technologies have accelerated the integration of transportation industry and people's livelihood industry, changed the construction mode of transportation infrastructure, expedited new travel modes and new management modes, and promoted the subversive development of society as a whole.

In contrast, there are still some problems in the intelligent development of China's transportation, such as the improvement of the intelligence of urban transportation system, the overall intelligence of comprehensive three-dimensional transportation network, the intelligence of transportation resource adjustment and guidance, the intelligence of urban transportation infrastructure construction, and the intelligence of transportation management. The development of China's intelligent transportation needs to learn from the experience of developed countries in the aspects of innovative new travel modes, construction of development of vehicle road collaboration, and construction of intelligent transportation industrial ecosphere, so as to speed up the development process of China's intelligent transportation.

The transportation industry in various countries has achieved qualitative change and development with the application of digital twin on the global scope, and the global intelligent transportation upgrade has become a major trend. Through the deep combination of digital twin and transportation, China has gradually built a new, extensive, and advanced transportation infrastructure, realizing the real-time perception of transportation. The intelligent upgrade of this development model has had a positive and far-reaching impact on China's transportation industry, reducing the substantial gap between China and the transportation-developed countries.

In 2021, after the full launch of China's "14th Five-Year Plan", the State Council issued the "Outline for the National Comprehensive Three-dimensional Transportation Network", which is another important document after the "Outline for the Construction of Strengthening Country with Strong Transportation Network". The document emphasizes the importance of digital and intelligent development of transportation and sets specific development goals, and moreover, it points out the important reference object for China to build a strong transportation country in the scheduled time.

If we say that the "Outline for the Construction of Strengthening Country with Strong Transportation Network" specifies the important tasks for the development of intelligent transportation in the next 10 years, then the "Outline for the National Comprehensive Three-dimensional Transportation Network" points out the specific content of the development of intelligent transportation in the next ten years. From the content of "promoting the level of intelligent development" in the "Outline for the National Comprehensive Three-dimensional Transportation Network", it can be seen that the focus of China's transportation intelligent development is the digital construction level of transportation technology facilities, and comprehensive digital infrastructure is an important foundation to support the digital upgrading of China's transportation. In the

"Outline for the National Comprehensive Three-dimensional Transportation Network", it also clearly states that China will complete the extensive advanced transportation information infrastructure, realize the comprehensive perception of the transportation network, and build an intelligent, efficient, and real-time transportation operation and maintenance management system by 2035. This is not only the development goal of the transportation industry but also the main way to accelerate the development of transportation in China.

On the basis of the direction guidance and construction content of few key documents such as "Outline for the Construction of Strengthening Country with Strong Transportation Network", "Outline for the Development Plan of Digital Transportation", and "Outline for the National Comprehensive Three-dimensional Transportation Network", it can be concluded that the comprehensive digitalization of China's transportation infrastructure in the future is the digital power can give more momentum to China's transportation development mainly through three approaches of the digitalization of the whole lifecycle of the transportation infrastructure, the enhancement of the perception of the transportation network, and the strengthening of intelligence and advancement of transportation technology, and then deeply excavating the values brought by transportation big data collection technology, transportation big data mining technology, and transportation big data integration technology so that the role of digital twin can be unlimited enlarged, and the collision between transportation data can be enhanced, so as to bring more help to the goal of China's transportation power.

The achievements of vehicle road collaboration, smart high-speed, and full network perception are highlighted in all directions looking at the transportation development planning and construction based on digital twin in China. Digital twin has made breakthrough progress in the process of improving the intelligent development of transportation. However, all of them are to enhance the intelligent attribute of transportation infrastructure when looking at the implementation of digital twin in various fields of China's transportation. Digital twin realizes the real-time interaction between China's transportation in digital space and the real world through digital improvement of transportation infrastructure.

After the State Council successively issued the "Outline for the Construction of Strengthening Country with Strong Transportation Network" and the "Outline for the National Comprehensive Three-dimensional Transportation Network", and the relevant requirements for the development of digital transportation during the "14th Five-Year Plan" period, China's transportation infrastructure will embark on the road of digital transformation and intelligent upgrading. In this process, digital twin technology gives full play to its own value and promotes the high-quality development of China's transportation infrastructure. The State Council of China issued the "Action Plan for New Infrastructure Construction in the Field of Transportation (2021–2025)", which clearly stated that "Based on the new development stage, implement the new development concept, build a new development pattern, take the promotion of high-quality transportation development as the theme, speed up the construction of a transportation power as the general goal, adhere to innovation driven and intelligent development, take digitalization, networking and intellectualization as the main line, organize and promote a number of new key transportation infrastructure projects, create an influential new transportation infrastructure model, and create innovative development environment, promote the development of new infrastructure from point to area, improve the efficiency, expand the functions and increase the momentum of transportation, and continuously enhance the sense of gain, happiness and security of the people".

In the future, China's new transportation infrastructure is a comprehensive and intelligent livelihood project, with a broad application prospect, and moreover, it produces a far-reaching impact on China's goal of becoming a strong country. Based on this situation, the author clearly sorted out the main ways of digital twin in helping the comprehensive development of China's new infrastructure, and found that the smart construction of transportation system, the smart construction of energy field, and the smart construction of transportation system are the key. Digital twin has gradually become the mainstream construction mode of new infrastructure in these three fields. It is believed that in the future development, China's intelligent transportation, new energy, warehousing and logistics, and other industries will be empowered by the digital twin, and China's basic strength and economic development will obtain more breakthroughs due to the digital twin.

4. Digital Twin Transportation Improves Transportation Technology Innovation

The development of human digital technology has brought new opportunities and challenges to intelligent transportation. Digital twin technology has opened up a new perspective window for the cause of intelligent transportation. By combining the texture of the application of this technology in transportation management, the whole society will have a clearer and more comprehensive view of its future development. In recent years, China's transportation industry has attached more and more importance to the application of digital twin and has successively held China Smart City Summit, Global Smart Logistics Summit, Digital Twin Technology Summit, and many other activities. HUAWEI, Xiaomi, Beijing 51 WORLD, and many other well-known enterprises have demonstrated the research results in areas such as the digital twin in logistics, warehousing, port management, and automatic driving.

In the process of the grand "clonal" banquet of transportation and digital twin, digital twin technology will make full use of data information such as physical model, sensor update, and operation history to conduct multi-dimensional simulation of transportation phenomenon described by different physical quantity, probability, and scale standards from different disciplines. Not only that but also the digital twin will completely map it to the virtual digital space, reflecting the whole cycle process of the transportation system. In simple terms, the digital cloning plan of future transportation development is that on the basis of advanced equipment or systems, a digital version of the transportation system is created.

Transportation system is created in reality, and its digital twin is created on the information platform. It is because of its "clonal" nature of transportation reality that it can complete dynamic simulation of the realization of transportation system and become a real "dynamic" information system.

The "movement" of the digital twin comes from the components of the main model of the transportation system, instead of occurring autonomously, which mainly includes the data fed back by the sensors on various ontologies in the transportation system, as well as the historical data of the operation of the ontologies. This means that the real-time status of each vehicle and road in the transportation system, as well as the changes in the relevant external environmental conditions, will be restored to its digital "clonal body".

The combination of digital twin and transportation system reflects the technical characteristics of global, real time, and two way. From the date of its birth, digital twin technology has continuously strengthened these characteristics, and people are able to have a deep insight into the future transportation development trend. 5G technology plays an important role in the construction and application of digital twin intelligent transportation system. There is real-time massive data interaction between physical entities and their digital twin in the intelligent transportation system. Just like the interaction between a healthy human body and the external environment, where various kinds of perception and control information are transmitted through the nervous network in this process to maintain the normal operation of the human body. Similarly, communication network plays an irreplaceable similar role in the digital twin of intelligent transportation system.

5G, an advanced network technology, has not only become a strong outlet in the field of implementing digital twin in transportation but also will actively carry out dynamic switching and adjustment under different branch application scenarios of intelligent transportation digital twin system to achieve continuous empowerment.

In recent years, 5G technology has gradually developed into the focus of science and technology in China, and a number of policies to strengthen 5G development have been introduced, such as the "Regulations on Promoting 5G Network Construction and Development" issued by the Xinjiang Uygur Autonomous Region in China; Anhui Province issued the "Outline of 5G Development Plan of Anhui Province (2019–2022)". In the process of guiding 5G development by these policies, 5G technology has also begun to gather with digital twin to help the transportation industry achieve digital transformation.

Mobile communication is closely related to transportation itself. The new mobile communication technology can also change the operation mode and mechanism of the entire transportation industry, instead of only improving the operation efficiency of the transportation industry. The service model of communication operators must be changed again when the era of digital twin landing in the transportation field comes. Communication operators can no longer set a fixed network capacity according to the usage peak value every day as in the past but need to realize intelligent operation together with application enterprises:

(1) 5G technology realizes efficient interaction between physical transportation and digital transportation. The development of communication and transportation has always been closely related. The development of digital transportation is neither a castle in the air nor a simple transfer of traditional information work to the Internet. It is under the integration of 5G technology and implementation of digital

twin in transportation field that new business forms such as vehicle road collaboration, automatic driving, and smart high speed derive a new development trend of efficient interaction between physical and digital transportation.

(2) 5G technology supports the multi-dimensional dynamic mapping of digital transportation. With the in-depth application of 5G technology in intelligent transportation, its advantages in supporting digital transportation are revealed. The technology can fully integrate virtual reality, artificial intelligence, and other technologies into the operation of digital transportation, making the digital twin intelligent transportation a qualitative leap. Before the integration of 5G technology into the operation of intelligent transportation, intelligent transportation has made some progress in relying on cloud computing and big data analysis, but the technical requirements put forward by the digital twin construction goal are getting higher and higher. Only 5G can replace some of the original backward information technology and accelerate the digital and diversified dynamic mapping.

(3) 5G technology accelerates the formation and application of digital twin. By giving full play to its advantages, 5G technology realizes the collection and supervision of transportation big data, solves many problems in the development of intelligent transportation, speeds up the formation and application of digital twin to a large extent, and also promotes the development of China's intelligent transportation in the digital twin intelligent transportation system.

After the comprehensive and deep integration of 5G technology and digital twin, the visual management of intelligent transportation in China is smoother. Digital twin first constructed the intelligent transportation ecological network, then promoted the construction of digital infrastructure, and built a three-dimensional transportation map with simulation characteristics. Finally, it built a transportation "cloud road" to realize the dynamic perception of human, material, and road data. In a word, through the texture of intelligent transportation, we can see that digital twin is the key direction and application scenario for the development of the transportation industry and also the key means to complete the visual presentation of intelligent transportation. Digital twin technology will realize dynamic virtual scenes at multiple levels through various intelligent applications, which can not only synchronize the real transportation but also make full use of the constructed information model to complete scientific research, analysis, and prediction of the real transportation.

In February 2021, the State Council issued the "Outline for the National Comprehensive Three-dimensional Transportation Network", in which the development goal of China's comprehensive three-dimensional transportation network is clearly put forward: "by 2035, basically build a modern and high-quality national comprehensive three-dimensional transportation network that is convenient, smooth, economical, efficient, green, intensive, intelligent, advanced, safe and reliable; realize international and domestic interconnection, three-dimensional access to major cities in China and effective coverage of county-level nodes, and strongly support the "National 123 Travel Traffic Circle" (1-hour commuting in metropolitan areas, 2-hour access in urban agglomerations and 3-hour coverage in major cities in China) and the "Global 123 Express Cargo Flow Circle" (1-day delivery in China, 2-day delivery in neighboring countries and 3-day delivery in major cities in the world). The quality, intelligent and green level of transportation infrastructure ranks top in the world. The transportation fully meets the people's growing good life needs, effectively protects national security and supports the basic realization of socialist modernization in China".

From the perspective of the construction results of China's comprehensive three-dimensional transportation network, each development focus mentioned in the "Outline for the National Comprehensive Three-dimensional Transportation Network" can be directly related to digital twin. For example, the completion of the "global 123 fast cargo logistics circle" requires not only the digital twin to build a digital model to enhance the intelligence of the logistics system, but also the digital twin to perform the optimal scheme of the logistics system operation, so as to improve the logistics efficiency.

In addition, the value of digital twin is not only highlighted in the field of real transportation but also more powerful in virtual space. Through data collection in the real world, digital twin has completed the design and testing of various advanced transportation equipment in the virtual space, which has improved the upgrading speed of modern transportation equipment and saved a lot of research and development costs for the transportation departments of various countries.

First of all, digital twin, combined with automobile industry, can effectively improve the research and development capacity and production efficiency of automobile industry, reduce production loss, and predict the possibility of error in advance. The "visualization + simulation + prediction" is used to support the full element simulation of automobile R&D so that the R&D, production, and testing of virtual industry chain company in the automobile industry can be fully simulated, and the "three-in-one development of automobile, road and cloud" in the automobile industry can be realized.

Second, the digital twin revived the Boeing aircraft, spawned the intelligent airport, clarified the development direction of aerospace and aviation equipment, and became an important guarantee for the solution of modern aerospace transportation equipment to the present and the future.

Third, digital twin enables high-quality development of rail transit through the linkage between physical reality and virtual space. It has constructed the real twin base of rail transit, promoted the integration of virtual and real elements of rail transit, built the three-dimensional twin model of rail transit, and directly solved the security problem of 17.59 billion passengers.

Finally, the digital twin also provides a new path for the transformation of the traditional transportation industry. The combination of digital twin and water conservancy industry will also form an important driving force for the reform of water conservancy industry as a carrier of digitalization and intelligence.

With the deepening of the development of digital and information technology in the field of transportation in China, the digital twin has started to empower the development of smart cities in China after the upgrade of the enabling transportation equipment.

In March 2021, the "Outline of the 14th Five-Year Plan for National Economic and Social Development of the People's Republic of China and the Vision for 2035", which was adopted at the fourth session of the 13th National People's Congress, clearly involved in the construction of smart cities in a number of places, and clearly defined the planning and deployment of digital infrastructure, digital economy, digital society, digital government, and digital ecology.

In the process of digital upgrading of China's society, the intelligent technology based on digital twin in urban transportation has been applied more and more widely, and the urban transportation system has ushered in the key stage of digital transformation. Virtual figures are constantly integrated with reality in the development of urban transportation system in China. As the cutting-edge technology of intelligent transportation, the key task of digital twin is how to find the appropriate application scenarios in combination with the current practical needs, so as to enable urban transportation more efficiently. Therefore, digital twin is carrying out a new round of research and development to promote urban transportation from intelligent to intelligent, so as to meet the needs of the existing urban transportation development.

As for the intelligent development of urban transportation in China, the formation of urban intelligent brain is closely related to digital twin. The implementation and application of digital twin in urban transportation has greatly improved urban transportation. It uses Internet of Things, 3D modeling, cloud computing, and other technologies to replicate the complex transportation network system in reality and constructs a digital twin transportation system through real-time data transmission, achieving the effect of optimizing urban transportation and ensuring transportation safety.

At present, China's digital twin transportation system has been piloted in major cities by way of hierarchical zoning, for example, in 2021, China's Guiyang City piloted the digital twin transportation system first; Qingdao launched the "Digital Qingdao 2021 Action Plan" and strived to build a "new scene" of digital transportation around multiple projects such as ETC intelligent parking and cruise taxi vehicle intelligent terminals, so as to improve the operation efficiency of Qingdao transportation.

In addition, the digital twin also clearly explains the difference between intelligent transportation and intelligent transportation. Intelligent transportation and intelligent transportation are different stages of the twin digital transportation system. Although the concepts of the two are similar, a clear distinction between the two is made, intelligent transportation is conducive to a better understanding of the hierarchical structure and technical connotation of the digital twin technology in the transportation system, to prevent the difficulties in promoting due to the excessive demand for preliminary development, and to avoid the lack of forward-looking and limited iteration of the system. In a narrow sense, intelligent transportation is biased towards the iteration of technology and pursues more high-tech applications. Intelligent transportation is an upgraded version of intelligent transportation, which is closely related to "people" and solves the problem of human-machine system coordination.

It can be said that the promotion of digital twin to smart city is all-round. It accelerates the infrastructure construction of smart city, makes the city management achieve the effect of "take in everything in a glance, be able to rally multitudes at one's call and be clear at a glance", and improves the urban transportation safety, innovates the new concept of transportation safety management, and deepens the connotation and extension of urban transportation with digital twin infrastructure, which has forged a intelligent transportation safety road.

Driven by the digital twin, the cutting-edge forces in the digital era have extended from transportation infrastructure to smart cities, and then led the transportation technology of land, sea, and air.

First of all, digital twin has "intelligently" created a new era of aerospace technology, which has triggered digital changes in the aerospace field. In December 2020, China Aerospace Science and Technology Corporation announced: "Aerospace Science and Technology Corporation has always been guided by the thought of socialism with Chinese characteristics, implemented the guiding spirit of network power and digital China, studied and implemented the decision and deployment of the Central Committee of the Communist Party of China and the State Council on promoting the integrated development of the digital economy and the real economy, regarded digital transformation as an important means to transform and enhance traditional kinetic energy and cultivate new kinetic energy for development, and comprehensively promoted the construction of digital aerospace".

Since the formulation of this strategy, China Aerospace Science and Technology Corporation has begun to strengthen the application of digital technology, in which the application effect of digital twin is extremely prominent. It opens the metaverse channel, with virtual spaceflight to accelerate the pace of development of the times. At the same time, digital twin upgrades aerospace intelligence brain while new energy is replacing the wings of human spaceflight; digital twin has helped aviation production and nurtured the development of high-end technology industry. The future of "mixed virtual and real, human-computer coordination" aviation is clearly presented to the aviation public.

All in all, as a key technology to solve the physical integration of intelligent manufacturing information and to practice the concept and objective of intelligent manufacturing, digital twin plays a great role in promoting the multi-dimensional design and practice of aerospace tools through the mutual mapping of physical entities and digital space, and makes the aerospace industry enter the "intelligent" manufacturing era.

Second, digital twin has "intelligently" created stronger surface transportation. In February 2020, 11 national ministries and commissions including the National Development and Reform Commission and the Ministry of Industry and Information Technology of the People's Republic of China jointly issued the "Innovation and Development Strategy of Intelligent Vehicles", proposing that by 2025, the technological innovation, industrial ecology, infrastructure, regulatory standards, product supervision, and network security system of China's standard smart vehicles will be basically formed. By 2035, China's standard intelligent vehicle system will be fully established. China has started to accelerate the development of the unmanned driving industry. In October 2021, the policy pilot area of Beijing intelligent connected vehicle officially opened the unmanned road test. Baidu and Pony obtained the first batch of 15 licenses for the unmanned road test. In the L4 level automatic driving test in this unmanned road test area, the driver safety officer can sit in the copilot, which represents that China's unmanned driving technology has taken a big step forward.

The booming development of unmanned driving vehicles is also supported by digital twin technology. The modeling and simulation of digital twin make it possible for unmanned driving vehicles to be online forever; digital twin technology has also entered the field of automatic navigation logistics robot, which makes modern logistics present the characteristics of "digitalization, automation, and intelligence"; digital twin has also built a "mirror system" of smart public transport in virtual cities, running out of the optimal solution of urban public transport management in virtual space, enabling urban public transport to enter the optimal operation state.

In the end, the digital twin depicts the "unity of water and sky". Digital twin creates flexible management of smart river channel by flexible simulation; the whole line management and control is used to innovate the segmented management method of water transportation; through the analysis of all factors, it controlled the river operation research in the digital space, thereby supporting China to complete the digital upgrade of intelligent water affairs, and realized the refined, intelligent, and standardized management of water resources.

5. Digital Twin Transportation Changes Transportation LifeCycle

Since the digital twin was born in the US Air Force Research Laboratory, the real world and virtual space began to merge. In recent years, with the development of VR, AR, and other technologies, the distance between the virtual world and reality has been further shortened. Digital twin technology is bound to usher in greater changes, and the complete integration of digital space and the real world will no longer be a fantasy. In September 2020, the State-owned Assets Supervision and Administration Commission (SASAC) of China issued the "Notice on Accelerating the Digital Transformation of State-owned Enterprises", which clearly proposed to accelerate the promotion of digital twin and accelerate the application of digital twin in various fields.

Digital twin technology has become one of the largest technology development trends in the future. Not only that, but the digital twin is also building the infrastructure of future technology.

Based on the real-time monitoring, simulation, and prediction of digital model, most of the practical applications of technology need to be based on the digital twin technology. For example, the logistics transportation robot, logistics sorting robot, logistics stacking robot, and logistics operation robot, which are widely used in intelligent logistics, are put into use on the basis of the twin integration of logistics network, AI, and digital. In addition, the construction of smart city and intelligent transportation also requires the integration of GIS, GPS, Internet of Things, and digital twin technology.

Digital twin can bring virtual presentation of physical entities, and not only the surface of physical entities but also their internal structure and operation will be presented in real time through a high simulation model. It also benefits from the support of big data, cloud computing, and other technologies. More importantly, digital twin technology is still at the stage of development and can continue to evolve. Digital twin evolution will also promote the development of other technologies. Therefore, digital twin technology is not only the infrastructure of science and technology but also the innovation pillar of future science and technology.

After the digital twin opens the door to connect the real world and digital space, the future technological infrastructure and the future development direction of virtual and real world will gradually be clear. The collaborative interaction between real space and virtual space is inevitable. The limitations of the real world make people have greater expectations for virtual space.

However, the unrestricted nature of virtual digital space may cause people to deviate from the basic moral standards of real society, thus promoting the generation of malice. In addition, in the future digital space, everyone's information and data will be stored in cloud. At that time, it is also a bottleneck of the development of digital twin as for whether the security of people's own information can be guaranteed. Therefore, there has been a heated discussion on whether the digital twin technology should be used to build a virtual space.

The issues that must be considered in the process of digitalization are as follows: Whether the digital twin technology will develop "good" or "evil", how the virtual space constructed by it will impact the real society, how to balance the relationship between the real world and the virtual space, and so on.

In terms of the implementation and application of various emerging technologies in China, the application of technology in the field of benefiting the country and the people is good; if it is used in the field of calamity to the country and the people, it is evil. There is no good or evil in science and technology itself, but the use of technology affects people's evaluation of technology. Therefore, we need to actively balance the relationship between technology and human, and society and nature, in the process of using digital twin technology, and promote the development of science and technology in a favorable direction. Digital twin technology will not fail us, and we can trust it, as the law and time will prove.

In September 2021, the Ministry of Transport of China issued the "Action Plan for New Infrastructure Construction in the Field of Transportation (2021–2025)", in which the plan clearly states, "With the overall goal of accelerating the construction of a strong transportation country, adhering to innovation driven and intelligent development, we will take digitalization, networking and intellectualization as the main line, organize and promote a number of key projects of new transportation infrastructure, create an influential model of new transportation infrastructure, create an innovative development environment, promote the development of new infrastructure from point to area, improve the efficiency, expand functions increase momentum of transportation, and continuously enhance the sense of gain, sense of happiness and sense of safety of the people". The plan also mentions that China's transportation construction focuses on the intelligence of roads, channels, ports, and hubs and deeply integrates technologies, such as digital twin, AI, big data, and cloud computing.

It can be seen from the content of the action plan that modern technology is applied to build a transportation network with car road coordination, which helps the transportation industry to achieve digital transformation, so as to obtain broad development space and support from a number of policies.

Future transportation is the intersection and integration of the real world and the virtual space. By acquiring real-time transportation data in the real world and digitalizing and modeling it in the virtual space, we can open up the digital twin space of the physical world and the cloud to extrapolate and solve the transportation problems encountered in the real world.

With the support of digital technology, for example, China has completed the construction of a number of highly difficult transportation equipment projects. With the emergence of the Pearl River Mouth tunnel, the first undersea high-speed railway in China, it means that China has made another major breakthrough in the field of transportation equipment.

The Pearl River Estuary tunnel is part of the high-speed railway built in connection with the Belt and Road Initiative. The Shenzhen–Jiangmen high-speed railway connects Shenzhen and Jiangmen and passes seven stations, including Xili, Shenzhen airport, Dongguan Binhai Bay, Nansha, Zhongshan north, Henglan, and Jiangmen. Among them, the Pearl River Estuary tunnel is located at the estuary of the Pearl River between Dongguan and Guangzhou, which is the only way for the Shenzhen River high-speed railway. It is a great problem for the high-speed railway to cross the Pearl River Estuary as the geographical environment near the Pearl River Estuary is very complex and the water corrosiveness is relatively serious.

With the advantage of digital technology, China Railway Engineering Equipment Group Co., Ltd. and China Railway Tunnel Group jointly developed the "No. Greater Bay Area" super large diameter slurry balance shield machine, which solved the problem. With the help of "No. Greater Bay Area", the project completed the construction of the Pearl River Estuary tunnel with a maximum underwater depth of 115 meters and a maximum water pressure of 1.06 MPa.

Under the guidance of advanced technologies such as digital twin, China has also made major breakthroughs in the construction of high-speed trains, heavy loaded trains, urban rail trains, and other transportation equipment fields, enabling the transportation mode to truly achieve the omnipotent "go to the sky and enter the sea".

In the future, society will also develop in an all-round digital way. Digital transportation is an important part of the digital economy. Therefore, the whole society should make full use of the advantages of digital twin technology, especially in the transportation equipment manufacturing industry, and actively integrate with the Internet, Internet of

Things, artificial intelligence, big data, and other technologies, accelerate the digital and intelligent construction of China's transportation equipment, and promote the transformation and upgrading of the transportation equipment field.

The key to the deep application of digital twin lies in its three major features, namely whole lifecycle, real-time mapping, and two-way interconnection.

The whole lifecycle refers to the whole process of an organism from birth to death in biology. The whole lifecycle in the field of transportation equipment refers to the whole process of design, research and development, manufacturing, testing, operation, and maintenance of transportation equipment covered by digital twin.

Real-time mapping refers to the comprehensive real-time connection between physical entities and digital twin. Two-way communication refers to the two-way flow of data and information between physical entities and their digital twin. During the whole lifecycle operation, the digital model will interact with each link of transportation equipment manufacturing in real time and continuously through sensors, artificial intelligence, operation history data, etc. and complete accurate mapping of physical entities in digital space so as to improve its own accuracy. At the same time, it forms a data flow closed-loop between the digital space and the physical world. In the end, real-time monitoring in various aspects, directions, and scales is carried out by use of digital space model, providing scientific reference for each link of operation, and further optimizing and improving the work of all stages of the whole lifecycle of transportation equipment.

In view of the three characteristics of digital twin, digital twin has been applied in many aspects in the field of transportation equipment. For example, Pci Technology has built a set of digital twin systems in a large city, which is one-by-one corresponding and mutually mapping with the "whole time, whole range and whole volume" information of urban pavement transportation. In addition, at the "2021 Zhongguancun Forum", Sun Chitian, technical director of Tencent's autopilot simulation, shared the digital twin technology to accelerate the landing of autopilot, and said that it can solve the problems of intelligent Internet connection, autopilot test, efficiency, and safety in the development of intelligent transportation through the characteristics of digital twin.

In the three links of transportation equipment manufacturing, the production line level involves a number of simple and complex processes, which is a very important part. However, due to the limitations of the traditional production line, such as relatively independent processes, data transmission delay, and machine accuracy, the production efficiency and cost, and product quality cannot be guaranteed continuously and stably. For this reason, digital twin provides managers with multi-perspective status monitoring and visual display of various data, which strongly promotes the construction of Intelligent Manufacturing in China.

Enterprises can further use digital twin technology to build factory-level in digital space on the basis of equipment-level and production line-level, centralize and integrate all control systems of each link of the whole transportation equipment factory, and conduct comprehensive control by a unified platform to ensure the efficient operation of production and manufacturing work in each link.

In addition, the digital twin successfully created the intelligent transportation visual decision-making system and constructed the "digital + intelligent" integrated three-dimensional transportation network. It is not only because of its unique talent and outstanding characteristics but also because of its good interaction.

Digital twin integrates not only 5G technology but also AI technology, cloud computing and other emerging technologies. Future transportation is the intersection and integration of the real world and virtual space. The closer the combination of digital twin and emerging technology is, the higher the speed and accuracy of its model construction in the virtual world, and the stronger the ability to deduce and solve real transportation problems.

For example, AI has been used more strongly on the ground with the support of 5G with the commercialization of 5G network, and face recognition, voice recognition, data analysis, unmanned driving, intelligent robot, and other technologies have been improved so that urban transportation problems have obtained better solutions. By integrating "5G + AI" into the model of intelligent transportation, the intelligence of transportation will be greatly upgraded.

In the new era of science and technology, 5G, the Internet of Things, and Beidou satellite, combined with earth observation, high-precision map, and other technologies, make the digital space more refined, real-time, and simulated. At the same time, cross-regional and cross-industry multi-source data are also constantly gathering to form a macro circulation transportation management model. This has brought convenience to the construction of the "digital + intelligent"-integrated three-dimensional transportation network. With the research and development and application of artificial intelligence + 3D digital automatic modeling and other technologies, the implementation efficiency of digital twin will be greatly improved, and the visual operating system of intelligent transportation will be realized. Moreover, it will also enable the industrial Internet to upgrade from 2D content to 3D, or even the promotion of 4D real-scene 3D in the future, and give human vision the leap from 2D to 4D digital everything.

From the key policies issued by the Ministry of Transport of China, such as the "Outline for the Construction of Strengthening Country with Strong Transportation Network", "Opinions on Accelerating the Construction of a Transportation Power Driven by Scientific and Technological Innovation", "Action Plan for New Infrastructure Construction in the Field of Transport 2021–2025", and "Sustainable Development of Transportation in China" white papers, it can be seen that the next ten years will be a blowout period for the development of China's transportation digital infrastructure. The field of transportation infrastructure in China has begun to experience the reconstruction, switching, and migration of digital infrastructure, so as to ensure the intelligent technology community represented by Internet of Things, cloud computing, edge computing, artificial intelligence, mobility, and digital twin to provide the technology base with high economy, high availability, and high reliability for the future economic development in the continuous integration, overlapping, and iterative upgrading. In the past 10 years, urban transportation has been involved in more and more fields. In the early stage of reform and opening up, transportation management was only a matter of concern to a few departments, such as the transportation police. Transportation planning is dominated by urban planning departments, and there is no special transportation planning unit. Today, the urban transportation industry is the focus of attention from the government, enterprises, and all walks of life of society, instead of just the work of a certain management department, and will become the top priority of future smart city construction.

6. Digital Twin Transportation Enhances the Level of Intelligence in Transportation

The core of intelligent transportation is "intelligence", which usually refers to the ability of people to quickly and accurately understand, handle and solve things and problems. "Intelligence" is the concept of spiritual level sublimation, the core of which is to emphasize the spiritual, savvy and talent innovation. In a word, "intelligence" emphasizes a state of unconsciousness or subconsciousness, without much external intervention, mainly due to its own subjective initiative.

Intelligent transportation is to install a "brain" for urban transportation, so that it can see, hear, smell relevant information in a timely manner, and make timely response, fundamentally solving problems such as urban transportation congestion, resource waste, frequent safety accidents, and difficulty in real-time control of the situation so that the development of urban transportation is on the track of sound development.

China is already a big transportation country worthy of the name, but it is still not a country with strong transportation network (transportation power). The reason is that, in terms of service, service is the essential attribute of transportation. One is safety and the other is convenience in terms of transportation services, which has always been a good wish for people to travel and live, and also the main purpose of intelligent transportation construction.

According to the planning objectives in the "Outline for the Construction of Strengthening Country with Strong Transportation Network", China will become a country with strong transportation network in an all-round way by the middle of the 21st century. "China Sustainable Transport Development Report" issued by the Ministry of Transport of China also clearly states, "Based on the new development stage, China's transportation will fully, accurately and comprehensively implement the new development concept, serve to build a new development pattern, accelerate the solution of imbalance and inadequacy in development, strive to achieve higher-quality, more efficient, more fair, more sustainable and safer development, and contribute China's intelligence to promoting global sustainable development and contributing Chinese intelligence and strength to the building of a community with a shared future for mankind".

The development of digital, intelligent, and powerful transportation in China needs the help of cutting-edge technologies, such as the Internet, big data, artificial intelligence, and digital twin. These key technologies will provide strong technical support for the construction of intelligent transportation.

At the beginning of 2018, the new infrastructure wave guided by "accelerating the pace of 5G business and strengthening the construction of new infrastructure such as artificial intelligence, industrial Internet and

Internet of Things", especially the characteristics of digital infrastructure, requires top-level design and targeted layout across the country, which is also an important issue faced by intelligent transportation.

It is an important challenge for transportation management departments to make use of Internet high and new technologies such as machine intelligence technology, cloud computing technology, and big data integration technology to provide new means and new thinking for urban intelligent transportation governance and meet the new needs arising from the continuous progress of technology and the continuous improvement of people's quality of life. Starting from the transportation integrated command and dispatching platform, transportation signal control system, video monitoring resources, transportation guidance system, transportation flow information collection, and other practical business systems, the introduction of mainstream Internet information technology and the construction of precision business service scenarios for intelligent transportation management can effectively improve the data and information service capabilities in terms of application of intelligent transportation in data integration, video intelligent analysis algorithm, business innovation support, efficiency enhancement of transportation management business, and services for public travel.

The new mode of intelligent transportation under the "ubiquitous interconnection and comprehensive perception" adopts the Internet + intelligent transportation governance framework. Through the establishment of a perception "ubiquitous network" and the coordination of people, vehicles, ships, roads, and ports related perception equipment, the integrated transportation operation status can be seen, measured, and controlled; the perception network is a collection of information from various departments, users, processes, equipment, and systems. Meanwhile, the perception network serves for different applications, which greatly expands the range of information perceived by various applications.

In 2019, the Ministry of Transport of China formulated and issued the "Outline for the Development Plan of Digital Transportation", which proposed for the first time to use the strategy of digital power to achieve "strengthening country with strong transportation network". It also points out the specific direction for the comprehensive integration of intelligent transportation and digital twin.

By 2025, a new step will be made in the digital upgrading of all elements and full cycle of transportation infrastructure and transportation equipment, and the digital acquisition system and network transmission system will be basically formed. The public network such as the fifth generation of mobile communication (5G) and the new generation of satellite communication system have been initially applied in the industry. The application level of transportation big data has been greatly improved, travel information services have been fully covered, the logistics service platform and integration have entered a new stage. The transportation infrastructure will be digitalized in all elements and cycles, and the transportation control network integrating heaven and earth will be basically formed by 2035. The on-demand instant travel service will be widely used. Moreover, China will become a major international standard setter or participant in the field of digital transportation, and the overall competitiveness of the digital transportation industry will be leading in the world". In order to achieve this wonderful prospect, innovation must become the first driving force for the future development of digital transportation, and the new concept of "travel as a service" will be fully popularized. Data is used to link travel demand and service resources, so that travel becomes an on-demand real-time service and makes travel easier. It is emphasized to build a digital collection system, a network transmission system, and an intelligent application system with the "data chain" as the main line, so as to accelerate the development of transportation informatization to digitalization, networking and intelligence, and provide support for the construction of "strengthening country with strong transportation network". Digital transportation mainly consists of seven systems, including digital collection system, network transmission system, intelligent application system, industrial ecosystem, network and data security system, standard system, support and guarantee system, etc. It can be predicted that in the near future, China will take data as the key element, fully empower transportation and related industries, improve the qual+ity of travel and logistics services, and create a more happy and harmonious social living environment.

References

An Xiaopeng. *Digital Infrastructure: The Migration Road to the Digital Twin World*. Beijing: Electronic Industry Press, 2021.

Chen Gen. *Digital Twin: An Important Application Scenario in the 5G Era, the Cornerstone of Future Real Industry*. Beijing: Electronic Industry Press, 2020.

Chen Gen. Intelligent manufacturing, digital twin and strategic scenario modeling. *Science & Technology and Economy*, 2019, 32(7): 1–5.

Christian Manzai and Linus Sloyepner. *Industry 4.0 in the Global Context*. Changsha: Hunan Science and Technology Press, 2021.

Dong Hua and Di Xiaofeng. The road thinking of transportation information construction under the background of big data era. *Scientific and Technological Innovation*, 2019, (23): 139–140.

Du Mingfang and Xing Chunxiao. *Digital Twin City: Research on Urban Smart Governance in the New Infrastructure Era*. Beijing: China Construction Industry Press, 2021.

Feng Bo. Design of transportation light system based on Siemens PLC1200. *Information System Engineering*, 2019 (12): 89–90.

Gao Yanli and Chen Cai. *Digital Twin City: The Integration of the Virtual and the Real Opens the Door of Wisdom*. Beijing: Posts and Telecom Press, 2019.

Geng Yanbin, Sun Peng, Tang Pengcheng, *et al.* International benchmarking research on the modernization development of transportation in the Yangtze River Delta. *Integrated Transportation*, 2019, 41(2): 10–15.

Hu Xiaoli. Thinking on information technology construction and transportation management. *Science and Wealth*, 2020, 12(1): 246.

Huang Haisong, Chen Qipeng, Li Yiting, *et al.* Overview of the development and application of digital twin technology in intelligent manufacturing. *Journal of Guizhou University (Natural Science Edition)*, 2020 (5).

Li Zhe. Discussion on the management ideas of road transportation informatization. *TranspoWorld*, 2019, (16): 160–161.

Luo Yu. Thoughts on the information construction of transportation management. *China Small & Medium Enterprises*, 2019, (11): 89–90.

Tao Fei, Liu Weiran, and Liu Jianhua. Digital twin and its application. *Computer Integrated Manufacturing System*, 2018, 24(1): 4–21.

Wang Qiang. Digital twin technology can simulate the optimal solution of urban transportation. *Auto Parts*, 2020, (2): 36.

Wu Fubiao. Digital twin empowers new infrastructure. *China Construction Informatization*, 2019, (6): 56–59.

Wu Fubiao. Opportunities for "digital twin" China's "new infrastructure". *China Survey and Design*, 2020, (12): 72–74.

Wu Kewei. Transportation digital twin application based on video intelligent analysis technology. The Fifteenth China Intelligent Transportation Annual Conference, 2020.

Wu Zhaohui, Liu Zhenzheng, Shi Ke, *et al.* Research on transportation scene digital twin construction and application of the integration of virtual and real. *Journal of System Simulation*, 2021, 33(2): 295–305.

Xiao Jinghua, Xie Kang, and Chi Jiayu. Intelligent manufacturing, digital twin and strategic scenario modeling. *Journal of Beijing Jiaotong University (Social Sciences Edition)*, 2019, 18(2): 69–77.

Zeng Yue and Zhu Liangcai. Research and industrialization of large-scale robot scheduling system under industry 4.0. *China Science and Technology Achievements*, 2020, (18): 4–7.

Zhang Jingtao. Trends and suggestions of digital twin technology in the application of intelligent transportation. *Information and Communications Technology and Policy*, 2020, 309(3): 27–31.

Zhang Long. Looking at digital twin from the development of intelligent manufacturing. *Software and Integrated Circuit*, 2018, (9): 59–62.

Zhang Yanli, Yuan Lei, Wang Yining, *et al.* Future learning under the integration of digital twin and hologram technology: New connotation, new prospect and new field. *Journal of Distance Education*, 2020, 38(5): 36–44.

Zheng Guorong, Liu Xiaoming, Shen Hui, *et al.* Research on collaborative optimization framework of multi transportation mode scheduling of urban comprehensive transportation hub driven by digital twin. *Automation Panorama*, 2020, 37(2): 52–56.

Zhu Guozhen. Germany and the United Kingdom transportation management work inspection and experience reference. *Road Transportation Management*, 2019, (9): 121–128.

Index

E

O

P

Q

R

S

U

V

W

www.ingramcontent.com/pod-product-compliance
Lightning Source LLC
LaVergne TN
LVHW011656200926
848484LV00004B/953